The Role of Phosphonates in Living Systems

Editor
Richard L. Hilderbrand, Ph.D., LCDR
Biochemist
Naval Biosciences Laboratory
Oakland, California

CRC Press
Taylor & Francis Group
Boca Raton London New York

CRC Press is an imprint of the
Taylor & Francis Group, an **informa** business

PREFACE

Phosphorus is required for growth, health, and reproduction in all forms of plants and animals. Since derivatives of orthophosphate occur naturally and occupy such important roles in life processes, it is apparent that an understanding of these roles might allow outside manipulation to facilitate, as in the therapeutic use of drugs, or to interfere with, as in the use of pesticides, normal life processes. Because of the unique character of the carbon-phosphorus bond and its effect on the phosphorus moiety, the phosphonates are of increasing importance to the facilitation of, or interference with, normal life processes, and to industrial processes using phosphate derivatives. Although all organic esters of phosphate are loosely classified as organophosphates, it is the phosphonate class of organophosphates, which contain a carbon-phosphorus covalent bond, that consists of "true" organophosphates.

Throughout the last three decades, a large amount of information on phosphonates has accumulated, but has not been compiled. At this time of the increasing importance of the phosphonate class of molecules, this information needs to be drawn together to make a manageable volume for use as a reference text and as a learning text for interested persons. In compiling a volume of the size planned, discussion must be held to a minimum on the peripheral aspects of phosphonates. Thus, some information has not received the discussion that it may well deserve. For the unintentional oversight of significant ideas, I assume full responsibility. It is desirable that in the future there be more thorough discussion of all aspects of the subject.

The U.S. Navy (the Naval Medical Research and Development Command and the Naval Medical Research Institute) has been supportive in this effort and I acknowledge that with gratitude. Despite that support, however, the opinions and assertions contained herein are the private views of the authors and are not to be construed as either official statements or as reflecting the views of the U.S. Navy.

I would like to thank CRC Press, Inc. for publishing this manuscript, the individual authors for their cooperation and contributions, Mrs. P. J. Gergely for her superb typing, and Mrs. Gale E. Crain for her editorial assistance. Then there are those, to whom my debt is more long-standing but no less real, who have been mentors over the years. My thanks to Dr. Thomas O. Henderson and Dr. Terrill C. Myers of the University of Illinois, Medical Center Campus, for introducing me to scientific research and, specifically, to studies of the role phosphonates play in living systems. Lastly, I dedicate this work to Heidi and Tim, my two children.

Richard L. Hilderbrand, Ph.D.
1982

THE EDITOR

Richard L. Hilderbrand, Ph.D., is currently a biochemist at the Naval Biosciences Laboratory, Oakland, Calif. In 1968, he graduated from Southwest Baptist College, Boliver, Mo. with a major in biology and a minor in chemistry. His graduate work was in biological chemistry and was completed in 1972 at the Graduate School, University of Illinois, Medical Center Campus, Chicago. Immediately upon receiving his degree, Dr. Hilderbrand entered active duty with the U.S. Navy and is currently a Lieutenant Commander in the Medical Service Corps. Previously, he has been assigned duty at the Naval Medical Research Unit #1, Berkeley, Calif., the Naval Health Research Center, San Diego, Calif., and the Naval Medical Research Institute/Toxicology Detachment, Wright Patterson Air Force Base, Ohio. This work was completed while he was at the latter laboratory. In addition, he has done research and taught part-time at San Diego State University, San Diego; is currently an Assistant Clinical Professor of Pharmacology and Toxicology at Wright State University School of Medicine, Dayton, Ohio; and has worked collaboratively with the Bob Hipple Laboratory for Cancer Research of the Wright State University School of Medicine.

Dr. Hilderbrand is a member of the New York Academy of Sciences and the American Association for the Advancement of Science. His current major research interests are the effects of toxic chemicals on hematopoietic, renal, and neural tissue. Past research has included rapid identification of infectious organisms, biochemical changes associated with human endurance performance, binding of prostaglandins to receptors on adipocytes, and studies of the natural occurrence of phosphonates.

CONTRIBUTORS

Timothy A. Calamari, Jr.
Supervisor Research Chemist
Southern Regional Research Center
New Orleans, Louisiana

George L. Drake, Jr.
Supervisor Research Chemist
Southern Regional Research Center
New Orleans, Louisiana

Robert Engel
Professor of Chemistry and Biochemistry
Department of Chemistry
Queens College
The City University of New York
Flushing, New York

Marion D. Francis
Senior Scientist
Miami Valley Laboratories
Procter and Gamble Company
Cincinnati, Ohio

Thomas O. Henderson
Department of Biological Chemistry
University of Illinois Medical Center
Chicago, Illinois

Richard L. Hilderbrand
Lieutenant Commander
Medical Service Corps
U.S. Navy
Biochemist
Naval Biosciences Laboratory
Oakland, California

Raymond R. Martodam
Staff Scientist
Miami Valley Laboratories
Procter and Gamble Company
Cincinnati, Ohio

Joseph Donald Smith
Assistant Professor of Chemistry
Department of Chemistry
Southeastern Massachusetts University
North Dartmouth, Massachusetts

TABLE OF CONTENTS

Chapter 1

FOREWORD

Richard L. Hilderbrand*

TABLE OF CONTENTS

I. FOREWORD

Organic compounds of phosphorus play an integral role in the biochemical processes of all living systems. These life processes require orthophosphate ($PO_4^{\equiv}$) as a primary constituent of nucleic acids, phospholipids, and phosphorylated proteins and carbohydrates. Of equal importance is the use of phosphate to generate high energy phosphate bonds whose energy is either used for synthetic processes or for production of other types of energy. In addition, phosphate provides buffering capacity, imparts solubility in aqueous solutions to organic molecules, precipitates with calcium to form the insoluble hydroxylapatites required in bone, and can provide a high concentration of negative charge within a given molecular dimension.[1] Phosphate esters are relatively stable in aqueous solutions at physiological pH but will readily hydrolyze when appropriate enzymes are present. Thus, phosphate is uniquely suited to provide the properties necessary to maintain the mosaic of biochemical reactions occurring within a living organism.

All phosphorus accessible to living organisms occurs in phosphate minerals as orthophosphates. The phosphorus atom occurs at the +5 level of oxidation with four oxygen atoms bonded to the phosphorus in a tetrahedral structure. The organophosphates which occur naturally in a living system are usually oxygen esters, diesters, or anhydrides of phosphoric acid. There are rather infrequent exceptions to this bonding of phosphates in living systems. One exception is the natural occurrence of the carbon to phosphorus (C–P) bond in the phosphonate class of organophosphates. Although phosphate had long been recognized as crucial to life processes, it was not until 1947 that Chavane,[2] a chemist involved in synthesizing phosphonates, observed that, since the C–P bond was stable, there was a possibility for the occurrence of phosphonates in nature. The actual identification of a naturally occurring phosphonate finally came in 1959 by Horiguchi and Kandatsu,[3] who identified 2-aminoethylphosphonic acid (AEP) in an amino acid extract from an hydrolysate of rumen protozoal lipid. Phosphonates have subsequently been shown to occur naturally in a variety of organisms and several metabolic processes involving phosphonates have been elucidated.

The study of synthetic organophosphorus molecules was initiated by Lassaigne in 1820 with the esterification of alcohols and phosphoric acid. In the late 19th and early 20th centuries, A. E. Arbuzov conducted research on the chemistry of trivalent phosphorus compounds. From this work came the well-known Michaelis-Arbuzov reaction, for the synthesis of the C–P bond, and a number of compounds which were forerunners of the insecticides in use today.[4]

The chemistry of phosphonates has developed more slowly than that of phosphate because of the difficulty of synthesizing the C–P bond and the wide availability of, and information on, orthophosphate and phosphate esters. With recognition of the unique chemical properties of phosphonates, commercial interest is increasing markedly.[5] The types of phosphonates of potential or actual industrial interest include catalysts, lubricant additives, flame retardants, surfactants, corrosion inhibitors, and plasticizers. In addition, commercial interest has developed in the use of phosphonates as insecticides, plant growth effectors, and health care products.

The chemical properties of the phosphonates, mechanisms of synthesis, and various analytical methodologies will be presented as required for each topic. The initial chapters include a summary of the natural occurrence of phosphonate and the information available on the metabolism of those compounds. The latter chapters of the book are descriptions of specific biological effects and uses of various types of synthetic phosphonates, including industrial materials. The information on industrial processes is included to show the variety of ways in which phosphonates can affect not only life processes, but our life style. In addition, the industrial applications can indirectly affect our life processes by inadvertent environmental or occupational exposure, thus generating the necessity of our knowing how phosphonates are used industrially. Information on both naturally occurring organophosphonates and synthetically produced organophosphonates will be included.

REFERENCES

1. **Katchman, B. J.,** Phosphates in life processes, in *Phosphorus and Its Compounds*, Vol. 2, Van Wazer, J. R., Ed., Wiley-Interscience, New York, 1961, chap. 20.
2. **Chavane, M. V.,** Synthese de quelques acides phosphoniques amines de formule generale H_3N^+-$(CH_2)_n$-PO_3^-H, *Compt. Rend.*, 224, 406, 1947.
3. **Horiguchi, M. and Kandatsu, M.,** Isolation of 2-aminoethane phosphonic acid from rumen protozoa, *Nature (London)*, 184, 901, 1959.
4. **O'Brien, R. D.,** *Toxic Phosphorus Esters*, Academic Press, New York, 1960, chap. 1.
5. **Chadwick, D. H. and Watt, R. S.,** Manufacture of phosphate esters and organic phosphorus compounds, in *Phosphorus and Its Compounds*, Vol. 2, Van Wazer, J. R., Ed., Wiley-Interscience, New York, 1961, 1272.

Chapter 2

PHOSPHONIC ACIDS IN NATURE

Richard L. Hilderbrand* and Thomas O. Henderson

TABLE OF CONTENTS

* Dr. Hilderbrand is an officer or employee of the U.S. Government. This work was prepared as part of his official duties. Under 17 U.S.C. 105.

I. INTRODUCTION

In 1947, a chemist involved in synthesizing organic molecules containing a carbon-phosphorus (C–P) covalent bond observed that the bond was stable and, therefore, might exist naturally in analogs of aminocarboxylic acids.[1] These biological materials went undetected until 1959 when an ether-ethanol soluble material from a mixed culture of rumen protozoa was hydrolyzed and an unknown ninhydrin-positive substance was found on a paper chromatogram. The material was subsequently purified and identified as 2-aminoethylphosphonic acid (NH_2-CH_2-CH_2-PO_3H_2, AEP).[2,3] The discovery of this compound has led to investigations on the distribution, metabolic pathways, chemical properites, and biological significance of naturally occurring phosphonates.

AEP (sometimes called ciliatine) has since been found in many species in a variety of phyla in the animal kingdom and in lower plants. AEP has been found as a free molecule, in lipids, and in macromolecular material. Phosphonate molecules other than AEP that are unbound have been identified, but apparently do not occur with the frequency that AEP does. These are primarily antibiotics but may be amino acids such as *N*-methylated-AEP. The lipids containing AEP are analogs of phospholipids and are both glycerophosphonolipids and sphingophosphonolipids. In addition, the usually nonphosphorylated cerebrosides have been detected as derivatives containing AEP. Although the C–P bond could theoretically occur at the position of either of the two phosphate ester bonds in the phospholipids, the known naturally occurring materials contain the C–P bond only in the position it is found in AEP. The macromolecular materials are proteoglycan type materials and may contain fatty acids. These phosphonate molecules may make up a significant proportion of the phospholipids which occur in a given organism or tissue. The answers to many of the questions concerning phosphonates have proven to be elusive. Of increasing importance are the questions concerning the role of these unusual materials in the intact organism and the teleological explanation for their existence.

Although studies on the metabolic pathways are intimately related to the information on the natural occurrence of phosphonates, metabolism is covered in a separate chapter of this volume (see Chapter 3). This may lead to some duplication of information or to the more critical problem of pertinent information being inadvertently neglected. The intent of this work is to provide a comprehensive tabular listing of the known natural occurrence of phosphonates, with discussion of certain aspects of the information. Literature through July 1981 is included in this survey.

II. DISTRIBUTION IN NATURE

Over the last 20 years phosphonates have been unequivocally identified from over 80 species that are members of 8 phyla in the animal kingdom. In addition, some bacteria and plants (such as fungi and dinoflagellates) have been found to contain or produce phosphonates. However, these animals (and plants) represent an almost insignificant fraction of even the animal kingdom, which alone consists of 1 to 2 $\times$ 10^6 living species.[4] The projection of true distribution is a difficult matter since the natural occurrence of phosphonates has not been thoroughly investigated.

One factor contributing to the limited availability of information on the distribution has been succinctly stated by Quin.[5] Quin points out that, of four groups that investigated the chemical nature of the lipids of the sea anemone *Anthopleura elegantissima,* three groups failed to find a phosphonate derivative, while the other unequivocally identified ceramide-AEP as 9.3% of the lipid fraction. Thus, it is apparent that some systems for identification of phospholipids fail to differentiate the phosphonate- from phosphate-derivatives and that AEP, as a component of natural materials, may have been overlooked on numerous occasions.

Another factor contributing to the lack of information is that the species being investigated are frequently associated with microorganisms. The microorganisms may contribute materials which are assimilated by the host or which contaminate the material being assayed, thus creating doubt concerning the origin of the C–P bond.[6]

Table 1 is a compilation of the species from which phosphonates have been isolated. Although some duplicate references may have been omitted, the listing includes all phosphonate-containing species of which we are aware, and other information, when available. Genus and species designations that appear in Table 1 are those that were assigned by the author of the paper identifying the phosphonate in that organism. Where no classification was available, the classification is as listed for that species by *Invertebrate Zoology*.[7]

A. Phosphonates as Free Molecules in Living Systems

AEP has been isolated as a free molecule and has been implicated in the anabolic processes of both the C–P bond and the phospholipids themselves. The free AEP molecule has no apparent biological function (e.g., as a storage mechanism for phosphate) in the organisms which synthesize it. The metabolism of the C–P bond and incorporation of AEP into biological molecules will be discussed further in Chapter 3.

One finding of particular interest is that free AEP has been reported in bovine brain.[8] In this case, the concentration of free AEP is twice that of AEP present in the lipid fraction. This would indicate either a limited ability to incorporate the AEP into the abundant phospholipids of the brain or a need for the AEP in the free form (for which there is no evidence thus far).

Another class of phosphonate molecules which has been found free and is of biological significance is that of the phosphonate antibiotics. The first antibiotic to be isolated and characterized was fosfomycin (formerly phosphonomycin, Structure 1).[9] Fosfomycin is produced by streptomycetes

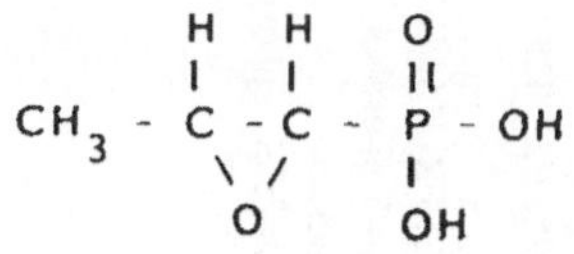

(-) (1R, 2S) - 1, 2 - EPOXYPROPYLPHOSPHONIC ACID

1. (Fosfomycin)

and is effective in mice against a number of Gram-negative and -positive microorganisms. The action of this antibiotic is the inhibition of bacterial cell wall synthesis. Recent work on the toxicological,[10-14] microbiological,[15-20] and clinical[21-24] aspects of fosfomycin has been carried out predominantly by investigators in West Germany, Italy, and Japan.

More recently, several other phosphonate containing antibiotics have been isolated from various species of the genus *Streptomyces*. The initial compound of this phosphonate antibotic series, 3-(*N*-acetyl-*N*-hydroxyamino) propylphosphonic acid (Structure 2), was initially isolated[25,26]

Table 1

Classification	Common name	Monomer	Biochemical fraction	Tissue	Ref.
I. Lower organisms					
A. Plants					
Bdellovibrio bacteriovorus	Bacterium	AEP	Phospholipid	Whole organism	63
Mycobacterium sp.	Bacterium	AEP	Phospholipid	Whole organism	102
Streptomyces sp. ATCC 21096, 21239 and 21240	Bacterium	Phosphonomycin	Free	Extracellular	9
S. rubellomurinus	Bacterium	FR-900098	Free	Extracellular	25,29,103,104
S. rubellomurinus (Subsp. indigoferus)	Bacterium	FR-900098, FR-33289	Free	Extracellular	29,103,104,105
S. lavendulae	Bacterium	FR-32863, FR-31564	Free	Extracellular	29,103,105
S. plumbens (nov sp.)	Bacterium	D-2-amino-5-phosphono-3-pentenoic acid	Tripeptide	Extracellular	106
Streptomyces	Bacterium	Phosphinothricin (phosphinic acid)	Tripeptide	Extracellular	107
Pythium prolatum	Fungus	AEP	Ceramide	Mycelium	64
Monochrysis lutheri	Phytoplankton	AEP	Lipid, residue	Whole organism	108
Amphidinium carteri	Phytoplankton	AEP	Lipid, residue	Whole organism	98
Peridinium trochoidem	Phytoplankton	AEP	Lipid, residue	Whole organism	98
Exuviella cassubica	Phytoplankton	AEP	Hydrolysate	Whole organism	98
Coccolithus huxleyi	Phytoplankton	AEP	Hydrolysate	Whole organism	98
Syracosphaera elongata	Phytoplankton	AEP	Hydrolysate	Whole organism	98
B. Animals					
Acanthamoeba castellanii	Soil amoeba	AEP, 1-hydroxy-2-AEP	Lipophosphonoglycan	Cell membrane	78,79
Tetrahymena pyriformis	Protozoan	AEP	Lipid, free, residue	Whole organism	35,36,68,109,110,111
		AEP	Phospholipid	Cilia	38,39,101
		AEP	Phosphonoprotein	Whole organism	75
		α-NH_2-β-phosphonopropionic acid (PALA)	Hydrolysate of ethanol extract	Whole organism	94,101

Mixed cultures	Rumen protozoa	AEP	Lipid hydrolysates	Whole organism	2,3,37,112
Isotricha sp. and *Entodinium* sp.	Protozoa	AEP	Hydrolysate	Whole organism	112
E. caudatum	Protozoan	AEP	Phospholipid	Whole organism	37
Amonyx nugax	Zooplankton	AEP	Hydrolysate	Whole organism	98
II. Coelenterata					
A. Anthozoans					
Anthopleura elegantissima	Anemone	AEP,*N*-Methyl-AEP,*N,N,N*-Trimethyl-AEP	Phospholipid	Whole animal	34,101
		AEP	Phospholipid	Whole animal	6,51,52
A. kurogane	Anemone	AEP	Lipid hydrolysate	Whole animal	101
A. pacifica	Anemone	AEP	Lipid hydrolysate	Whole animal	101
A. xanthogrammica	Anemone	*N*-Methyl-AEP, *N,N*-Dimethyl-AEP, *N,N,N*-Trimethyl-AEP	Lipid	Whole animal	33,101
		AEP,*N*-Methyl-AEP	Residue, protein	Whole animal	72,113
Aiptasia pallida	Anemone	AEP	Residue	Whole animal	114
Bunadosama cavernata	Anemone	AEP	Lipid, residue	Whole animal	114
Bunadosama sp.	Anemone	AEP	Lipid, water soluble, residue	Whole animal	115
Cerianthus americanus	Anemone	AEP	Residue	Whole animal	114
Paractus rapiformis	Anemone	N.I.	Residue	Whole animal	114
Metridium dianthus	Anemone	AEP	Lipid, residue	Whole animal	69
		AEP	Residue (protein)	Whole animal	67, 70,71
Actinia equina	Anemone	N.I.	Residue	Whole animal	114
Metridium senile	Anemone	AEP	Ceramide	Whole animal	53,95
		AEP	Lipid hydrolysate	Whole animal	101
		AEP	Glycoprotein	Whole animal	72,73,74,76,82,116
		AEP	Proteases	Gastric tissue	83,84
Tealia felina	Anemone	AEP	Lipid, residue	Whole animal	69,101
Leptogorgia virgulata	Soft coral	N.I.	Lipid, residue	Rind, axial rod	114
Zoanthus sociatus	Coral	PALA	Lipid hydrolysate	Whole animal	94,101
Renilla reniformis	Sea pansy	N.I.	Lipid, residue	Whole animal	114
Harenactis sp.	Anemone	AEP	Lipid hydrolysate	Whole animal	101
Acropora pulchra	Coral	AEP	Lipid hydrolysate	Whole animal	101
Stylophora pistillata	Coral	AEP	Lipid hydrolysate	Whole animal	101

Table 1 (continued)

Classification	Common name	Monomer	Biochemical fraction	Tissue	Ref.
Fungia fungites	Coral	AEP	Lipid hydrolysate	Whole animal	101
Favia specioso	Coral	AEP	Lipid hydrolysate	Whole animal	101
Tubipora musica	Coral	AEP, PALA	Lipid hydrolysate	Whole animal	101
Muriciopsis flavida	Coral	AEP, PALA	Lipid hydrolysate	Whole animal	101
Eucinea succinea	Coral	AEP, PALA	Lipid hydrolysate	Whole animal	101
Briarium asbestinum	Coral	AEP	Lipid hydrolysate	Whole animal	101
Microspicularia sp.	Coral	AEP, PALA	Lipid hydrolysate	Whole animal	101
B. Hydrozoa					
Physalia physalius	Portuguese Man-of-War	N.I.	Insoluble residue	Float, tentacles	114
C. Scyphozoa					
Dactylometra quinquecirrha	Jellyfish	N.I.	Lipid	Whole animal	114
Pelagia colorata	Jellyfish	AEP	Hydrolysate of lipid	Whole animal	101
III. Mollusca					
A. Pelecypoda					
Mytilus edulis	Blue mussel	AEP	Ethanol extract	Whole animal	69,117
Lucina radians	Scallop	N.I.	Lipid	Whole animal	114
Hinnites gigantum	Scallop	AEP	Ceramide	Whole animal	124
Dinocardium robustum	Cockle	N.I.	Lipid	Whole animal	136
Crassostrea virginica	Oyster	N.I.	Phospholipids	Adductor, whole animal	96,114
C. gigas	Oyster	AEP	Lipid	Whole animal	118
Ostrea gigas	Oyster	AEP	Ceramide	Adductor, gills, mantle, viscera	117,119,120
Oyster sp.	Oyster	AEP	Ceramide	Gill	55
Pinctada martensii	Oyster	AEP	Ceramide	Whole animal	55
Corbicula leana	Clam	AEP	Ceramide	Whole animal	55
C. sandai	Clam	*N,N*-acyl methyl-AEP and *N*-acyl-AEP	Ceramide	Whole animal	54,55
		AEP	Ceramide	Whole animal	121
C. japonica	Clam	AEP	Ceramide	Whole animal	121

Unio biwae	Clam	AEP	Ceramide	Whole animal	121
Inversidens hirase	Clam	AEP	Ceramide	Whole animal	121
Anodonta lauta (rostrata)	Clam	AEP	Ceramide	Whole animal	121
Cristaria plicata	Clam	AEP	Ceramide	Whole animal	121
Hyriopsis schlegelii	Clam	AEP	Ceramide	Whole animal	122
Spisula sp.	Clam	AEP	Protein	Hepatopancreas and gill	74
Venus mercenaria	Clam	AEP	Ethanol extract	Whole animal	69
Macrocallista nimbosa	Clam	N.I.	Lipid	Whole animal	136
B. Gastropoda					
Turbo cornutus	Turbo	*N*-Methyl-AEP	Ceramide	Viscera	56,57,58
		*N*ethyl-AEP, AEP	Cerebroside	Muscle and viscera	55,59,60,61,117
Tegula argyrostoma	Snail	AEP	Ceramide	Whole animal	55
T. lischkei	Snail	AEP, *N*-Methyl-AEP	Ceramide	Whole animal	117
T. pfeifferi	Snail	AEP	Ceramide	Whole animal	55
Monodonta labio	Snail	*N*-Methyl-AEP, AEP	Cerebroside (or ceramide)	Whole animal	47,55,62,117
Haliotis midae	Abalone	AEP	Lipid	Whole animal	123
H. corrugata	Abalone	AEP	Ceramide	Whole animal	124
Chlorostoma argirostoma	Abalone	*N*-Methyl-AEP	Cerebroside		62
Sinotaia histrica	Snail	AEP	Ceramide	Whole animal	55
Heterogen longispira	Snail	AEP	Ceramide	Whole animal	55
Purpura bronni	Snail	AEP	Ceramide	Whole animal	55
Busycon canaliculatum	Whelk	N.I.	Lipid and residue	Whole animal	69
Busycon sp.	Whelk	AEP	Protein (mucus)	Hypobranchial gland	74
Buccinum undatum	Whelk	AEP	Peptidoglycan	Hypobranchial gland	74
Aplysia kurodai	Sea Hare	AEP	Ceramide	Nervous tissue	66
Lehmannia poireiri	Slug	AEP	Free, lipid, residue	Whole animal	125
Conomurex luhuanus		AEP	Ceramide	Whole animal	117
Celluna eucosmia		AEP, *N*-Methyl-AEP	Ceramide	Whole animal	117
Limus flavus	Slug	AEP	Free, lipid, residue	Whole animal	125
Archidoris sp.	Nudibranch	AEP	Residue	Whole animal	114
Lymnaea stagnalis	Snail	AEP	Ceramide	Whole animal	126
Helisoma sp.	Snail	AEP	Phosphonoglycan	Egg masses, adult	85,86
Biomphalaria glabrata	Snail	AEP	N.I.	Egg masses	85
Helix aspersa	Snail	AEP	Free, lipid, residue	Whole animal	125
Helix lactea	Snail	AEP	Ceramide	Whole animal	127

Table 1 (continued)

Classification	Common name	Monomer	Biochemical fraction	Tissue	Ref.
C. Cephalopoda					
Loligo edulis	Squid	AEP	Ceramide	Whole animal	55
L. pealia	Squid	N.I.	Lipid	Whole animal	114
Polypus vulgaris		AEP	Ceramide	Whole animal	55
D. Loricata					
Liolophira japonica		AEP, *N*-Methyl-AEP	Ceramide	Whole animal	117
IV. Echinodermata					
A. Echinoidea					
Arbacia punctulata	Sea urchin	N.I.	Lipid	Whole animal	114
B. Asteroides					
Asterias forbesi	Starfish	N.I.	Lipid	Whole animal	69,114
V. Annelida					
A. *Chaetopterus variopedatus*	Segmented worm	N.I.	Lipid	Whole animal	114
VI. Arthropoda					
A. Crustacea					
Cyclograpsus punctatus	Salt-water crab	AEP	Phospholipid	Whole animal	128
Callinectes sapidus	Blue crab	N.I.	Lipid	Whole animal	114
Minippe mercenaria	Stone crab	N.I.	Lipid	Muscle and organs	114
Peneus duorarum	Shrimp	N.I.	Lipid	Muscle and organs	114
Euphausia superba	Antarctic krill	AEP	Lipid	Whole animal	129
VII. Porifera					
A. *Hymeniacidon heliophila*	Sponge	N.I.	Lipid and residue	Whole animal	114
VIII. Chordata					
A. Mammalia					
Homo sapien	Man	AEP	Lipid and nonlipid	Brain, liver, heart and skeletal muscle	130,131
		N,N,N-Trimethyl-AEP	Lipid	Aorta	43

Bos sp.	Cow	AEP, *N,N,N*-Trimethyl-AEP	Lipid	Bile	42
		AEP	Bile acid conjugate	Bile	132,133
		AEP	Lipid	Liver	40,41
		AEP	TCA extract	Brain	8
		N.I.	Lipid	Blood and sperm	134
Capra sp.	Goat	AEP	Free	Liver	135

Note: N.I. = not identified.

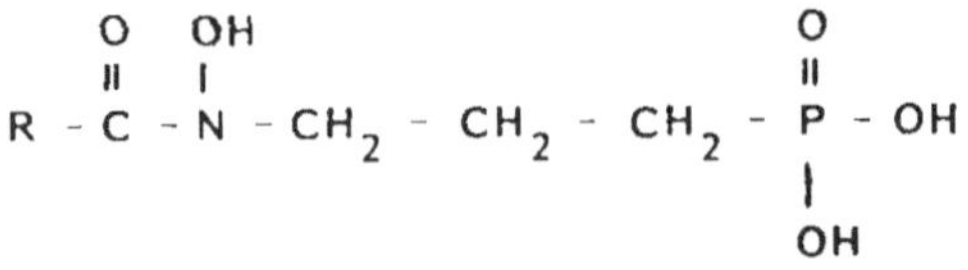

3 - (N - ACYL - N - HYDROXYAMINO) PROPYLPHOSPHONIC ACID

2. R = CH_3 (FR - 900098)

3. R = H (FR - 31564)

from *S. rubellomurinus* and the *N*-formyl congener synthesized.[27,28] The *N*-formyl (Structure 3) and the *N*-formyl dehydro (Structure 4) congeners were then found

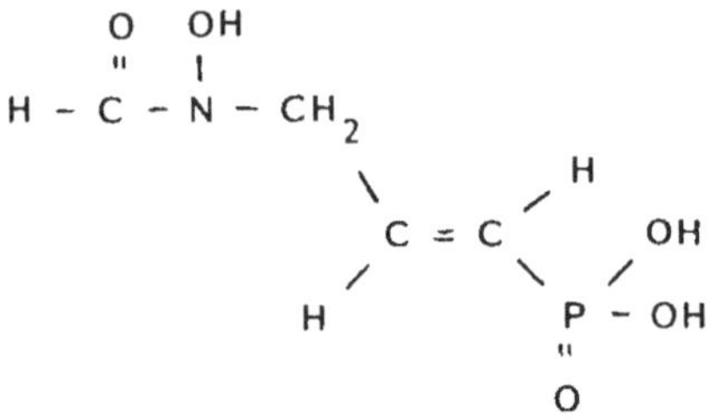

3 - (N - FORMYL - N - HYDROXYAMINO) - 1 - trans - PROPENYLPHOSPHONIC ACID

4. (FR - 32863)

to occur naturally in cultures of *S. lavendulae* and were found to have much stronger antibacterial activity than antibiotic 2.[29] Another antibiotic, 3-(*N*-acetyl-*N*-hydroxyamino)-2(R)-hydroxypropylphosphonic acid (Structure 5), was also isolated from *S. rubellomurinus* subsp. *indigoferis.*

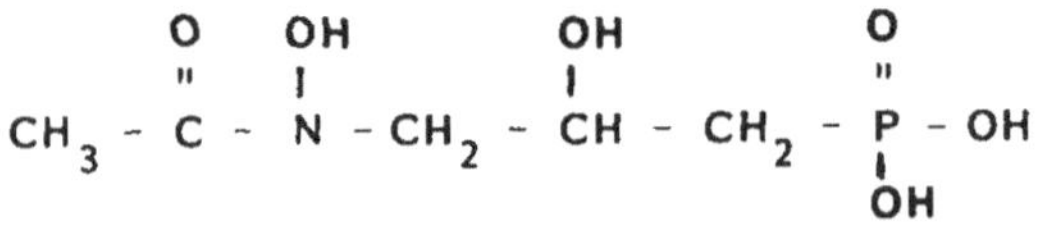

3 - (N - ACETYL - N - HYDROXYAMINO) - 2 (R)

- HYDROXYPROPYLPHOSPHONIC ACID

5. (FR - 33289)

This antibiotic had a potency similar to the initial antibiotic (Structure 2) in the series but was not as efficacious as were the *N*-formyl congeners.[29] The antibiotics in this series are similar in activity to fosfomycin; when resistance is observed to one there is generally cross-resistance to the other antibiotics. This resistance is apparently due to the loss of an active system to transport the antibiotic through the cell wall. The transport system involved may be that for L-α-glycerophosphate, although exclusion of other systems is not possible at this time.[30]

The 3-(*N*-formyl-*N*-hydroxyamino)propylphosphonate (Structure 3) exhibited the greatest antibacterial activity and was active toward most Gram-negative bacteria (including *Pseudomonas aeruginosa).*[31] *Serratia marcescens* and glucose-nonfermenting Gram-negative rods (except *P. aeruginosa)* were exceptions and were resistant to the antibiotic. This antibiotic was, in general, more effective than was fosfomycin, although the antibiotics differ in antibacterial spectrum. Antibacterial activity is closely related to transport as mentioned above.[30] In addition, transport of the *N*-formyl antibiotic is more rapid than for fosfomycin in organisms that are more susceptible to it than to the fosfomycin. In vivo activity was also greater for the *N*-formyl congener than for fosfomycin.[31]

One other phosphonate compound that is effective as an antibacterial agent is L-1-aminoethylphosphonic acid (Structure 6). This synthetic material is

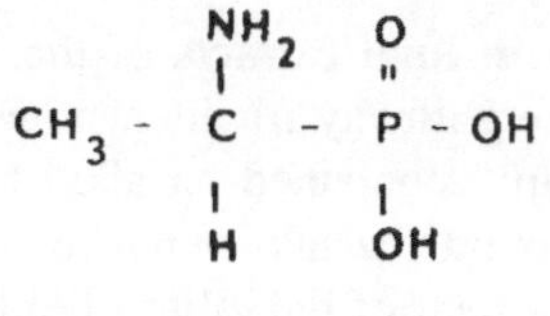

L - 1 - AMINOETHYLPHOSPHONIC ACID

6.

administered as a dipeptide (Alafosfalin) with alanine. This compound and other synthetic phosphonate antibiotics are discussed in Chapter 6.

B. Glycerophosphonolipids

A group of phosphonolipids that are analogs of glycerophospholipids has been identified. AEP was found to be bound to glycerol in lipids from *A. elegantissima.*[6] Although detailed structure proofing was not carried out, glycerol and AEP occurred in equimolar amounts in the hydrolysate of a lipid fraction isolated by ion exchange chromatography; the lipid was presumed to be the glycerol ester of AEP. Kittredge et al.[33] isolated and characterized 2-*N*-methyl-AEP and 2-*N,N,N*-trimethyl-AEP (Structure 7)

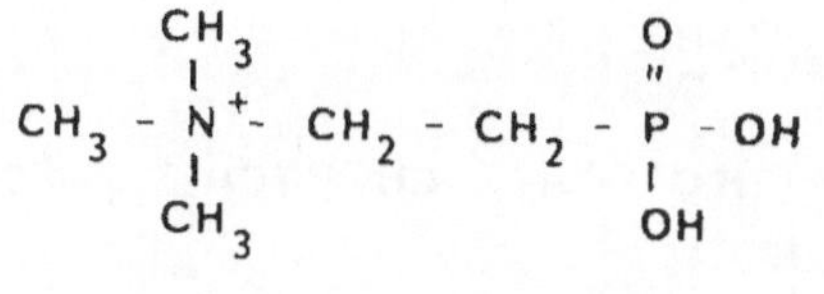

2 - (N,N,N - TRIMETHYL) AMINOETHYLPHOSPHONIC ACID

7.

from *A. xanthogrammica*. These derivatives were then identified in phospholipid fractions other than sphingolipids, which would presumably indicate their presence as glycerophosphonolipids.[34]

Liang and Rosenberg[35] identified diacylglyceryl-AEP (Structure 8), the AEP

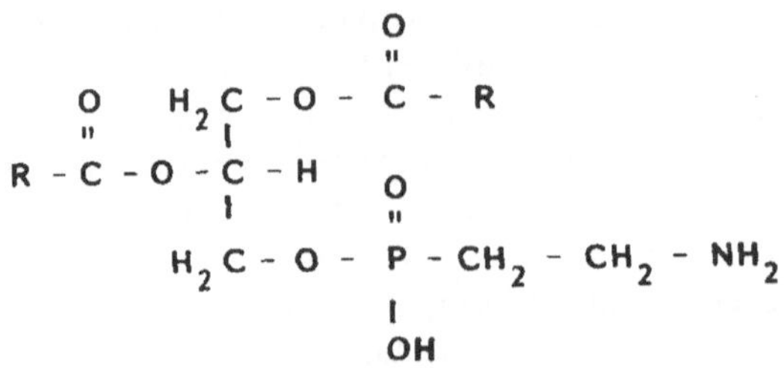

DIACYLGLYCERYL - AEP

8.

analog of phosphatidylethanolamine, in lipid extracts of the protozoan *Tetrahymena pyriformis* and presented evidence for the pathway for its synthesis. Thompson[36] found that a large percentage of this phosphonolipid contained an alkyl-ether bond at carbon 1 of the glycerol and Dawson and Kemp[37] found the alk-1-enyl analog in rumen protozoa. Later work has shown that the major component of the ciliary lipids of *T. pyriformis* is glycerophosphonolipids. These phosphonate-containing analogs seem to be enriched in ciliary lipids over the remainder of the protozoal lipids.[38,39] Since that time diacylglyceryl-AEP and diacylglyceryl-*N,N,N*-trimethyl-AEP, which is the phosphonate analog of phosphatidylcholine, have been isolated from bovine liver[40,41] and bile.[42] Also, Alam and Bishop[43] detected the phosphonate analog of choline, but not AEP, in human aortic tissue.

Synthetic analogs of phosphatidylethanolamine[44,45] and phosphatidylcholine[46] were available before the natural materials were actually isolated and thus were helpful in the characterization of the natural phosphonolipids. The chemical characteristics of the phosphate and phosphonate analogs are so similar that only specific chromatographic procedures[47-50] allow adequate separation for unequivocal identification.[38]

C. Sphingophosphonolipids

Ceramide-AEP was first identified as a component of the lipids of the sea anemone *A. elegantissima* by Rouser et al.[51] Simon and Rouser[52] then elucidated the structure (9) of the ceramide-AEP and found the

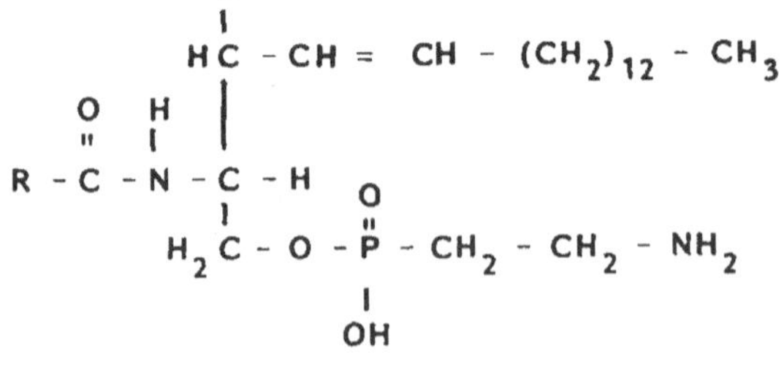

CERAMIDE - AEP

9.

base to be sphingosine with an AEP esterified at the first hydroxyl and an *N*-acyl group at the second carbon. The fatty acids were predominantly saturated and had chain lengths from 14 to 18 carbons. Longer chain fatty acids usually found in animal ceramides were absent, as was sphingomyelin.[52]

Ceramide-AEP has been identified in a variety of other species (see tables). These ceramides contain sphingosine or closely related derivatives as the long-chain bases. The bases of ceramide from the anemone *Metridium senile* were found to be D-*erythro*-1,3-dihydroxy-2-amino-*trans*-4,*trans*-8-octadecadiene (95%) (Structure 10) and sphingosine (5%).[53]

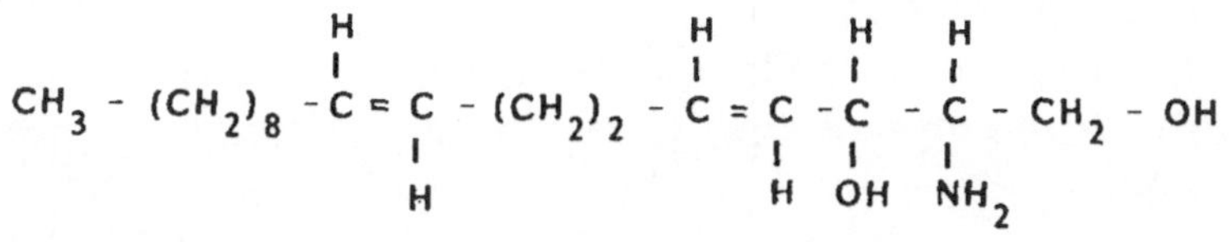

D - erythro - 1,3 - DIHYDROXY - 2 - AMINO - trans - 4, trans -8- OCTADECADIENE

10.

Again the fatty acids were found to be moderate in length (14 to 18 carbons) with hexadecanoic acid making up 52% of the fatty acid fraction. The fatty acid composition was unusual in that branched-chain acids made up 35% of the fatty acids; branched-chain acids or bases are only rarely found as constituents of ceramide.[53]

One ceramide-AEP has been isolated that has an unusual structure and is of special interest. The amino group of the AEP apparently exists as *N,N*-acylmethyl-AEP (Structure 11) or *N*-acyl-AEP (Structure 12).[54] These

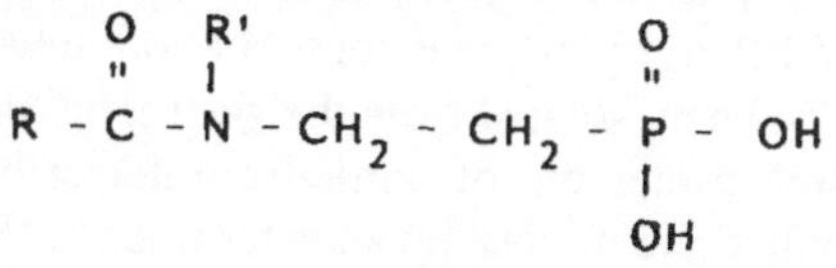

11. R' = CH_3 N-ACYL - N - METHYL - AEP

12. R' = H N-ACYL - AEP

structures provide a precedent for the occurrence of *N*-acyl-AEP in macromolecular materials (see Section II.D).

Sphingophosphonolipids from *Turbo cornutus* have been extensively characterized. As opposed to other snails which contain both ceramide aminoethylphosphate and ceramide-AEP, the turbo does not contain the phosphate derivative.[55] Hayashi et al.[56] first identified ceramide-*N*-methyl-AEP (ceramide-MAEP) in lipids from the viscera of *T. cornutus* and found two molecular species types in the long-chain base and fatty acid composition.[57] The first type contains predominantly *N*-hexadecanoic acids and C_{16}-, C_{17}-, C_{18}-, and branched chain C_{18}-sphingosine, along with C_{18}- and C_{22}-sphingodienine.[57] The second type contains predominantly a 2-hydroxy derivative of hexadecanoic acid and C_{18}- and C_{19}-phytosphingosine (Structure 13) and C_{22}-dehydrophytosphingosine.[58] The phytosphingosine derivatives

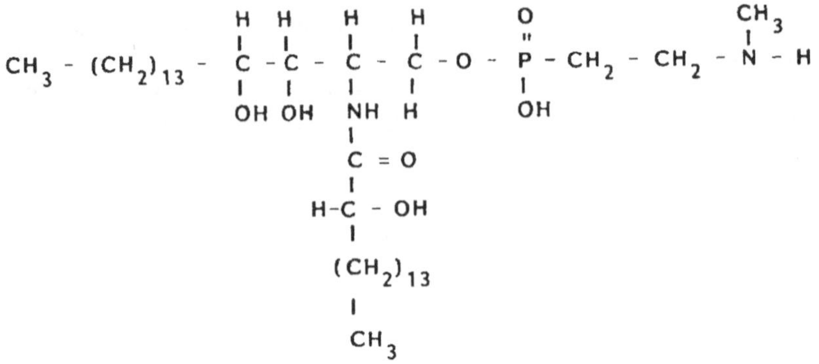

2 - HYDROXYHEXADECANOYL C_{18}- PHYTOSPHINGOSINE
(with 2 -N- methyl - AEP)

13.

contain two additional hydroxy groups per molecule than do the ceramides in the first group. It is interesting that the hydroxy fatty acids are in combination with phytosphingosine rather than sphingosine.

More complex sphingolipids, containing both a phosphonate and a carbohydrate moiety, have also been characterized. Cerebrosides containing *N*-methyl-AEP (or AEP), palmitic (or 2-hydroxypalmitic) acid, galactose, and a long-chain base have been isolated from viscera and muscle of *T. cornutus*.[59-61] The bases were predominately mono- or diunsaturated, dihydroxy bases 18 or 22 carbons long. The AEP derivative of these cerebrosides was found only in muscle, while the *N*-methyl-AEP derivative was found in both muscle and viscera.[60] Cerebroside-AEP compounds have also been isolated from the gastropod *Monodonta labio*.[62] These phosphonyl derivatives of cerebrosides are of particular interest because the alkylphosphonic acid is bound to the carbohydrate moiety by an ester linkage (Structure 14) (See Section II.D)[60-62] and may establish a precedent

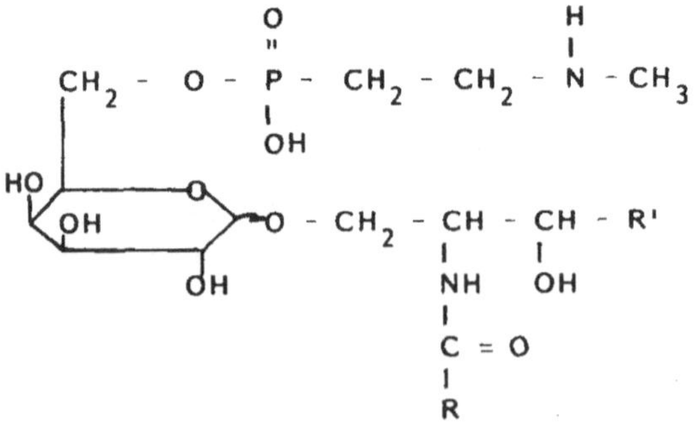

1-O- [6' - O - (N - METHYLAMINOETHYLPHOSPHONYL) GALACTOPYRANOSYL]

CERAMIDE

14.

for the linkage of AEP to macromolecular materials.

The cerebrosides from the muscle of *T. cornutus* contain primarily *N*-methyl-AEP, with some AEP, and the cerebrosides from the viscera contain only the *N*-methyl-AEP derivative; the long-chain base and fatty acid composition is similar in the cerebrosides of both muscle and viscera.[60,61] The visceral cerebrosides contain more hydroxylated fatty acids than do the muscle cerebrosides. A notable feature is that a molecular species of the cerebrosides from viscera contains normal (dihydroxy) long-chain bases and hydroxylated fatty acids. This does not occur in the ceramide-*N*-methyl-AEP and is a characteristic in common with galactosylceramide and may show a common biosynthetic pathway for the phosphonyl cerebroside and the galactosylceramide.[61]

Sphingophosphonolipids also occur in rumen protozoa and have been found in two parasitic microorganisms, *Bdellovibrio bacteriovorus*[63] and *Pythium prolatum*.[64]

The phosphonate incorporated into the cerebrosides of *M. labio* is mainly the *N*-methyl derivative, although the shellfish contains both AEP and *N*-methyl-AEP, indicating that there is some specificity for the incorporation of the alkylphosphonate.[62] The significance of the presence of the alkylphosphonate to the molecular structure differences found in these sphingolipids has yet to be determined. The incorporation of the specific base may be the organism's response to the availability of fatty acids for base synthesis.[53,63] In turn, the molecular structure of the fatty acids may be related to environmental considerations such as the ambient temperature of the sea water.[53,65] Watanabe et al.[65] reported that the fluidity of phosphonolipids increased within 10 hr in *T. pyriformis* organisms undergoing a temperature decrease and the degree of unsaturation of the fatty acid component increased. They proposed that the ratio of diacyl- to alkyl-acyl-phosphatidylcholine may be involved in thermal adaptation by controlling the available fatty acid acceptor sites.[65]

Simon and Rouser[52] proposed that, in the species above, the ceramide-AEP or derivatives of ceramide-AEP substitute for the sphingomyelin present in most nervous tissue. Further work is consistent with this possibility. Komai et al.[66] found that ceramide-AEP occurred in the nervous tissue (ganglion and fiber) of *Aplysia kurodai* and that sphingomyelin was absent. Cerebrosides were not detected and the ceramide-AEP from liver had a different fatty acid composition than did the ceramide-AEP from nervous tissue in that organism.[66]

D. Aminophosphonates in Macromolecular Materials

AEP was first identified as a component of proteinaceous material following its isolation from a residue of the sea anemone *Metridium dianthus*. Following removal of lipid and complete hydrolysis, the AEP was purified by ion exchange chromatography and found to make up 1.1% of the insoluble material. Following partial hydrolysis, AEP was found in material precipitated by trichloroacetic acid (TCA), providing evidence that the AEP was truly associated with protein.[67] Simultaneously, Rosenberg[68] noted that nonlipid residual material from *T. pyriformis* contained AEP and suspected that it made up a structural component of the cell. In addition, he noted that the residue was resistant to proteolysis by Pronase and trypsin, implying that an unusual protein structure was involved. Further studies on *M. dianthus* indicated that the AEP was associated with protein and that a peptide containing 4.4% AEP could be solubilized by pepsin and separated by gel filtration chromatography.[69]

Hilderbrand et al.[70] found that a proteinaceous extract from *M. dianthus* residue, following exhaustive lipid extraction, contained 50% of its phosphorus in the phosphonate form. Gel filtration chromatography on Sephadex® G-25 removed most of the orthophosphate; the phosphonate to orthophosphate ratio increased to 20:1. Initial characterization of the material[71] showed that AEP is associated with a complex biological polymer. Further purification by ion exchange chromatography and gel filtration provided a high molecular weight fraction made up of 22% neutral sugars, 4% hexosamines, and 40% amino acids with AEP making up 28% of the total ninhydrin positive material as determined by amino acid analysis. Only phosphonate-phosphorus (as AEP) was detectable, and glucosamine and galactosamine made up 10% of the ninhydrin reactive material. The remainder of the material was unidentified.

The isoelectric point of the phosphonoprotein was 3.7. This low value could not be accounted for by the presence of aspartic or glutamic acid, since the first pKs for these acids are 3.8 and 4.2, respectively. The presence of the phosphonate with at least one unesterified position with a pK_1 of 2.1 could account for the low isoelectric point.[71] This observation is not consistent with a significant quantity of AEP being present as a polyphosphonate type structure, since this would be an anhydride and the material would not have such a low pI. Anomalous behavior prevented accurate molecular weight determination by ultracentrifugation or gel filtration, with the molecular weight estimated to be in the midrange of 1.5×10^5 to 1.5×10^6 daltons.[71] Data from gel filtration chromatography of the phosphonoprotein suggested that the Stokes radius was much greater than the molecular weight indicated that it should be.[71,72] During ultracentrifugation the molecule behaved as if it were smaller in molecular weight than was expected from the gel filtration data. These observations are consistent with and indicate that the macromolecule is very widely dispersed as a result of a high density of negative charge. This phenomenon has been seen with other proteins such as sulfated proteins and acidic glycoproteins.[71] Since the usual acidic components of other glycoproteins (sialic and glucuronic acid, and sulfated sugars) are absent, the AEP may be providing the acidic function necessary to maintain viscoelastic structural properties.[71,72]

Phosphonate containing macromolecules have been found in other species as well. Kirkpatrick and Bishop[72,73] isolated a similar type of material from *M. senile*. The material from *M. senile* and *A. xanthogrammica* contained AEP (0.8 μmol/mg) and hexosamine (0.36 μmol/mg), was made up of a least two different components, and had an apparent molecular weight of 3.0×10^5 daltons. Phosphonoprotein isolated from *A. xanthogrammica* contained *N*-methyl-AEP (0.03 μmol/mg) and AEP (0.37 μmol/mg) and had a molecular weight of 3.0×10^5 daltons.[72] Bunde et al.[74] have isolated proteins containing AEP from other marine organisms and have isolated a highly sulfated polysaccharide containing AEP from hypobranchial gland mucous of *Buccinum undatum*. In addition, Lepak and Smith[75] recently isolated a phosphonoglycoprotein, from *T. pyriformis*, which contains about 8% AEP and 32% carbohydrate. Anomalous behavior of this material on gel exclusion column chromatography was observed and was similar to that of the above mentioned phosphonoproteins.

The AEP has been assumed to be covalently linked to the phosphonoproteins because rigorous conditions have failed to separate it from the other constituents. The structure of that linkage has not been determined and is the primary question concerning these unusual biological molecules. 1-Fluoro-2,4-dinitrobenzene treatment failed to produce the 2,4-dinitrophenyl-AEP derivative which would be formed with a primary amino group on the AEP.[5,71] Furthermore, Kirkpatrick and Bishop[72] could not form a dansyl chloride derivative without acid hydrolysis of the protein. Thus, the amino group of the AEP may be in a peptide linkage or *N*-acylated. The absence of fatty acid in the phosphonoprotein indicates that the amino group is not present as a long-chain acyl derivative, although an acetyl derivative is possible since acetic acid would not have been detected in the fatty acid assay.[72] A precedent exists for this structure with AEP since Hori and Arakawa[54] have identified an acylated ceramide-AEP.

A possible relationship between concentrations of neutral carbohydrate and AEP and between hexosamine and AEP has been noted.[71,72] Hurley et al.[76] have isolated a high molecular weight glycoprotein from *M. senile* which contains 66% amino acids, 30% carbohydrate, and 220 residues of AEP per 1000 amino acid residues. The partial hydrolysis of the phosphonoprotein using 0.5 *M* alkali produced no free AEP; the bound AEP then chromatographed with the hexosamine containing oligosaccharides rather than with peptidic material. This alkaline hydrolysis should destroy any serine or threonine residues and the corresponding *O*-glycosidic or *O*-phosphoryl linkages. Thus the AEP would appear to be attached to the oligosaccharide side chain.[76] In support of this observation is the finding that no appropriate amino acid (or acids) exists in sufficient quantity in the glycoprotein to allow AEP to be linked to the protein as a monoester.[76]

Phosphonimide bonding of the AEP into the macromolecule is not likely because the nitrogen to phosphorus bond is relatively labile to hydrolysis by alkali. The presence of a phosphonimide would have resulted in hydrolysis and free AEP rather than the products that were observed upon partial hydrolysis. In addition, a phosphonimide bond would be readily detectable by ^{31}P-NMR and was not seen in available spectra of the phosphonate material.[71] Thus, it is possible that the AEP is attached as a monoester to a carbohydrate and the amino group of the AEP is acetylated. The recent isolation of a phosphonomonoesterase from *M. senile* indicates that the enzymatic pathways may exist for the inclusion of AEP as a monoester. The biological substrate for this enzyme is not definitely known but AEP was a competitive inhibitor for the hydrolysis of 4-nitrophenyl(phenyl) phosphonic acid in vitro.[77] This is the type of phosphonomonoester linkage reported in a sphingoglycolipid by Matsuura.[62]

Korn et al.[78,79] purified to apparent homogeneity a lipophosphonoglycan from the soil amoeba *Acanthamoeba castellanii*. This structural material contains both AEP and 1-hydroxy-2-aminoethylphosphonic acid and makes up about 30% of the plasma membrane of the amoeba. The mass of the isolated material is neutral sugars (26%), hexosamine (3.3%), aminophosphonates (10%), phosphate (3.2%), fatty acids (14%),[79] inositol (8%), C_{24}- and C_{25}-phytosphingosines (13%), and unknown (22%).[80] Of the 14% fatty acids, 8.4% was made up of C-22 to C-28 normal and branched, saturated 2-hydroxy fatty acids.[81] This material is distinguished from the previously mentioned phosphonoprotein by the presence of fatty acids.

A working model for the structure of the two components of the lipophosphonoglycan has been proposed by Dearborn et al.[80] on the basis of the composition and the constituents which fractionate together following partial hydrolysis. The fatty acids and phytosphingosine bases are proposed to be in ceramides, and the ceramides linked through acid hydrolyzable phosphate to inositol. The aminophosphonates may be involved in the linkage of the lipids, perhaps through inositol, and galactosamine may also be linked to inositol. One segment of both components may be an oligomeric lipid containing three inositol moieties and six aminophosphonates, with the sugar constituents varying for each component of the lipophosphonoglycan.[80]

Korn et al.[79] have proposed that the materials from *M. dianthus* contain undetected fatty acids; however, a second attempt to identify fatty acids in the phosphonoproteins gave negative results.[76] There are several other differences besides composition between the phosphonoglycoprotein from *M. dianthus* and the lipophosphonoglycan from *A. castellanii*. The lipophosphonoglycan has two homogenous components and a lower molecular weight than the phosphonoprotein. In addition, the lipophosphonoglycan will bind to concanavalin A and precipitate while the phosphonoprotein will not.[76]

In addition to being present in sphingophosphonolipids and macromolecular material of *M. senile*, AEP has been found in an acid mucopolysaccharide-like material which covers the entire body and tentacles of that anemone.[82] This material may be responsible for the anthozoans feeding by trapping suspended marine organisms in a mucus on their body surface and tentacles and transporting it to their stomadium.[82] In addition, the material may function in the intracellular matrix. The apparent lack of sulfated sugars, uronic acids, and sialic acid in the mucus may be accounted for by the presence of AEP in a monoester linkage.[82] The mucus is probably produced by columnar epithelial cells of the epidermis. Of interest here is the apparent absence of any phosphonoprotein in the collagen from the mesoglea of the same organism.[82]

Bunde et al.,[82] using histochemical techniques, also observed an acid mucopolysaccharide-like material in the gastrodermal portion of *M. senile*. This observation is consistent with those of Gibson and Dixon[83] and Stevenson et al.[84] that the AEP is located in two proteolytic enzymes from the digestive tract of *M. senile*. Small amounts of AEP and hexosamine are present in these two proteases with chymotryptic-like activity.

Miceli et al.[85,86] have published information on the function of phosphonate-phosphorus in the pulmonate snail *Helisoma* sp. Phosphorus-31 nuclear magnetic resonance studies on this snail have shown that in freshly laid egg masses essentially all phosphorus is present as phosphonate-phosphorus and 85% of the phosphonate-phosphorus is AEP. The phosphonate is not present in lipids, but is present in high molecular weight molecules consisting mainly of carbohydrate. During embryonic development, the phosphonate-phosphorus was gradually converted to hydrolyzable phosphate with no increase in total phosphorus. When the eggs hatched the infant snails contained mostly hydrolyzable phosphorus and 87% of the phosphorus remaining in the empty cases was phosphonate-phosphorus in the form of AEP.[85] When eggs were laid following administration of [^{33}P]-H_3PO_4 to adult snails, the eggs contained most of the ^{33}P in AEP and an unidentified phosphonate. During embryonic development, the [^{33}P]-AEP was hydrolyzed to [^{33}P]-phosphate. Thus, in *Helisoma* sp. and other planorbid snails, the adults can synthesize AEP and other phosphonates from H_3PO_4, transfer the C–P by way of the egg masses, and the embryonic snails can break the C–P bond.[85] Phosphonate-phosphorus obviously plays an integral role in the development of the snail and provides an excellent model for metabolic studies. It is possible that the rate of hydrolysis of AEP is a controlling factor in the developmental processes. This work may lead to a mechanism to control snail proliferation, especially in light of the importance of the planorbid snail *Biomphalaria glabrata* as a host for the human pathogenic trematode *Schistosoma mansoni*.[85]

III. DISCUSSION

A. The Role of Naturally Occurring Phosphonates

Since there is no documented role for free AEP, the role of the aminophosphonates in vivo consists of two interrelated aspects: first, the role of the aminophosphonates in the structure and/or function of the molecule into which it is incorporated and, second, the role of that macromolecule in the structure and/or function of the organism in which it is found.

The first aspect must be addressed by examining the aminophosphonates and the molecules into which they are incorporated naturally. The molecular properties involved are discussed in detail in Chapter 5 and will only briefly be mentioned here; however, they include effects on the pI, solubility, and steric relationships. Secondary effects such as changes in structural conformation of macromolecules may result from the primary effects. All of these factors have the capability of imposing metabolic or functional restrictions on the phosphonate-containing molecules or of imparting a unique function to the materials.

Many phosphonate-containing molecules have been observed to be resistant to enzymatic hydrolysis (see Chapter 5, Section IV.C).[39,70,72,87] This property may be observed only at the C–P bond or at other parts of the molecule as well. An example of this is the resistance of ceramide-AEP to hydrolysis by phospholipase C from *Clostridium welchii*,[62] an enzyme that ordinarily cleaves the phosphate ester bond between ethanolaminephosphate and sphingosine. Rosenthal and Pousada[88] have investigated the effects of phosphonates on phospholipases. The enzymes most effectively inhibited by phosphonates are those that would hydrolyze an ester at the location of the C–P bond. Although others have found that phospholipase C will hydrolyze the phosphate ester bond,[53,58] there is a difference in the rate of hydrolysis which may account for the discrepancy in the results. Rosenthal and Pousada[88] found that the phosphonate analog of lecithin, with the C–P bond in the nitrogenous moiety, was a weak inhibitor of phospholipase C from *C. perfringens*. Thompson[49] has also noted that the presence of incorporated AEP prevents the degradation of phosphonolipids by phospholipases. Thus, the presence of a C–P bond apparently can impart resistance to a variety of hydrolytic enzymes.

In addition to the resistance to enzymatic hydrolysis, there are some structural considerations of importance. The presence of an alkyl ether in the glycerophosphonolipids imparts an inert character to the molecule but does not disturb the normal lipid interaction (see Chapter 5, Section V). The presence of AEP in the spingosine may provide a molecule of even more inert character in the hydrocarbon backbone than the glycerophosphonolipids, but which display the normal physico-chemical behavior of nonphosphonate lipids in membranes. One might speculate that if the C–P bond provides resistance to oxidation, a class of compounds may have developed that not only contains the C–P bond but a linkage of C–C–C of the ceramides rather than C–O–C of the plasmalogens.

Of interest here also, is the work on molecular species differences in the phosphonolipids from the shellfish *T. cornutus* and *M. labio* (Section II.C). Fatty acids with a hydroxyl at the 2 position have been found in a number of sphingophosphonolipids. These fatty acids make ceramides resistant to enzymatic hydrolysis by bacterial enzymes.[89] Thus, the phosphonate containing molecules are apparently inert and resistant to hydrolysis not only because of the C–P bond but because of other components of the molecules as well. (Other biological materials from marine organisms have been found to have unusual biological and metabolic properties.[90-93])

The second aspect of the role which the phosphonate molecule plays in the organism is directly related to the properties of the molecule itself. The inert character of the phosphonate-containing molecules may be used to generate membranes or structural components which are resistant to oxidation, hydrolysis, or other metabolic processes. Whether these resistant structures are serendipitous or if there is an actual biological requirement is not known. The specificity of incorporation of *N*-methyl-AEP into sphingolipids noted by Matsuura[62] is indicative of either a special function or a unique metabolic pathway for incorporation into that material. The nonrandom incorporation of AEP into anatomic structures as seen in *T. pyriformis*,[38,39] M. senile,[72] M. dianthus,[71] and other organisms is also indicative of a unique function. A number of authors have made proposals concerning the physiological function of the phosphonate molecules. Phosphonates have been proposed to provide buffering or cationic binding capacity.[94] An extension of this thought has been offered by Mason[95] who found the membrane of the sea anemone to be relatively impermeable to elemental nutrients. He proposed that the highly ionic lipid ceramide-AEP plays a direct role in the facilitated transport of small ions from the aqueous environment into the anemone.

Rosenberg[68] proposed that in *T. pyriformis,* the AEP was incorporated into structural material and conferred resistance to enzymatic hydrolysis on the cell membrane. This AEP containing material is apparently not metabolized further, is necessary for structural integrity, and is synthesized only during cell growth. This is consistent with the enrichment of phosphonates in ciliary lipids noted by Smith et al.[38] and Kennedy and Thompson.[39]

A similar observation was made from work with the oyster *Crassostrea virginica* by Swift.[96] Phosphonolipids were conserved at the expense of phosphodiester bonds in starved oysters. Thus, phosphonates may be a storage mechanism for phosphorus in a phosphorus deficient environment. This phosphonate-phosphorus may then be utilized as required for the structural integrity of membranes,[39,87] since it is apparently not available metabolically. (How this relates to the recent finding in the planorbid snail that phosphonate-phosphorus can be utilized by the embryonic snail is not known [see Section II.D]. In the case of the planorbid snail [Section II.D] there is some residual phosphonate following hatching that may be structural material; however, most of the phosphonate present in the early embryonic stages seems to be available for metabolic processes.)

It is apparent that structures and functions involving phosphonates are intimately related and that much of the above discussion presents hypothesis based on available data. The aminophosphonates have unique biochemical properties and impart unusual characteristics to macromolecules. From the data on the resistance of phosphonate-containing molecules to enzymatic hydrolysis that have accumulated, to this time, the proposal that phosphonates confer an important enzymatic resistance to membrane seems to be highly credible.

Another potential role for the unique structures of the aminophosphonates is in interspecies (or intraspecies) communication or recognition processes. An example of this type of communication is given in a recent report[97] of a survival enhancing relationship between the sea anemone, *A. xanthogrammica,* and the mussel, *Mytilus californianus*. The anemone reproduces by epidemic spawning of gametes which develop into feeding (planktotrophic) larvae. After several weeks the larvae locate, recognize, and settle into the bed of *M. californianus,* which acts as a refuge and nursery. As the juvenile anemones grow, they migrate downward and establish a position to capture dislodged mussels and grow to adult size. Thus, there is a unique recognition pattern for the anemone larvae for the mussel beds. *M. californianus* has not been reported specifically to have metabolic capability to synthesize the C–P bond; however, another mussel, *M. edulis,* does have that capability and opens the possibility that *M. californianus* is the same.

B. The Metabolic Origin of the C–P Bond

Since organisms which are at the base of the food chain have been found to synthesize the C–P bond, questions arise as to whether organisms higher in the food chain are synthesizing phosphonates, if they are incorporating exogenous phosphonates or if a contaminating microorganism is present (as in the case of zooxanethellae and *A. elegantissima*).[6] In the case of mammals, there is little evidence to support any endogenous synthesis of phosphonates. The reported occurrences apparently result from dietary intake of dairy products, beef, or bivalves or from the synthesis by rumen protozoa followed by assimilation of phosphonate from the exogenous source. In cases where an organism exists in a symbiotic relationship with other organisms (especially microorganisms) the determination of natural occurrence may be difficult. The determination must be based on quantities present, rates of particular metabolic processes, or presence in uncontaminated tissues or subcellular fractions.

In other cases both the food source organism and the consuming organism have the capability to synthesize the C–P bond and utilize the phosphonate products. Both plankton[98] and a species of mussel[69] are known to synthesize phosphonates and the mussel is a filter feeder which consumes plankton[98] for its own survival. If one would care to add in the potential use of mussels as a food source for the anemone *A. xanthogrammica,*[97] there are three tiers in the food chain with a common metabolic capability. The role of the food source in the development of the metabolic capability is not known.

IV. SUMMARY

In this chapter, the distribution and molecular structures of naturally occurring phosphonates have been examined. The primary natural molecule which contains the C–P bond is AEP which is incorporated into lipid and proteinaceous residues. The chemical properties of the monomeric aminophosphonates impart unique characteristics to molecules and membranes into which they are incorporated. The C–P bond is known to occur only in a relatively few species. However, when it does occur, it may involve a significant fraction of the phosphorus of those species. This indicates a possible functional requirement for the phosphonate-phosphorus rather than the phosphonates just being end products of metabolism. Possible functional roles include the ability to provide cationic buffering capacity or facilitated transport of essential ions, the ability to provide resistance to hydrolytic enzymes, and the ability to provide inter- or intraspecies communication or recognition processes. The true distribution of the phosphonates and the molecular structure of the proteinaceous phosphonates remains to be determined as does an explanation for their existence. Several reviews on naturally occurring phosphonates are available.[5,87,89,99,100,101]

REFERENCES

1. **Chavane, M. V.,** Synthese de quelques acides phosphoniques amines de formule generale H_3N^+-$(CH_2)_n$-PO_3^-H, *Compt. Rend.*, 224, 406, 1947.
2. **Horiguchi, M. and Kandatsu, M.,** Isolation of 2-aminoethane phosphonic acid from rumen protozoa, *Nature (London)*, 184, 901, 1959.
3. **Horiguchi, M. and Kandatsu, M.,** Ciliatine: a new aminophosphonic acid contained in rumen Ciliate protozoa, *Bull. Agric. Chem. Soc. Jpn.*, 24, 565, 1960.
4. **Storer, T. I., Usinger, R. L., Stebbens, R. C., and Nybakken, J. W.,** *General Zoology*, 6th ed., McGraw-Hill, New York, 1979, 303.
5. **Quin, L. D.,** The natural occurrence of compounds with the carbon-phosphorus bond, in *Topics in Phosphorus Chemistry*, Vol. 4, Grayson, M. and Griffith, E., Eds., Wiley-Interscience, New York, 1967, 23.
6. **Kittredge, J. S., Roberts, E., and Simonsen, D. G.,** The occurrence of free 2-aminoethylphosphonic acid in the sea anemone, *Anthopleura elegantissima*, *Biochemistry*, 1, 624, 1962.
7. **Barnes, R. D.,** *Invertebrate Zoology*, 4th ed., Sanders College, Philadelphia, 1980.
8. **Shimizu, H., Kakimoto, Y., Nakajima, T., Kanazawa, A., and Sano, I.,** Isolation and identification of 2-aminoethylphosphonic acid from bovine brain, *Nature (London)*, 207, 1197, 1965.
9. **Hendlin, D., Stapley, E. O., Jackson, M., Wallick, H., Miller, A. K., Wolf, F. J., Miller, T. W., Chaiet, L., Kahan, F. M., Foltz, E. L., Woodruff, H. B., Mata, J. M., Hernandez, S., and Mochales, S.,** Phosphonomycin, a new antibiotic produced by strains of streptomyces, *Science*, 166, 122, 1969.
10. **Lebere, A., Marre, R., and Sack, K.,** Fosfomycin: animal experiments on nephrotoxicity, pharmacokinetics and therapeutic efficacy (author's transl.), *Immun. Infekt.*, 8, 101, 1980.
11. **Koeda, T., Moriguchi, M., and Hata, T.,** Effect of fosfomycin-calcium on reproductive performance of rats. III. Peri- and post-natal examination (author's transl.), *Jpn. J. Antibiot.*, 33, 733, 1980.
12. **Koeda, T., Moriguchi, M., and Hata, T.,** Effect of fosfomycin-calcium on reproductive performance of rats. II. Fertility test (author's transl.), *Jpn. J. Antibiot.*, 33, 613, 1980.
13. **Viotte, G., Morin, J. P., Bendirdjian, J. P., Godin, M., and Fillastre, J. P.,** Functional, histological, ultrastructural and biochemical study of rat kidney treated with fosfomycin and gentamycin, administered separately or together, *C. R. Soc. Biol.*, 173, 1152, 1980.
14. **Koeda, T., Odaki, M., Sasaki, H., Yokota, M., Watanabe, H., Kawaoto, H., Ito, T., and Ishiwatari, N.,** Toxicological studies on fosfomycin sodium. I. Acute toxicity in mice and rats, *Jpn. J. Antibiot.*, 32, 61, 1979.
15. **Ceruti, T., Biazzi, E., D'Auossa, I., Pennacchio, A., and Santini, F.,** *In vitro* activity of fosfomycin, *Arch. Sci. Med.*, 137, 221, 1980.
16. **Pessione, E., Martinetto, P., Gaffodio, A. M., Angeretti, A., and Biunta, C.,** Influence of culture media composition on the interpretation of susceptibility tests to fosfomycin, *G. Ital. Chemioter.*, 26, 195, 1979.
17. **Mendoza, C., Garcia, J. M., Claneza, J., Mendez, F. J., Hardisson, C., and Ortiz, J. M.,** Plasmid-determined resistance to fosfomycin in *Serratia marcesens*, *Antimicrob. Agents Chemother.*, 18, 215, 1980.
18. **Geveudan, M. J., Mallet, M. N., Charrel, J., Palmas, N., Gulian, C., and Gevaudan, P.,** Antibacterial kinetics of fosfomycin (author's transl.), *Ann. Biol. Clin.*, 37, 327, 1979.
19. **Ullmann, U. and Lindemann, B.,** *In vitro* investigations on the action fosfomycin alone and in combination with other antibiotics on *Pseudomonas aeruginosa* and *Serratia marcescens*, *Arzneim. Forsch.*, 30, 1247, 1980.
20. **Lindenschmidt, E. G. and Schassan, H. H.,** Fosfomycin, a new antibiotic: *in vitro* activity compared with mezlocillin, cefuroxime and gentamicin, *Immun. Infekt.*, 8, 121, 1980.
21. **Rossi, R., Colantonio, G., Fraticelli, M., Giura, C., Quarenghi, M. I., Polli, O., Ferradini, M. A., and Nonihi, M. O.,** Use of fosfomycin in antibiotic therapy. Its effectiveness in patients with acute renal insufficiency or on hemodialysis treatment, *G. Ital. Chemioter.*, 26, 315, 1979.
22. **Peters, H. J. and Eicher, R.,** Clinical experience with fosfomycin in urinary tract infections (author's transl.), *Muench. Med. Wochenschr.*, 123, 748, 1981.
23. **Hirschl, A., Stanek, G., and Rotter, M.,** Improvement of therapeutic efficacy of fosfomycin by addition of glucose-6-phosphate in the treatment of intraperitoneally infected mice (author's transl.), *Zentralbl. Bakteriol., Parasitenkd., Infektionskr., Hyg., Abt. 1: Reihe A*, 246, 562, 1980.
24. **Rizzo, M., Nicoletti, P., Mazzei, T., and Carini, M.,** Fosfomycin chemotherapy of urological infections. Bacteriological and clinical considerations, *Drugs Exp. Clin. Res.*, VII, 37, 1981.
25. **Okuhara, M., Kuroda, Y., Goto, T., Okamoto, M., Terano, H., Kohsaka, M., Aoki, H., and Imanaka, H.,** Studies on new phosphonic acid antibiotics. I. FR-900098, isolation and characterization, *J. Antibiot.*, 33, 13, 1980.
26. **Kamiya, T., Hemmi, K., Takeno, H., and Hashimoto, M.,** Studies on phosphonic acid antibiotics. I. Structure and synthesis of 3-(*N*-acetyl-*N*-hydroxyamino)propylphosphonic acid (FR-900098) and its *N*-formyl analogue (FR-31564), *Tetrahedron Lett.*, 21, 95, 1980.

27. **Hashimoto, M., Hemmi, K., Takeno, H., and Kamiya, T.,** Studies on phosphonic acid antibiotics. II. Synthesis of 3-(*N*-acetyl-*N*-hydroxyamino)-2(R)-hydroxypropylphosphonic acid (FR-33289) and 3-(*N*-formyl-*N*-hydroxyamino)-1-*trans*-propenylphosphonic acid (FR-32863), *Tetrahedron Lett.*, 21, 99, 1980.
28. **Kamiya, T., Hemmi, K., Takeno, H., and Hashimoto, M.,** Studies on phosphonic acid antibiotics. I. Structure and synthesis of 3-(*N*-acetyl-*N*-hydroxyamino) propylphosphonic acid (FR-900098) and its *N*-formyl analogue (FR-31564), *Tetrahedron Lett.*, 21, 95, 1980.
29. **Okuhara, M., Kuroda, Y., Goto, T., Okamoto, M., Terano, H., Kohsaka, M., Aoki, H., and Imanaka, H.,** Studies on new phosphonic acid antibiotics. III. Isolation and characterization of FR-31564, FR-32863, and FR-33289, *J. Antibiot.*, 33, 24, 1980.
30. **Kojo, H., Shigi, J., and Nishida, M.,** FR-31564, a new phosphonic acid antibiotic: bacterial resistance and membrane permeability, *J. Antibiot.*, 33, 44, 1980.
31. **Mine, Y., Kamimura, T., Nonoyama, S., and Nishida, M.,** *In vitro* and *in vivo* antibacterial activities of FR-31564, a new phosphonic acid antibiotic, *J. Antibiot.*, 33, 36, 1980.
32. **Atherton, F. R., Hall, M. J., Hassall, C. H., Lambert, R. W., Lloyd, W. J., and Ringrose, P. S.,** Phosphonopeptides as antibacterial agents: mechanism of action of alaphosphin, *Antimicrob. Agents Chemother.*, 15, 696, 1979.
33. **Kittredge, J. S., Isbell, A. F., and Hughes, R. R.,** Isolation and characterization of the *N*-methyl derivatives of 2-aminoethylphosphonic acid from the sea anemone, *Anthopleura xanthogrammica, Biochemistry*, 6, 289, 1967.
34. **Kittredge, J. S.,** *N*-methyl derivatives of 2-aminoethylphosphonic acid in the lipids of the sea anemone, *Anthopleura elegantissima, 7th Int. Congr. Biochem. (Jpn.)*, 453, 1967.
35. **Liang, C. R. and Rosenberg, H.,** The biosynthesis of the phosphonic analogue of cephalin in *Tetrahymena, Biochim. Biophys. Acta*, 125, 548, 1966.
36. **Thompson, G. A., Jr.,** Studies of membrane formation in *Tetrahymena pyriformis*. I. Rates of phospholipid biosynthesis, *Biochemistry*, 6, 2015, 1967.
37. **Dawson, R. M. C. and Kemp, P.,** The aminoethylphosphonate-containing lipids of rumen protozoa, *Biochem. J.*, 105, 837, 1967.
38. **Smith, J. D., Snyder, W. R., and Law, J. H.,** Phosphonolipids in *Tetrahymena* cilia, *Biochem. Biophys. Res. Commun.*, 39, 1163, 1970.
39. **Kennedy, K. E. and Thompson, G. A., Jr.,** Phosphonolipids: localization in surface membranes of *Tetrahymena, Science*, 168, 989, 1970.
40. **Hasegawa, S., Tamari, M., and Kametaka, M.,** Isolation of diacylglyceryl-2-aminoethylphosphonate from bovine liver, *J. Biochem.*, 80, 531, 1976.
41. **Hasegawa, S., Tamari, M., and Kametaka, M.,** The distribution of ciliatine (2-aminoethylphosphonic acid) in bovine liver, *Agric. Biol. Chem.*, 40, 2097, 1976.
42. **Tamari, M., Ogawa, M., Hasegawa, S., and Kametaka, M.,** Etudes sur les phosphonolipides de la bile de boeuf, *Agric. Biol. Chem.*, 40, 2057, 1976.
43. **Alam, A. U. and Bishop, S. H.,** Choline phosphonate in vascular tissue, *156th Meet. Am. Chem. Soc.*, American Chemical Society Abstracts of Papers, Abstr. 276, Atlantic City, 1968.
44. **Baer, E. and Stanacev, N. Z.,** Phosphonolipids. I. Synthesis of a phosphonic acid analogue of cephalin, *J. Biol. Chem.*, 239, 3209, 1964.
45. **Rosenthal, A. F. and Pousada, M.,** Phosphonate analogs of cephalin, *Recl. Trav. Chim. Pays-Bas*, 84, 833, 1965.
46. **Baer, E. and Stanacev, N. Z.,** Phosphonolipids. V. Synthesis of phosphonic acid analogues of L-α-lecithins, *J. Biol. Chem.*, 240, 3754, 1965.
47. **Hori, T., Sugita, M., and Itasaka, O.,** Biochemistry of shellfish lipids. X. Isolation of a sphingolipid containing 2-monomethylaminoethylphosphonic acid from shellfish, *J. Biochem.*, 65, 451, 1969.
48. **Kapoulas, V. M.,** The chromatographic separation of phosphonolipids from their phospholipid analogs, *Biochim. Biophys. Acta*, 176, 324, 1969.
49. **Thompson, G. A., Jr.,** The metabolism of 2-aminoethylphosphonate lipids in *Tetrahymena pyriformis, Biochim. Biophys. Acta*, 176, 330, 1969.
50. **Stillway, L. W. and Harmon, S. J.,** A procedure for detecting phosphonolipids on thin layer chromatograms, *J. Lipid Res.*, 21, 1141, 1980.
51. **Rouser, G., Kritchevsky, G., Heller, D., and Lieber, E.,** Lipid composition of beef brain, beef liver, and the sea anemone: two approaches to quantitative fractionation of complex lipid mixtures, *J. Am. Oil Chem. Soc.*, 40, 425, 1963.
52. **Simon, G. and Rouser, G.,** Phospholipids of the sea anemone: quantitative distribution; absence of carbon-phosphorus linkages in glycerol phospholipids; structural elucidation of ceramide aminoethylphosphonate, *Lipids*, 2, 55, 1967.
53. **Karlsson, K.-A. and Samuelsson, B. E.,** The structure of ceramide aminoethylphosphonate from the sea anemone, *Metridium senile, Biochim. Biophys. Acta*, 337, 204, 1974.

54. **Hori, T. and Arakawa, I.**, Isolation and characterization of new sphingolipids containing *N,N*-acylmethylaminoethylphosphonic acid and *N*-acylaminoethylphosphonic acid from the mussel, *Corbicula sandai*, *Biochim. Biophys. Acta*, 176, 898, 1969.
55. **Hori, T., Arakawa, I., and Sugita, M.**, Distribution of ceramide 2-aminoethylphosphonate and ceramide aminoethylphosphate (sphingoethanolamine) in some aquatic animals, *J. Biochem.*, 62, 67, 1967.
56. **Hayashi, A., Matsuura, F., and Matsubara, T.**, Isolation and characterization of a new sphingolipid containing 2-*N*-methylaminoethylphosphonic acid from the viscera of *Turbo cornutus*, *Biochim. Biophys. Acta*, 176, 208, 1969.
57. **Matsuura, F., Matsubara, T., and Hayashi, A.**, Identification of molecular species of ceramide 2-*N*-methylaminoethylphosphonates containing normal fatty acids and dihydroxy long chain bases from *Turbo cornutus*, *J. Biochem.*, 74, 49, 1973.
58. **Hayashi, A. and Matsuura, F.**, 2-Hydroxy fatty acid- and phytosphingosine-containing ceramide 2-*N*-methylaminoethylphosphonate from *Turbo cornutus*, *Chem. Phys. Lipids*, 10, 51, 1973.
59. **Hayashi, A. and Matsuura, F.**, Isolation of a new sphingophosphonolipid containing galactose from the viscera of *Turbo cornutus*, *Biochim. Biophys. Acta*, 248, 133, 1971.
60. **Hayashi, A. and Matsuura, F.**, Characterization of aminoalkylphosphonyl cerebrosides in muscle tissue of *Turbo cornutus*, *Chem. Phys. Lipids*, 22, 9, 1978.
61. **Matsuura, F.**, Phosphonosphingoglycolipid, a novel sphingolipid from the viscera of *Turbo cornutus*, *Chem. Phys. Lipids*, 19, 223, 1977.
62. **Matsuura, F.**, The identification of aminoalkylphosphonyl cerebrosides in the marine gastropod, *Monodonta labio*, *J. Biochem.*, 85, 433, 1979.
63. **Steiner, S., Conti, S. F., and Lester, R. L.**, Occurrence of phosphonosphingolipids in *Bdellovibrio bacteriovorus* strain UKi2, *J. Bacteriol.*, 116, 1199, 1973.
64. **Wassef, M. K. and Hendrix, J. W.**, Ceramide aminoethylphosphonate in the fungus *Pythium prolatum*, *Biochim. Biophys. Acta*, 486, 172, 1977.
65. **Watanabe, T., Fukushima, H., and Nozawa, Y.**, Studies on thermal adaptation in *Tetrahymena* membrane lipids. Positional distribution of fatty acid in diacyl- and alkyl-acyl-phosphatidylcholines and -(2-aminoethyl)phosphonolipids from cells grown at different temperatures, *Biochim. Biophys. Acta*, 620, 133, 1980.
66. **Komai, Y., Matsukawa, S., and Satake, M.**, Lipid composition of the nervous tissue of the invertebrates *Aplysia kurodai* (gastropod) and *Cambarus clarki* (arthropod), *Biochim. Biophys. Acta*, 316, 271, 1973.
67. **Quin, L. D.**, 2-Aminoethylphosphonic acid in insoluble protein of the sea anemone *Metridium dianthus*, *Science*, 144, 1133, 1964.
68. **Rosenberg, H.**, Distribution and fate of 2-aminoethylphosphonic acid in *Tetrahymena*, *Nature (London)*, 203, 299, 1964.
69. **Quin, L. D.**, The presence of compounds with a carbon-phosphorus bond in some marine invertebrates, *Biochemistry*, 4, 324, 1965.
70. **Hilderbrand, R. L., Henderson, T. O., Glonek, T., and Myers, T. C.**, Characterization of a phosphonate-rich macromolecular complex from *Metridium dianthus* utilizing ^{31}P NMR, *Fed. Proc. Abstr.*, 30, 1072, 1971.
71. **Hilderbrand, R. L., Henderson, T. O., Glonek, T., and Meyers, T. C.**, Isolation and characterization of a phosphonic acid rich glycoprotein preparation from *Metridium dianthus*, *Biochemistry*, 12, 4756, 1973.
72. **Kirkpatrick, D. S. and Bishop, S. H.**, Phosphonoprotein. Characterization of aminophosphonic acid rich glycoproteins from sea anemones, *Biochemistry*, 12, 2829, 1973.
73. **Kirkpatrick, D. S. and Bishop, S. H.**, Aminoethylphosphonic acid; constituent in glycopeptide, *Fed. Proc. Abstr.*, 31, 874, 1972.
74. **Bunde, T. A., Hurley, J. C., Seymour, F. R., Dell, J. C., and Bishop, S. H.**, Aminophosphonic acids in glycoproteins from marine animals, *Fed. Proc. Abstr.*, 35, 1446, 1976.
75. **Lepak, N. M. and Smith, J. D.**, Characterization of phosphonoglycoproteins from *Tetrahymena*, *Fed. Proc. Abstr.*, 40, 1845, 1981.
76. **Hurley, J. C., Bunde, T. A., Dell, J. C., Kirkpatrick, D. S., and Bishop, S. H.**, Phosphonoglycoprotein from *Metridium senile* — heterogeneity of glycoproteins containing aminoethylphosphonic acid, *Comp. Biochem. Physiol. B*, 58, 253, 1977.
77. **Burcham, J. M., Dearlove, G. E., and Bishop, S. H.**, Phosphonomonoesterase activity in *Metridium senile* L., *Comp. Biochem. Physiol. B*, 67, 147, 1980.
78. **Korn, E. D., Dearborn, D. G., Fales, H. M., and Sokoloski, E. A.**, Phosphonoglycan, *J. Biol. Chem.*, 248, 2257, 1973.
79. **Korn, E. D., Dearborn, D. G., and Wright, P. L.**, Lipophosphonoglycan of the plasma membrane of *Acanthamoeba castellanii*, *J. Biol. Chem.*, 249, 3335, 1974.
80. **Dearborn, D. G., Smith, S., and Korn, E. D.**, Lipophosphonoglycan of the plasma membrane of *Acanthamoeba castellanii*, Inositol and phytosphingosine content and general structural features, *J. Biol. Chem.*, 251, 2976, 1976.

81. **Dearborn, D. G. and Korn, E. D.,** Lipophosphonoglycan of the plasma membrane of *Acanthamoeba castellanii*, *J. Biol. Chem.*, 249, 3342, 1974.
82. **Bunde, T. A., Dearlove, G. E., and Bishop, S. H.,** Aminoethylphosphonic acid-containing glycoproteins: the acid mucopolysaccharide-like components in mucus from *Metridium senile* (L), *J. Exp. Zool.*, 206, 215, 1978.
83. **Gibson, D. and Dixon, G. H.,** Chymotrypsin-like proteases from the sea anemone, *Metridium senile*, *Nature (London)*, 222, 753, 1969.
84. **Stevenson, K. J., Gibson, D., and Dixon, G. H.,** Amino acid analyses of chymotrypsin-like proteases from the sea anemone *(Metridium senile)*, *Can. J. Biochem.*, 52, 93, 1974.
85. **Miceli, M. V., Henderson, T. O., and Myers, T. C.,** 2-Aminoethylphosphonic acid metabolism during embryonic development of the planorbid snail *Helisoma*, *Science*, 209, 1245, 1980.
86. **Miceli, M. V., Henderson, T. O., and Myers, T. C.,** A high molecular weight phosphonoglycan from the egg masses of the snail *Helisoma* sp., *11th Int. Congr. Biochem.*, Toronto, Canada, Abstr. 07-7-H94, 1979, 505.
87. **Kittredge, J. S. and Roberts, E.,** A carbon-phosphorus bond in nature, *Science*, 164, 37, 1969.
88. **Rosenthal, A. F. and Pousada, M.,** Inhibition of phospholipase C by phosphonate analogs of glycerophosphatides, *Biochim. Biophys. Acta*, 164, 226, 1968.
89. **Rosenberg, H.,** Phosphonolipids, in *Form and Function of Phospholipids*, Ansell, G. B., Hawthorne, J. N., and Dawson, R. M. C., Eds., Vol. 3, Elsevier, New York, 1973, 333.
90. **Pettit, G. R., Day, J. F., Hartwell, J. L., and Wood, H. B.,** Antineoplastic components of marine animals, *Nature (London)*, 227, 962, 1970.
91. **Shibata, S., Dunn, D. F., Kuchii, M., Kashiwagi, M., and Norton, T. R.,** Cardiac stimulant action of extracts of coelenterates on rat atria, *J. Pharm. Sci.*, 63, 1332, 1974.
92. **Quinn, R. J., Kashiwagi, M., Norton, T. R., Shibata, S., Kuchii, M., and Moore, R. E.,** Antitumor activity and cardiac stimulatory effects of constituents of *Anthopleura elegantissima*, *J. Pharm. Sci.*, 63, 1798, 1974.
93. **Mebs, D. and Gebauer, E.,** Isolation of proteinase inhibitory, toxic, and hemolytic polypeptides from a sea anemone, *Stoichactis* sp., *Toxicon*, 18, 97, 1980.
94. **Kittredge, J. S. and Hughes, R. R.,** The occurrence of α-amino-β-phosphonopropionic acid in the zonanthid, *Zoanthus sociatus*, and the ciliate, *Tetrahymena pyriformis*, *Biochemistry*, 3, 991, 1964.
95. **Mason, W. T.,** Isolation and characterization of the lipids of the sea anemone, *Metridium senile*, *Biochim. Biophys. Acta*, 280, 538, 1972.
96. **Swift, M. L.,** Phosphono-lipid content of the oyster, *Crassostrea virginica*, in three physiological conditions, *Lipids*, 12, 449, 1977.
97. **Sebens, K. P.,** Recruitment in a sea anemone population: juvenile substrate becomes adult prey, *Science*, 213, 785, 1981.
98. **Kittredge, J. S., Horiguchi, M., and Williams, P. M.,** Aminophosphonic acids: biosynthesis by marine phytoplankton, *Comp. Biochem. Physiol.*, 29, 859, 1969.
99. **Mastalerz, P.,** The biochemistry of carbon-phosphorus bond, *Postepy Biochem.*, 15, 151, 1969.
100. **Horiguchi, M.,** Natural carbon-phosphorus compounds, in *Chemical Analysis of Phosphorus Compounds*, Vol. 37, Elving, P. J. and Kolthoff, I. M., Eds., Wiley-Interscience, New York, 1972, 703.
101. **Kittredge, J. S.,** The Natural Occurrence of the Aminophosphonic Acids, β-Aminoethylphosphonic Acid and α-Amino-β-Phosphonopropionic Acid, Ph.D. thesis, University of California, San Diego, 1964.
102. **Sarma, G. R., Chandramouli, V., and Venkitasubramanian, T. A.,** Occurrence of phosphonolipids in mycobacteria, *Biochim. Biophys. Acta*, 218, 561, 1970.
103. **Iguchi, E., Okuhara, M., Kohsaka, M., Aoki, H., and Imanaka, H.,** Studies on new phosphonic acid antibiotics, II. Taxonomic studies on producing organisms of the phosphonic acid and related compounds, *J. Antibiot.*, 33, 13, 1980.
104. **Hemmi, K., Takeno, H., Hashimoto, M., and Kamiya, T.,** Studies on phosphonic acid antibiotics. III. Structure and synthesis of 3-(*N*-acetyl-*N*-hydroxyamino)-2(*R*)-hydroxypropylphosphonic acid (FR-33289), *Chem. Pharm. Bull.*, 29, 646, 1981.
105. **Kuroda, Y., Okuhara, M., Goto, T., Okamoto, M., Terano, H., Kohsaka, M., Aoki, H., and Imanaka, H.,** Studies on new phosphonic acid antibiotics. III. Structure determination of FR-33289, FR-31564, and FR-32863, *J. Antibiot.*, 33, 29, 1980.
106. **Park, B. K., Hirota, A., and Sakai, H.,** 2-amino-5-phosphono-3-pentenoic acid, a new amino acid from N-1409 substance, an antagonist of threonine, *Agric. Biol. Chem.*, 40, 1905, 1976.
107. **Bayer, E., Gugel, K. H., Hagele, K., Hagenmaier, H., Jessipow, S., Konig, W. A., and Zahner, H.,** Stoffwechsel produkte von mikroorganismen. Phosphinothricin und phosphinothricyl-alanyl-alanin, *Helv. Chim. Acta*, 55, 224, 1972.
108. **Baldwin, M. W. and Braven, J.,** 2-Aminoethylphosphonic acid in *Monochrysis*, *J. Mar. Biol. Ass. U. K.*, 48, 603, 1968.

109. **Kandatsu, M. and Horiguchi, M.,** Occurrence of ciliatine (2-aminoethylphosphonic acid) in *Tetrahymena, Agric. Biol. Chem.*, 26, 721, 1962.
110. **Chou, S. C. and Scherbaum, O. H.,** Ethanolamine and phosphonic acid complexes in heat-treated *Tetrahymena* cells, *Exp. Cell Res.*, 45, 31, 1966.
111. **Sugita, M. and Hori, T.,** Isolation of diacylglycerol-2-aminoethylphosphonate from *Tetrahymena pyriformis, J. Biochem.*, 69, 1149, 1971.
112. **Abou Akkada, A. R., Messmer, D. A., Fina, L. R., and Bartley, E. E.,** Distribution of 2-aminoethylphosphonic acid in some rumen microorganisms, *J. Dairy Sci.*, 51, 78, 1968.
113. **Shelburne, F. A. and Quin, L. D.,** Isolation of 2-(methylamino)ethylphosphonic acid from the proteinaceous residue of a sea anemone, *Biochim. Biophys. Acta*, 148, 595, 1967.
114. **Quin, L. D. and Shelburne, F. A.,** An examination of marine animals for the presence of carbon-bound phosphorus, *J. Mar. Res.*, 27, 73, 1969.
115. **Henderson, T. O., Glonek, T., Hilderbrand, R. L., and Myers, T. C.,** Phosphorus-31 nuclear magnetic resonance studies of the phosphonate and phosphate composition of the sea anemone, *Bunadosoma sp, Arch. Biochem. Biophys.*, 149, 484, 1972.
116. **Dearlove, G. E., Bishop, S. H., and Viles, J. M.,** Aminoethylphosphonic acid containing glycoproteins as acidic sea anemone mucus components, *Fed. Proc. Abstr.*, 38, 652, 1979.
117. **Matsubara, T.,** Distribution of the dienic long chain bases in shellfish sphingophosphonolipids, *Chem. Phys. Lipids*, 14, 247, 1975.
118. **Tamari, M. and Kametaka, M.,** Isolation and identification of ciliatine (2-aminoethylphosphonic acid) from phospholipids of the oyster, *Crassostrea gigas, Agric. Biol. Chem.*, 36, 1147, 1972.
119. **Matsubara, T.,** The structure and distribution of ceramide aminoethylphosphonates in the oyster *(Ostrea gigas), Biochim. Biophys. Acta*, 388, 353, 1975.
120. **Matsubara, T. and Hayashi, A.,** Identification of molecular species of ceramide aminoethylphosphonate from oyster adductor by gas-liquid chromatography-mass spectrometry, *Biochim. Biophys. Acta*, 296, 171, 1973.
121. **Hori, T., Itasaka, O., Inoue, H., and Yamada, K.,** Further study of structural components of the pyridine insoluble sphingolipid from corbicula, *Corbicula sandai*, and the distribution in other species, *Biochem. J.*, 56, 477, 1964.
122. **Higashi, S. and Hori, T.,** Studies on sphingolipids of fresh water mussel spermatozoa, *Biochim. Biophys. Acta*, 152, 568, 1968.
123. **DeKoning, A. J.,** Phosphonolipids of marine origin. IV. The abalone *(Haliotis midae), J. Sci. Food Agric.*, 17, 460, 1966.
124. **Simon, G. and Rouser, G.,** Species variations in phospholipid class distribution of organs. II. Heart and skeletal muscle, *Lipids*, 4, 607, 1969.
125. **Liang, C.-R. and Rosenberg, H.,** On the distribution and biosynthesis of 2-aminoethylphosphonate in two terrestrial molluscs, *Comp. Biochem. Physiol.*, 25, 673, 1968.
126. **Liang, C.-R. and Strickland, K. P.,** Phospholipid metabolism in the molluscs. I. Distribution of phospholipids in the water snail *Lymnaea stagnalis, Can. J. Biochem.*, 47, 85, 1969.
127. **Liang, C.-R., Segura, M., and Strickland, K. P.,** Phospholipid metabolism in the molluscs. II. Activities of choline kinase, ethanolamine kinase, and CTP: phosphorylethanolamine cytidyltransferase in the mollusc *Helix lactea, Can. J. Biochem.*, 48, 580, 1970.
128. **DeKoning, A. J.,** Detection of 2-aminoethylphosphonic acid in the phospholipids of the crab *(Cyclograpsus punctatus), Biochim. Biophys. Acta*, 202, 187, 1970.
129. **Tamari, M.,** Isolation and identification of ciliatine (2-aminoethylphosphonic acid) from lipids of edible antarctic krill, *Euphausia superba, Agric. Biol. Chem.*, 43, 651, 1979.
130. **Alhadeff, J. A. and Daves, G. D., Jr.,** Occurrence of 2-aminoethylphosphonic acid in human brain, *Biochemistry*, 9, 4866, 1970.
131. **Alhadeff, J. A. and Daves, G. D., Jr.,** 2-Aminoethylphosphonic acid: distribution in human tissues, *Biochim. Biophys. Acta*, 244, 211, 1971.
132. **Tamari, M. and Kametaka, M.,** Isolation of ciliatine (2-aminoethylphosphonic acid) from the bile of the bovine, *Agric. Biol. Chem.*, 37, 933, 1973.
133. **Tamari, M., Ogawa, M., and Kametaka, M.,** A new bile acid conjugate, ciliatocholic acid, from bovine gall bladder bile, *J. Biochem.*, 80, 371, 1976.
134. **Tamari, M. and Kametaka, M.,** Distribution de la phosphonates dans le sang et le sperme de boeuf, *Agric. Biol. Chem.*, 44, 1957, 1980.
135. **Kandatsu, M. and Horiguchi, M.,** The occurrence of ciliatine (2-aminoethylphosphonic acid) in the goat liver, *Agric. Biol. Chem.*, 29, 781, 1965.
136. **Burt, C. T. and Myers, T. C.,** Unpublished data, 1975.

Chapter 3

METABOLISM OF PHOSPHONATES

Joseph Donald Smith

TABLE OF CONTENTS

I. INTRODUCTION

Studies on the metabolism of phosphonates have encompassed a wide variety of compounds and have been carried out in a large number of different species. Use of phosphonates as analogues of carboxylates in the study of enzyme specificity predated the discovery of naturally occurring phosphonates,[1,2] but it is since the first natural phosphonate, 2-aminoethylphosphonate (AEP), was reported in 1959[3] that virtually all of the metabolic studies have been carried out.

Research in phosphonate metabolism has intensified in recent years not only because of the occurrence of phosphonates in nature but because of their use as antibiotics,[4,5] neurotransmitter antagonists,[6-11] and pesticides.[12] Metabolic studies may be grouped in three general areas: (1) biosynthesis of phosphonates, (2) metabolic utility and effects of both natural and unnatural phosphonates in an organism, and (3) catabolism of phosphonates, including cleavage of the carbon-phosphorus bond.

II. BIOSYNTHESIS OF PHOSPHONATES

A. 2-Aminoethylphosphonate and Related Compounds

1. General Studies

Of the three techniques for studying phosphonic acid synthesis — physical isolation and chemical characterization, ^{31}P-NMR (Nuclear Magnetic Resonance) and radioactive isotope incorporation studies — only the latter has been of great utility in helping to elucidate metabolic pathways.

Chemical characterization has indeed demonstrated the presence of phosphonic acids and their derivatives in many organisms[3,13-28] particularly in those grown on a medium which is clearly lacking phosphonates.[29-30] Yet it is clear that mammals, which do contain AEP, do not synthesize the molecule and its presence must be attributed either to ingestion, in the case of man,[21-23] or to absorbtion from the rumen, in the case of cow[16,24-28] or goat.[15]

NMR has been used to identify phosphonates, particularly in *Acanthamoeba*,[31] and to distinguish between phosphonate and phosphate-ester phosphorus.[32,33] Utilization of NMR to elucidate the biosynthesis of AEP or its breakdown will require use of the high-resolution superconducting instruments which are now coming into more general use.[34]

2. Incorporation of $^{32}P_i$ into Phosphonates

The use of radioactive isotopes provides direct evidence for the synthesis of phosphonates by various organisms.[35-38] However, in some cases ambiguity has resulted where a radioactive compound has not been completely characterized.[39]

a. Bacteria

Only one bacterium has been definitely shown to contain a phosphonate, *Bdellovibrio bacteriovorus*.[40] An unknown phosphonate has been isolated from the sphingolipids of *Bdellovibrio* grown on a medium containing $^{32}P_i$. The phosphonate is not AEP or any of its *N*-methyl derivatives. Following acid hydrolysis the compound is also resistant to alkaline phosphatase hydrolysis confirming its identification as a phosphonate. A suggestion that the compound might be 1-hydroxy-2-aminoethylphosphonate, which has been found in *Acanthamoeba,*[31] was made but no further evidence has been presented.

Several strains of mycobacteria, *Mycobacterium tuberculosis* $H_{37}Rv$, *M. phlei* and *M.* 607 have been reported to contain AEP.[41] The compound was isolated from the lipids by hydrolysis in 6 *N* HCl and ion-exchange chromatography, followed by comparison with standard AEP in a paper chromatographic system. Only one chromatographic system was used and it is not clear whether that one can separate AEP from ethanolamine phosphate, traces of which may remain even after exhaustive acid hydrolysis.[42] It is noteworthy that in further studies on the lipids of these organisms using $^{32}P_i$, no further mention of AEP or phosphonolipids is made.[43]

The presence of very low amounts of phosphonates in several other bacteria was noted by Horiguchi[44] in a review article but has never been further elaborated.

b. Phytoplankton

Five species representing the two classes of dinoflagellates and coccolithoporphids have been shown to incorporate $^{32}P_i$ into phosphonates.[38] In all cases AEP is the predominant phosphonate found, with various *N*-methyl-AEP derivatives also present in four of the species. [^{32}P]-Phosphonoalanine is found in significant amounts, relative to AEP, in *Amphidinium carteri, Coccolithus huxleyi,* and *Syracosphaerc elongata* with smaller quantities in *Exuviella cassubica.*[38] The fifth species *Peridinium trochoidem* lacks phosphonoalanine.[38]

c. Protozoa

The free-living ciliate *Tetrahymena pyriformis* has been the organism of choice for most of the detailed studies on phosphonate synthesis because of its high phosphonate content (13% of the total phosphorus of the organism) and its ease of cultivation.[29] Different strains of *Tetrahymena* have been used by different investigators but the results appear to be comparable. The rumen ciliate *Entodinium caudatum* also incorporates $^{32}P_i$ into AEP.[45]

Kittredge and Hughes[35] first demonstrated the incorporation of $^{32}P_i$ into both AEP and phosphonoalanine in growing cultures of *Tetrahymena.* They reported a substantial difference in the specific activities of the two compounds, that of AEP being almost three times that of the phosphonoalanine, implying that there could be two different pathways for the synthesis of the C–P bond. However, the specific activity of the AEP, which is present in much higher concentrations in the cell, was determined after crystallization of the compound, while that of the phosphonoalanine was determined on a column eluate (for which only the radioactivity profile is shown), so that additional ninhydrin-positive material could have been present. Indeed, so little [^{32}P]phosphonoalanine was present, to prove the identity of phosphonoalanine by crystallization to constant specific activity, synthetic material had to be added. Further studies both in vivo and in vitro with ^{32}P- and ^{14}C-labeled precursors (Section II.A.4) clearly indicate a common pathway for the C–P bond formation for both AEP and phosphonoalanine.

Rosenberg[36] demonstrated that AEP is formed only during logarithmic growth of *Tetrahymena* not during the stationary phase. This is consistent with the postulated structural role for AEP in the phospholipids and structural proteins of the cell which would be primarily synthesized during cell growth and division. In cells which were grown in the presence of $^{32}P_i$ for 1 hr, then placed in cold medium for an additional 3 hr (during which time the cell population had approximately doubled), there was no loss of [^{32}P]AEP from either lipid or protein-bound AEP indicating a lack of turnover of these structural elements.[36]

In labeling experiments in the continuous presence of $^{32}P_i$, Rosenberg found that 1 hr after $^{32}P_i$ addition to a growing culture the lipid-bound AEP had a higher specific activity (19,000 cpm/μmol P) than did the protein-bound AEP (11,000 cpm/μmol P) which was in turn higher than the free AEP (4400 cpm/μmol P).[36] The same pattern of the specific activities obtained at 6 hr after $^{32}P_i$ addition. Only at 23 hr had the protein-associated AEP attained the same specific activity as the lipid-bound AEP; the specific activity of the free AEP was still lower. The interpretation presented is that the AEP is synthesized at the lipid level using a phosphate from a phospholipid, with the AEP subsequently transferred to the AEP-containing protein and finally released to the soluble pool. This explanation is inconsistent with the observation that the [^{32}P]AEP is not released from either lipid or protein during the "chase" experiment.[36]

An alternative explanation would be that there may be several pools of free AEP within the cell. Newly synthesized AEP, containing $^{32}P_i$, would be directly incorporated into the macromolecules and would penetrate a second free pool much more slowly. In view of the work of Blum and co-workers demonstrating the existence of two metabolic pools of phosphoenolpyruvate, the presumed precursor of the phosphonates (Section II.A.4),[46,47] the possible existence of several pools for AEP becomes significant. The precursor-product relationship between phosphatidylcholine and diglyceride-AEP was made more attractive by the report that they both contained very similar amounts of glycerylether (chimyl alcohol) and had similar fatty acid profiles, quite distinct from that of phosphatidylethanolamine.[48] However, the observation by Liang and Rosenberg[49] that in a 2 hr labeling period with $^{32}P_i$ the phospholipid-bound AEP had a 30% higher specific activity than did phospholipid-ester-phosphorus would again suggest that phospholipid ester phosphorus is not converted to AEP phosphorus.

While AEP has been found at the lipid and protein level as well as in soluble form,[36] phosphonoalanine has been found only in the proteins[35,50] and in free form.[35,50] Despite extensive efforts to detect phosphonoalanine in the lipids,[35,50,51] where its decarboxylation could directly give diglyceride-AEP in the manner of the conversion of phosphatidylserine to phosphatidylethanolamine,[52] no such evidence has been obtained.

d. Invertebrates

Several organisms of this class have been shown to incorporate radioactive P_i into AEP either in vivo or in vitro. The fresh water mussel *Hyriopsis schlegelii,*[53] the snails *Helix aspera*[37] and *Helisoma* sp.,[54] and the garden slugs *Lehmannia poirieri*[37] and *Limus flavus*[37] are all capable of AEP biosynthesis. Miceli et al.[54] have demonstrated that when adult *Helisoma* are exposed to $^{33}P_i$ for 24 hr, newly-laid egg masses contain $^{33}P_i$ of which 85% is in AEP.

Liang and Rosenberg demonstrated that the pattern of ^{32}P-labeling of lipid-bound, protein-bound, and soluble AEP in *Lehmannia poirieri* was opposite to that observed in *Tetrahymena;* free AEP had a higher specific activity than did protein-bound AEP which in turn was higher than lipid-bound AEP.[37] This pattern obtained during a 9 hr labeling period. It was concluded that the synthesis of AEP takes place to give the free AEP which is then incorporated into proteins and lipids. Since the lipid-associated AEP in *Lehmannia*, as well as in most invertebrates, is in ceramide-AEP[19,20,53] (only in a few organisms is diglyceride-AEP present,[30,55]) it is apparent that the AEP synthesis cannot be at the phospholipid level if all organisms synthesize phosphonates in the same manner.

e. Vertebrates

There have been several studies examining the possibility of AEP synthesis in rats. Alhadeff et al.[56] incubated rat liver slices with $^{32}P_i$, [U-^{14}C]glucose or [3-^{14}C]pyruvate for 2 hr. The tissues were then hydrolyzed in HBr and AEP isolated by column chromatography. Although the liver was found to contain about 1.3 mg AEP/g wet tissue, there was no incorporation of any of the radioactivity into the AEP. Similar results were obtained when [U-^{14}C]glucose was injected into rats. After 12 days, the AEP isolated from several tissues was lacking in radioactivity.[56]

A study by Marmouyet et al.,[39] in which AEP was isolated from $^{32}P_i$-injected rats, demonstrated that the amount of radioactivity in AEP in normal rats was about 20 times that in "germ-free" rats, implying that most of the AEP was synthesized by the intestinal-microbes rather than by the animal itself. The traces of labeled AEP may be attributed either to the animals not being totally germ-free or to the "AEP" not being treated with phosphatase to remove traces of any remaining phosphate ester.[39]

One would conclude that, in fact, rats and presumably other mammals, lack the capacity to synthesize AEP.

3. *Incorporation of [^{14}C]-Compounds into Phosphonates*

a. *In Vivo Studies*

Both AEP and phosphonoalanine are labeled when *Tetrahymena* is cultured on media containing various ^{14}C-labeled compounds.[49,57-59] The results obtained are consistent with phosphoenolpyruvate being the precursor of the carbon skeletons of AEP and phosphonoalanine.

Those labeled precursors whose label could give rise to carbon-3 of phosphoenolpyruvate, such as [1- or 6-^{14}C]glucose, are found to label AEP in carbon 1, the phosphonate carbon.[49,57] Compounds such as [2-^{14}C]acetate,[58,59] [3-^{14}C]pyruvate,[49,58,59] [2,3-^{14}C]succinate,[59] [2,3-^{14}C]fumarate,[59] [2,3-^{14}C]malate,[59] and [3,4-^{14}C]glutamate,[59] which give rise to phosphoenolpyruvate via gluconeogenesis, are converted to AEP labeled equally in both carbons. Pyruvate and acetate can be converted to succinate through the glyoxylate pathway which is very active in *Tetrahymena*.[60] [2-^{14}C]Glucose labels carbon 2 of AEP almost exclusively[57] while [3,4-^{14}C]glucose, which would label the carboxyl carbon of phosphoenolpyruvate,[58] labels the carboxyl group of phosphonoalanine but does not label AEP.[58] The results obtained have been interpreted to mean that phosphoenolpyruvate is converted to phosphonopyruvate, either via intramolecular rearrangement or using another phosphate source. The poor conversion of [1,2-^{14}C]ethanolamine to AEP,[45,49,59] eliminates as a mechanism the phosphoramidic rearrangement proposed by Segal.[61]

Although the labeling patterns of phosphonoalanine and AEP suggested the former as precursor of the latter,[50,57,58] subsequent examination of this conversion in vitro (Section II.A.3.b) has led to the conclusion that the phosphonoalanine must be reconverted to phosphonopyruvate or to phosphonoacetaldehyde before AEP may be formed.[51,62-64]

The results of similar, although more limited, labeling experiments in the slug *Lehmannia poirieri*[37] and the protozoan *Entodinium caudatum*[45] are consistent with those obtained in *Tetrahymena*, suggesting a common pathway for AEP formation in these, if not in all, organisms.

b. *In Vitro Studies*

Warren[58] and Geike[65] first developed cell-free systems from *Tetrahymena* capable of phosphonate synthesis. Warren used a cell homogenate which could convert [^{32}P]phosphoenolpyruvate into both AEP and phosphonoalanine.[58] [^{32}P]Phosphonoalanine was also converted to AEP by this system, but $^{32}P_i$ was not incorporated into the phosphonates.[58] Geike found that glutathione was an absolute requirement for [U-^{14}C]glucose to be incorporated into AEP in his cell-free system (40 m*M* P_i, pH 7.6; 10 m*M* $MgCl_2$).[65]

Horiguchi developed a cell-free system capable of incorporating $^{32}P_i$ into AEP.[62] The cells were homogenized in 50 m*M* glycine-NaCl-NaOH, pH 10 with 2% albumin and 1 m*M* ethylenediaminetetraacetic acid (EDTA) and the homogenate was diluted 1:1 with 100 m*M* Tris-HCl, pH 7.7 containing 5 m*M* L-alanine, 1 m*M* pyridoxal phosphate, 0.1 m*M* ATP, 20 m*M* oxaloacetate, and 1.6% albumin. This system was totally dependent on oxaloacetate. If phosphoenolpyruvate replaced oxaloacetate, no radioactive AEP was formed, indicating trapping of [^{32}P]-phosphoenolpyruvate by the cold material,[62] again implicating phosphoen-

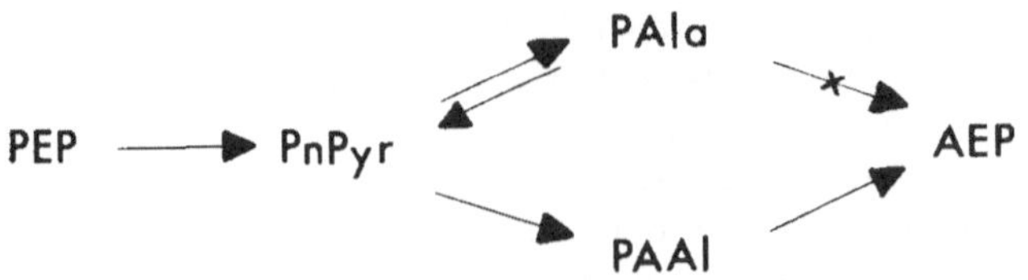

FIGURE 1. Pathway for the biosynthesis of AEP. The "x" between PAla and AEP indicates that this reaction does not take place. Abbreviations: PEP, phosphoenolpyruvate; PnPyr, phosphonopyruvate; PAla, β-phosphonoalanine; PAAl, phosphonoacetaldehyde.

olpyruvate as the probable precursor of AEP. Virtually all of the AEP formed in this system was found to be in free form, only 0.5% of the [^{32}P]AEP formed was in the lipid fraction and 1.5% in the protein fraction.[62]

Most striking of the results was the observation that phosphonoalanine added to the system had only minimal effect on the conversion of either $^{32}P_i$ or [3-^{14}C]phosphoenolpyruvate into AEP,[62] while addition of phosphonoacetaldehyde almost totally blocked the incorporation of the radioactive compounds into AEP.[62,63] Oxaloacetate also had minimal effect on the formation of [^{14}C]AEP from [3-^{14}C]phosphoenolpyruvate.

These results clearly demonstrate not only that phosphoenolpyruvate is the immediate precursor of phosphonates in *Tetrahymena,* but that phosphonoacetaldehyde is a direct precursor of AEP.[62,63] Phosphonoalanine is a side product in the synthesis and apparently is not directly converted to AEP.[64]

Subsequent experiments by Horiguchi and Rosenberg have demonstrated that doubly-labeled phosphoenolpyruvate ([^{32}P]phosphoenol[3-^{14}C]pyruvate) is incorporated into both AEP and phosphonoalanine in a cell-free system with the two products having the same ^{14}C/^{32}P ratio as the starting phosphoenolpyruvate.[51] Phosphonopyruvate and phosphonoacetaldehyde were not observed directly but the 2,4-dinitrophenylhydrazones corresponding to the two compounds were isolated. Subsequent hydrogenation of the dinitrophenylhydrazones gave AEP and phosphonoalanine demonstrating the formation of the two carbonyl-containing phosphonates in the biosynthetic pathway. Again, virtually none of the phosphonates formed in this system was found at the lipid level.[51]

It has also been recently demonstrated that the conversion of [3-^{14}C]phosphonoalanine into AEP proceeds through phosphonopyruvate. The latter compound added to a cell-free system traps radioactivity from phosphonoalanine and no radioactive AEP is formed.[64]

4. *Summary of the Biosynthesis of 2-Aminoethylphosphonate*

The pathway of biosynthesis of AEP as it has been elucidated is shown in Figure 1.

While most of the definitive work has been done using *Tetrahymena,* the pathway is consistent with all of the observations made with other organisms. Phosphonoalanine does not occur in many of the organisms which synthesize AEP and is thus a side product. Since AEP in most of the organisms is found as ceramide-AEP rather than diglyceride-AEP, the rapid appearance of ^{32}P-AEP in diglyceride-AEP rather than free AEP in *Tetrahymena*[36] may be attributed to metabolic channeling of newly-formed AEP into the lipids.[178]

No studies have been reported on the formation of the *N*-methyl-,[18,20,38,67] dimethyl-,[18,38] or trimethyl-AEP[18,38] which have been found in many organisms. The AEP in the glycerolipids is not methylated,[66] so that methylation of AEP probably occurs after its incorporation into sphingolipids.

5. *Regulation of Biosynthesis of 2-Aminoethylphosphonate*

AEP is synthesized only during growth of *Tetrahymena.*[36] When a stationary culture is incubated with $^{32}P_i$, virtually no radioactivity is incorporated into AEP.[36]

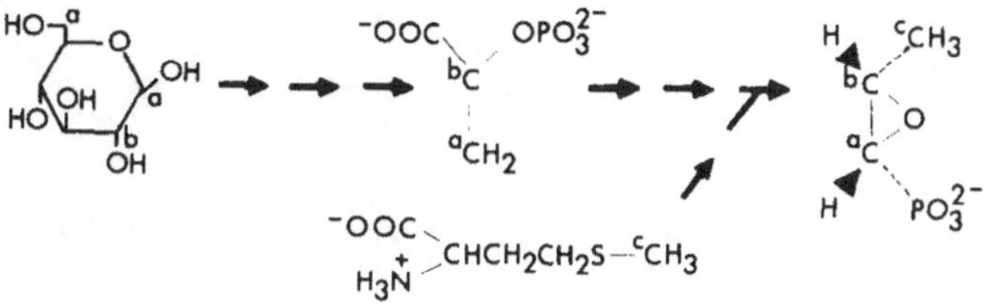

FIGURE 2. Biosynthesis of fosfomycin showing the labeling pattern of fosfomycin from glucose and methionine.

When AEP is present in the growth medium of *Tetrahymena* along with $^{32}P_i$, the amount of [^{32}P]AEP formed is decreased in proportion to the amount of AEP present, indicating feedback control on AEP synthesis.[68] The inhibitor may be either AEP itself or the diglyceride-AEP whose content in the cell increases with the amount of AEP present in the medium. This inhibition is specific to AEP since 3-aminopropylphosphonate (the isosteric analogue of ethanolamine phosphate, see Chapter 5) under the same experimental conditions has no effect on [^{32}P]AEP formation.[69] Since the actual amount of AEP formed in cell-free systems is so small, the decrease in radioactive AEP formed (Section II.A.3.b) when phosphonopyruvate[64] or phosphonoacetaldehyde[62,63] is added, may be due to the trapping of the intermediates rather than actual inhibition of the pathway.

B. Phosphonate-Containing Antibiotics

The only phosphonate-containing antibiotic whose biosynthesis has been investigated is fosfomycin (formerly phosphonomycin) which was originally isolated from used culture medium of *Streptomyces fradiae*.[4] Fosfomycin, (−)*1R,2S*-cis-1,2-Epoxypropylphosphonic acid, is a secondary metabolite of the organism and is produced in significant quantities only when the growth of the organism is slowing from logarithmic to stationary phase.[70] The compound is clearly made as some two-carbon unit to which the methyl group is subsequently added.[70] [^{14}C-Methyl]-methionine and [U-^{14}C]glycine provide the methyl group of fosfomycin.[70,71] Methionine is also required as an inducer of fosfomycin synthesis since it has to be present in the original medium for optimum synthesis and cannot be replaced by other methyl group sources such as choline, betaine, or serine. Carbon sources which give increased yields of fosfomycin are, in order of the activity, malate, α-ketoglutarate and glutamate, oxaloacetate, and succinate. These are all intermediates or sources for intermediates in the citric acid cycle and gluconeogenesis.[70]

Labeling of carbons 1 and 2 of fosfomycin is provided best by [1,6-^{14}C]glucose which primarily labels carbon 1, [2-^{14}C]glucose which labels carbon 2 and [2-^{14}C]acetate which labels both carbons 1 and 2 of fosfomycin in equal amounts.[70] [1-^{14}C]Acetate gives very poor incorporation of radioactivity. From these data, phosphoenolpyruvate was suggested as the most likely precursor of carbons 1 and 2 of fosfomycin[70] (Figure 2).

The labeling of fosfomycin by acetate may be explained by invoking the glyoxylate cycle, through which carbon 2 of acetate can label carbons 2 and 3 of oxaloacetate and hence phosphoenolpyruvate and be incorporated into fosfomycin, while the carboxyl carbon of acetate would be lost in the final product.

Several organisms, among them several species of *Penicillium*,[72] oxidize cis-1-propenylphosphonate to fosfomycin (Figure 3) but it is not known if this reaction is involved in the biosynthesis of the antibiotic by *Streptomyces*.

A number of other phosphonic[73-78] and phosphinic[79] acid-containing antibiotics, synthesized by various strains of *Streptomyces*, have been characterized but no biosynthetic studies have been reported. Several of these occur as tripeptides,[73,79] which may be transported more readily by bacteria than are the free phosphonates (Section V.A.1).[80]

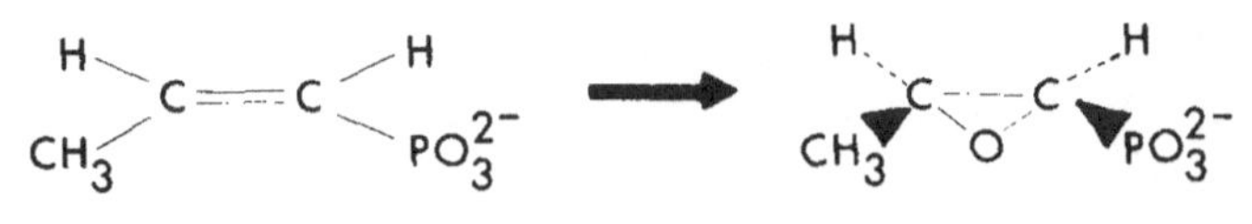

FIGURE 3. Epoxidation of cis-1-propenylphosphonate to fosfomycin by *Penicillium*.

III. BIOSYNTHESIS OF PHOSPHONOLIPIDS

A. Sphingophosphonolipids

The major sphingophosphonolipid found is ceramide-AEP.[17] Most of the other lipids of this class may be considered as derivatives of this compound with the modifications, *N*-methylation[19,20,67,81] or *N*-acylation[81] probably occurring after the ceramide-AEP is synthesized.

There have been no direct studies on the biosynthesis of ceramide-AEP. The lipid has thus far been found in all organisms which synthesize AEP. Many of these same organisms lack the glycerophosphonolipids, so that formation of ceramide-AEP by direct base transfer from phospholipid to ceramide as occurs for sphingomyelin[82] or ceramidephosphoinositol[83] formation may be considered unlikely.

Liang et al.[84] found that cell-free extracts of the snail *Helix lactea* could not form CMP-AEP from CTP and AEP under the same conditions in which CDP-ethanolamine was formed, suggesting that the AEP in sphingolipids may not come from CMP-AEP. They did not, however, examine this possibility. *Helix lactea* does not contain diglyceride-AEP.[84]

[^{14}C]AEP does label the sphingolipids of the sea anemone *Metridium senile* in vivo.[85,86] The most likely route of ceramide-AEP formation might be a base exchange reaction between free AEP and another sphingolipid.

B. Glycerylphosphonolipids

1. Bacteria

It is noteworthy that *Mycobacteria* are among the few bacteria which can incorporate free ethanolamine into phospholipids[87,88] by the cytidine nucleotide pathway (Figure 4). If the presence of AEP in these organisms can be definitely established (Section II.A.2.a) then a mechanism does exist in *Mycobacteria* by which glycerophosphonolipids may be synthesized.

In *Bacillus subtilis*[89] or *Escherichia coli*,[90,91] which do not utilize free ethanolamine for phosphatidylethanolamine synthesis, the only phosphonolipid formed is an analogue of phosphatidylglycerolphosphate in which the glycerol-3-phosphate which reacts with CDP-diglyceride is replaced by its isosteric analogue 3,4-dihydroxybutyl-1-phosphonate, when the phosphonate is added to the growth medium. The phosphonate is not acylated to the analogue of phosphatidic acid nor can 2,3-dihydroxylpropyl-1-phosphonate replace its homologue.

2. Tetrahymena

In *Tetrahymena* the glycerylphosphonolipids consist of a mixture of 1,2-diacylglyceryl-3-*O*-(2-aminoethylphosphonate)[30,48,92] and 1-*O*-alkyl-2-acylglyceryl-3-*O*-(2-aminoethylphosphonate),[48,92,177] in which the glycerylether is chimyl alcohol. Exogenous [^{14}C]AEP is incorporated into both the diacyl- and glycerylether phosphonolipid during growth of *Tetrahymena*.[50] When $^{32}P_i$ is used to label the phosphonolipids for 2 hr, AEP in the diacyl lipid is labeled about twice as much as that in the glycerylether phosphonolipid, perhaps reflecting specificity of the CMP-AEP:diglyceride AEP transferase for the diglyceride over the 2-acyl chimyl alcohol.[49]

CMP-AEP has been isolated from *Tetrahymena* and shown to donate AEP to diglyceride in a cell homogenate.[93] The enzymatic activity is highest in mitochondria. The formation

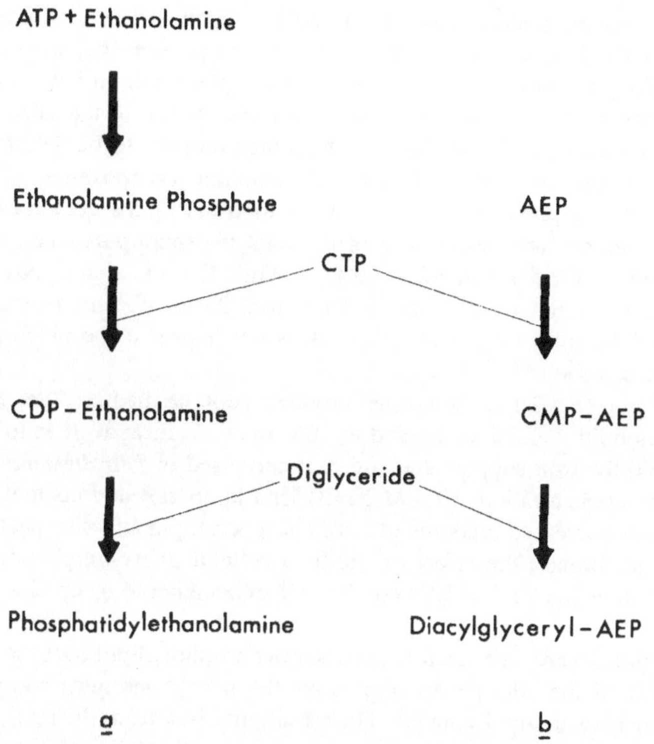

FIGURE 4. Cytidine nucleotide pathway for the formation of phospholipids. Phosphatidylcholine synthesis occurs from free choline in the same manner.

of CMP-AEP from CTP and AEP in a cell-free system from *Tetrahymena* has also been demonstrated.[93]

The phosphonolipid is not methylated by *S*-adenosylmethionine either in vivo or in vitro as is phosphatidylethanolamine, suggesting great specificity for the phosphatidylethanolamine:*S*-adenosylmethionine *N*-methyltransferase.[66] The *N*-methyl-AEP found in the sphingolipids of *Tetrahymena* is thus not derived from the glycerolipids.[67] The great specificity of the methyltransferase for phosphatidylethanolamine is also reflected in the observation that when ethanolamine phosphate is replaced by its isostere 3-aminopropylphosphonate in the phospholipids, the analogue is not methylated either.[69]

The level of the AEP-glycerides in *Tetrahymena* may be regulated by the amount of free AEP available to the cell.[68] When AEP is added to the growth medium in concentrations up to 10 m*M*, the level of the phosphonolipid rises from about 20% of the total cellular phospholipid to 40%. This is mirrored by an essentially equal decrease in the amount of phosphatidylethanolamine with minimal change in the level of phosphatidylcholine.[68] 3-Aminopropylphosphonate added to the medium almost totally replaces ethanolamine phosphate, not AEP, in the phospholipids but does not alter the relative amounts of the phospholipids.[69] This suggests that the enzymes responsible for the synthesis of AEP-glyceride and phosphatidylethanolamine are different with the level of AEP perhaps acting to control phospholipid levels.

A new strain of *T. pyriformis*, called NT-1, has been isolated which, unlike the most commonly used strain WH-14,[94] shows temperature-dependent alteration of the phospholipid composition of the organism.[95,96] On temperature shift from 39.5 to 15°C the phosphonolipid increases from 16 to 29% of the total phospholipids while phosphatidylethanolamine drops from 43 to 26%, the same pattern observed on AEP feeding. There is also an increase in the unsaturation of the fatty acids in response to the temperature decrease as a potential means of maintaining membrane fluidity.[96] Whether the phospholipid base change is also a means of altering membrane fluidity or is merely a consequence of the different fatty acid and glyceryl ether contents of the various phospholipids remains to be determined.[97]

The most striking environmental effect on phospholipid composition of *Tetrahymena* involves the decrease in phosphonolipid from 16 to 6% and the reciprocal increase in phosphatidylethanolamine from 39 to 46% of the total phospholipids when strain NT-1 is grown in the presence of 1.5% ethanol at 28°C.[98] While there is also a significant change in fatty acid content, a decrease of 16:1 + 16:2 from 23 to 5% and an increase of 18:2 from 14 to 25% of the total fatty acids, there does not appear to be much change in the overall level of unsaturation.[98]

The addition of iron ($FeCl_3$) to the normal proteose-peptone medium[99] leads to increased levels of phosphonolipid,[100] 26% compared to 20% in strain E, as well as to better growth characteristics[99] and the iron supplements are routinely used in *Tetrahymena* media.

High sodium ion concentrations (0.3 *M* NaCl) lead to an elevated content of phosphatidylethanolamine and decreased amounts of both phosphonolipid and phosphatidylcholine in the pellicle (cell membrane); the effect on the total cellular phospholipid composition was not reported.[101] Neither 0.1 *M* $CaCl_2$[101] nor 10 m*M* ethanolamine or choline[69] affect phospholipid levels.

The ciliary membranes are significantly enriched in phosphonolipid over the whole cell.[102-105] From 50 to 75% of the cilia phospholipids are the phosphonolipids compared to 20 to 25% of the cellular phospholipid content. The variation arises from differences in isolation procedure for the cilia as well as depending on the growth phase of *Tetrahymena*. The glycerylether-containing phosphonolipids tend to predominate in the cilia and also accumulate in stationary phase cells.[177] With two of the four normally-hydrolyzable bonds in the molecule resistant to lipolytic enzymes the phosphonolipid would tend to accumulate whereas hydrolyzable phospholipids would turn over.

3. *Invertebrates*

a. Insects

While phosphonates have not been found in insects, it has been shown that housefly larvae are capable of incorporating AEP and its methylated analogues into the glycerophospholipids.[106,107] Since choline is a dietary requirement in insects, they are capable of incorporating a wide variety of aminoalcohols into lipids via cytidine nucleotide intermediates[108,109] (Figure 4).

The added AEP apparently competes with ethanolamine phosphate, since the amount of phosphonolipid formed is approximately equal to the decrease in phosphatidylethanolamine, while there is little change in the amount of phosphatidylcholine.[106] All of the AEP incorporated into lipids is found in the glycerolipids, none in sphingolipids.

Both *N,N*-dimethyl-AEP and *N,N,N*-trimethyl-AEP replace choline in the lipids of housefly larvae to an equal extent and with no perturbation of the "lecithin" content of the organism. There is no conversion of the dimethyl-AEP to trimethyl-AEP.[107]

b. Marine Organisms

AEP has been found in the glycerophospholipids of only a few marine invertebrates. These include the sea anemone *Anthopleura elegantissima*, in which glyceryl-AEP was characterized from the saponifiable lipid fraction,[13] and the oyster *Crassostrea virginica*.[55,110] No biosynthetic studies have been reported.

In the oyster under starvation or postspawning conditions there is an increase of phosphonolipid relative to the total phospholipid of the organism.[110] Apparently the phosphonolipid is conserved at the expense of the ester phospholipid which is degraded.

4. *Vertebrates*

a. In Vivo Studies

Diacylglyceryl-AEP has been isolated both from bovine[111] and caprine[15] liver as well as from bovine bile.[25] As isolated from normal animals the amount of lipid is quite small, (e.g., in bovine liver the concentration was found to be 127 mg/kg liver)[111] but chronic exposure of an animal to AEP may have serious pathological consequences. Nair et al.[112] found that normal and cholesterol-fed rats, after 4 weeks of weekly, intraperitoneal injection of 100 mg AEP/kg, showed enhanced bilirubin:UDP-glucuronyl transferase activity and increased plasma cholesterol levels. Increased numbers of lipid-occlusion bodies in liver were found on sectioning and staining the tissue. Although the authors did not determine the amount of phosphonolipid formed, they speculate that the metabolic alterations may be due to replacement of phosphatidylcholine by diacylglyceryl-AEP which could lead to decreased lipid turnover.

This is in accord with the known metabolism of AEP in rats. When radioactive AEP is injected into rats it is found only as free AEP,[113-118] CMP-AEP,[115] and diacylglyceryl-AEP [113-118] with a small amount as ceramide-AEP.[116,117] In most studies,[113,114,118] no methylation of diacylglyceryl-AEP to the trimethyl form was observed; where it was noted,[115-117] the identification appears to be quite tenuous. When CMP-[^{14}C]AEP is injected,[117] the radioactivity is incorporated into lipids faster and to a greater extent than is free AEP, confirming the central role of CMP-AEP in phosphonolipid synthesis via the cytidine nucleotide pathway (Figure 4).

More than 90% of injected AEP is eliminated from rats within 48 hr.[114-116] What remains is predominantly in lipids, primarily in liver. There is no evidence for degradation of the compound in vivo.[114,116,118]

While trimethyl-AEP is not formed by rats, when it is injected the animals are capable of incorporating it into the "lecithin" fraction, primarily in liver.[119] Both trimethyl-AEP and *N,N,N*-trimethyl-3-aminopropylphosphonate, which is isosteric with choline phosphate, are incorporated into the lipids with no evidence of further metabolism.[119] Within 26 hr of injection, 80% of each compound is excreted.[119]

When [^{14}C]AEP is injected into vitamin A-deficient rats, there is an increase in phosphonolipid radioactivity relative to control animals over a 6-hr period.[120] The half-lives of the radioactive lipids do not differ significantly between control and experimental animals. The increased phosphonolipid levels may be due to more rapid synthesis or to increased uptake of AEP by the liver in the deficient animals.[120]

b. In Vitro Studies

Studies in several cell-free systems, rat liver,[115,116,121] mouse brain,[122,123] and chick intestinal mucosa,[124] have confirmed the in vivo observations on diacylglyceryl-AEP synthesis. AEP reacts with CTP to form CMP-AEP;[115,116] the AEP is then transferred to diglyceride to form the phosphonolipid.[116,124] AEP and 3-aminopropylphosphonate act both as substrates for and competitive inhibitors of ethanolaminephosphate:CTP cytidyl-transferase (E.C.2.7.7.14).[121] At pH 7.7, the K_i of 108×10^{-3} *M* for AEP is several orders of magnitude greater than the K_m of 1.33×10^{-3} *M* for ethanolamine phosphate indicating that substantial quantities of AEP must be present for there to be significant effects on this enzyme.[121]

In mouse brain slices, AEP inhibits the incorporation of $^{32}P_i$ into several phospholipids without apparently being incorporated itself.[122,123] While the 10 μM AEP used is well below the K_m for ethanolamine phosphate (1.33 m*M* for the rat liver enzyme) and thus is not incorporated, the effect may be on ethanolamine kinase or choline kinase or on [^{32}P]ATP formation rather than on the later steps in lipid synthesis.

Trimethyl-AEP is also converted to CMP-trimethyl-AEP by rat liver homogenates.[119] The K_m is 5.2×10^{-3} *M* compared to 0.35×10^{-3} *M* for choline phosphate indicating that large amounts of the phosphonate are needed if the phosphonolipid synthesis is to be significant.[119]

IV. PHOSPHONOGLYCANS AND PHOSPHONOPROTEINS

Phosphonate-containing macromolecules have been reported only in some of the organisms which contain phosphonates and have been characterized in even fewer species with metabolic studies being minimal. A "residue" fraction, characterized as protein in nature, which remains after aqueous and chloroform-methanol extraction of various organisms such as *Tetrahymena*,[36,50] sea anemone,[125-127] and bovine brain[114] contains covalently bound AEP.

A. Phosphonoglycans

Only *Acanthamoeba castellani* contains a phosphonoglycan in which AEP and 1-hydroxy-AEP are linked to a glycan, which also contains phytosphingosine, inositol, fatty acid, and a variety of sugars.[128] No protein is associated with the molecule. No metabolic or detailed structural studies have been reported.

A. Phosphonoproteins

1. Tetrahymena

Tetrahymena incorporate $^{32}P_i$ into residue-bound AEP. Rosenberg showed that this fraction releases amino acids but not AEP on proteolysis. The unhydrolyzed material still contains amino acids as well as AEP.[36]

Subsequent studies by Lepak and Smith[129] with *Tetrahymena* labeled with [^{14}C]AEP have shown the phosphonoprotein of the organism to be tightly bound in the cell membrane and absent from cilia and internal organelles. The phosphonoprotein requires strong detergent for solubilization and aggregates in the absence of detergent. The AEP-containing protein is included on a Sephacryl®-S200 gel filtration column in the presence of 0.1% sodium dodecyl sulfate (SDS) but is excluded from the column in the absence of detergent. The phosphonoprotein fraction contains 8% AEP as well as 32% carbohydrate, with the AEP bound to the carbohydrate not the polypeptide portion of the molecule.[129,179]

2. Sea Anemone

The anemones *Metridium senile*[85,86] and *M. dianthus*[130] incorporate [^{3}H]AEP into their protein fraction. Much of this appears to be soluble mucus glycoprotein which lacks other acidic components — sulfate, uronic acids, and sialic acids.[86] In addition, Stevenson et al.[131] have purified two chymotrypsin-like proteases which contain AEP. One contains 6 mol of AEP and 5 mol of galactosamine per 25,930 dalton molecule while the other lacks carbohydrate and contains 4 mol of AEP per mole of protein (24,585 daltons).

Quin[126] has shown that the AEP in the anemone glycoproteins does not react with 1-fluoro-2,4-dinitrobenzene indicating a blocked amino group. Shelburne and Quin[127] have also found *N*-methyl-AEP in the protein residue of *Anthopleura xanthogrammica*.

V. CATABOLISM OF PHOSPHONATES

Studies on phosphonate catabolism have been concerned with conversions of the carbon skeleton of phosphonates and with the cleavage of the carbon-phosphorus bond.

A. Bacteria

1. In Vivo Studies

Many bacteria are able to grow in a medium in which a phosphonate is the sole source of phosphorus (Table 1). While the range of utilization of phosphonates by a given organism is limited, it is significant that all bacteria which can utilize some phosphonate can use AEP. Indeed, Cook et al.[144] isolated 14 species of bacteria from sewage sludge by growing each isolate on a given phosphonate. Subsequently each organism was tested for growth on the

Table 1
BACTERIA ABLE TO UTILIZE PHOSPHONATE PHOSPHORUS AS SOLE PHOSPHORUS SOURCE

Organism	Ref.
Aerobacter cloacae	135
Bacillus cereus	136
Escherichia coli (various strains)	132,134,135,138,140,143
E. freundi	134
Klebsiella pneumoniae	135,145
Mycobacterium phlei[a]	134
Proteus mirabilis	135
Pseudomonas aeruginosa	133,137,139,141,142
P. putida	144
P. testosteroni	146,147
Salmonella typhimurium	135
Serratia marcescens	135
Shigella sonnei[b]	135

[a] Only in presence of glycerol.
[b] Low growth.

Table 2
PHOSPHONATES ABLE TO SUSTAIN GROWTH OF AT LEAST ONE BACTERIAL SPECIES WHEN THE PHOSPHONATE IS THE SOLE SOURCE OF PHOSPHORUS

Phosphonate	Organic product	Ref.
β-Alanyl-2-aminoethylphosphonate	β-Alanylaminoethane	141
1-Aminobutylphosphonate	N.D.[a]	144
2-Aminoethylphosphonate	Acetaldehyde	135,136,138—144
Aminomethylphosphonate	N.D.	135,143,144
2-Amino-4-phosphonobutyrate	N.D.	135,144
3-Aminopropylphosphonate	*n*-Propylamine	135,141,142
Chloromethylphosphonate	N.D.	138
2,3-Dihydroxypropylphosphonate	N.D.	134
Ethylphosphonate[b]	Ethane	132,134,144,145,147
Glycyl-2-aminoethylphosphonate	N.D.	141
Hydroxymethylphosphonate	N.D.	138
Methylphosphonate[b]	Methane	132,133,135,137,144, 145,147
β-Phosphonoalanine	N.D.	135,143,144
Phenylphosphonate	Benzene	138,145
Propylphosphonate[b]	Propane	147

[a] Not determined.
[b] Also several mono- and diesters.

other phosphonates. Each organism had a narrow spectrum of growth using primarily the compound on which it was isolated, and each used AEP. The ubiquitous degradation of AEP by bacteria would appear to be of potential importance in recycling phosphorus in nature.

The cleavage of the carbon-phosphorus bond apparently occurs generally by a hydrolytic process, with the elements of water incorporated into the products — OH on the phosphorus and H on the carbon (Table 2).

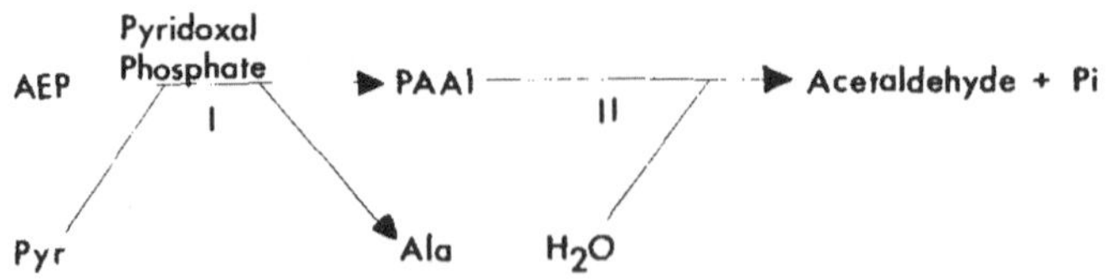

FIGURE 5. Enzymatic breakdown of AEP by *Bacillus cereus*. Abbreviations: Pyr, pyruvate; PAAl, phosphonoacetaldehyde; Ala, alanine.

In some organisms AEP can act as both a nitrogen and phosphorus source;[139,142] *Pseudomonas putida* can grow with AEP supplying carbon, nitrogen, and phosphorus.[144]

In no bacterium can a phosphonate be used if P_i is present.[136,140] Once the P_i has been depleted then, after a lag in growth, phosphonate uptake and utilization begins.[136,140] In general, the uptake and cleavage of the phosphonate appears to be correlated with phosphate utilization since little, if any, P_i appears in the medium, while the phosphonate does not accumulate in the cell.[134,135,138,141]

The transport systems for AEP and some other compounds appear to be specific and induceable.[136,142] Other phosphonates may use normal transport sytems of related compounds. Thus, a mutant of *Streptococcus faecalis* lacking the high affinity transport system for aspartate and glutamate cannot transport phosphonoalanine.[148] Similarly *E. coli*, which lacks the glycerol-3-phosphate transport system, cannot take up 3,4-dihydroxybutylphosphonate which the wild type organism does.[149,150] Likewise the uptake of the phosphonate analogues of phenylalanine and tyrosine by *S. typhimurium* is dependent on a functional aromatic amino acid permease.[151]

Phosphonopeptides appear to be transported into bacterial cells more readily than are free phosphonates.[80,152-154] Once inside the cell, the peptides are hydrolyzed to the free phosphonic acid.[152,154] For the compounds tested, the phosphonopeptide, for example, alafosphin (L-alanyl-L-1-aminoethylphosphonate) is a far more effective antibacterial agent than is free L-1-aminoethylphosphonate itself.[152-156] The phosphonate interferes with cell wall synthesis both by inhibiting alanine racemase[153,156] and by acting as a substrate for UDP-*N*-acetylmuramyl-L-alanine synthetase (E.C.6.3.2.8[b]).[133]

2. *In Vitro Studies*

The mechanism of breakdown of AEP by *Bacillus cereus* has been described by LaNauze and co-workers (Figure 5).[157-160]

Enzyme I, AEP transaminase, produces phosphonoacetaldehyde (PAA1) using alanine and pyridoxal phosphate.[157,158] Enzyme II, the phosphonatase (E.C.3.11.1.1),[156,160] hydrolyzes PAA1 to acetaldehyde and phosphate through an intermediate Schiff base, which may be trapped by borohydride reduction.[160] The enzyme is only inhibited by borohydride in the presence of either PAA1 or acetaldehyde.[160]

The phosphonatase contains two subunits of about 35,000 daltons each and is inhibited by cyanide, sulfite, and hydroxylamine which probably react with the imine of the intermediate.[160] Phosphite inhibits only in the presence of the substrate or product aldehydes, also by reacting with the imine.[159] No studies have been reported as to whether both enzymes are induced by growth on AEP or only one of the two is.

A different type of mechanism must obtain for the hydrolysis of the unsubstituted alkyl- and arylphosphonates; methyl-, ethyl-, propyl-, and phenylphosphonates yield methane,[133,146,147] ethane,[146,147] propane,[147] and benzene,[145] respectively. Indeed, in *Pseudomonas aeruginosa*, when AEP is substituted as in β-alanyl-2-aminoethylphosphonate the product obtained is β-alanylaminoethane.[141] No attempt to examine the enzymes responsible for these cleavages has been reported.

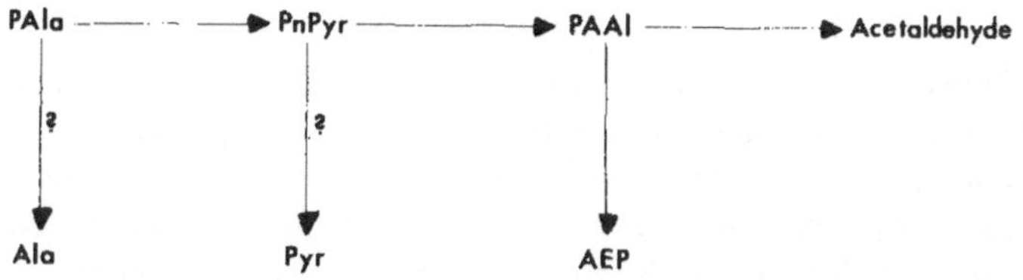

FIGURE 6. Metabolism of phosphonoalanine by *Tetrahymena*. Abbreviations: PAla, phosphonoalanine, PnPyr, phosphonopyruvate; PAAl, phosphonoacetaldehyde; Ala, alanine; Pyr, pyruvate.

Of a number of phosphonates tested in extracts of wild type *E. coli* as substrates for transaminases only 3-aminopropylphosphonate (44 μmol/hr/g protein), phosphonoalanine (105 μmol/hr/g) and 2-amino-4-phosphonobutyrate (1382 μmol/hr/g) were transaminated with α-ketoglutarate as acceptor.[161] Without specific induction of enzymes to utilize phosphonates the ability to use them is apparently minimal.

B. Tetrahymena

Tetrahymena take up AEP from the growth medium and incorporate it into lipids and proteins. There is no evidence for breakdown of the compound in vivo.[50] In cell homogenates however, there is some formation of phosphonoacetaldehyde by transamination.[161] The rate of 135 μmol/hr/g protein at a substrate concentration of 40 *mM* suggests that the reversal of the biosynthetic conversion of phosphonoacetaldehyde to AEP is insignificant, since the intracellular concentration of AEP must be well below that.

Phosphonoalanine is metabolized to a considerable extent in *Tetrahymena*. At least 30% of the radioactivity from exogenous [^{14}C]phosphonoalanine is found in other compounds — predominantly AEP and fatty acids.[50] Phosphonoalanine is not directly decarboxylated to AEP as had originally been postulated[58] but is degraded to phosphonopyruvate which is converted to phosphonoacetaldehyde and in turn to AEP.[64]

Studies on the metabolism of phosphonoalanine in a cell-free system have established this pathway and have also established the breakdown of phosphonoacetaldehyde to acetaldehyde[64] which is presumably oxidized to acetate and incorporated into fatty acids. There is also appearance of radioactivity from phosphonoalanine in both pyruvate and alanine (Figure 6). It has not been established whether phosphonoalanine is converted to alanine, phosphonopyruvate to pyruvate, or whether both may take place. The conversion of phosphonoalanine to products in the cell-free system is less than 1%.

There are several reports that nonenzymatic breakdown of phosphonoalanine in the presence of pyridoxal phosphate and a metal ion gives significant quantities of alanine.[162,163] To what extent this may have contributed to the formation of alanine in the in vitro studies is not known. In addition, nonenzymatic transamination of aminophosphonic acids takes place in the presence of pyridoxal phosphate and an acceptor α-ketocarboxylic acid.[164] This type of reaction could be involved in the low level of transamination of AEP seen in cell-free systems of *Tetrahymena* and other organisms.[161]

The phosphonoalanine used in both the in vivo and in vitro studies was the racemic mixture.[50,64] It may be that the D- and L-isomers are metabolized differently.

There is no conversion of phosphonoacetaldehyde to phosphonoacetic acid.[64] The latter compound is a potent antiviral agent whose mode of action is by inhibition of DNA polymerase.[5] It also inhibits the growth of *Tetrahymena* but is not metabolized itself.[165]

C. Invertebrates

Several snails, *Biomphalaria glabrata*[166] and *Helisoma* sp.[167] have been shown to degrade AEP during embryonic development and hatching. Freshly laid egg masses contain at least 95% of their phosphorus as AEP. The phosphonate content decreases to less than 50% of the total phosphorus during development.[166,167] The newly hatched larvae contain mostly esterified phosphorus as well as $P_{\equiv}$ while the residual egg cases contain phosphorus almost exclusively as an unidentified phosphonate.[167]

Extracts of the sea anemone *Anthopleura elegantissima* have a weak ability to transaminate AEP (107 μmol/hr/g protein) and some related compounds.[161]

D. Vertebrates

In the chicken, most AEP injected intraperitoneally is excreted in unchanged form.[168,169] Some is incorporated into lipids. The degradation which does occur appears to be due to bacterial action in the caecum.[168,169]

Likewise, rats given AEP either orally or intraperitoneally eliminate most of it in both feces and urine without metabolizing the AEP.[114,116,118,170] The breakdown which does occur on oral adminstration is attributed to intestinal bacterial action.[170]

Phosphonoalanine is metabolized by rats in a fashion similar to that in *Tetrahymena;*[171,172] 36 hr after injection of [^{14}C]phosphonoalanine, half of the radioactivity is found in the urine and one third as [^{14}C]CO_2. Of the material retained in the animal most is unchanged phosphonoalanine, some has been converted to AEP while the rest is found in a variety of organic acids.[171]

In vitro phosphonoalanine is metabolized to phosphonopyruvate, phosphonoacetaldehyde, acetaldehyde, and AEP.[172] There is no oxidation of phosphonoacetaldehyde to phosphonoacetate.[172] The latter is not metabolized to CO_2 by the rat, rabbit, dog, or monkey.[173]

The ability of rats to metabolize several other phosphonates has been examined.[174,175] Isopropyl methylphosphonate and methylphosphonate, hydrolysis products of the cholinesterase inhibitor Sarin (isopropyl methylphosphonofluoridate), are not metabolized by rats. Neither the ester nor phosphonate group is cleaved.[174]

Fosamine ammonium (ammonium ethyl phosphonoformamide, Du Pont® Krenite® Brush Control Agent, DPX-1108) is partially hydrolyzed to phosphonoformamide upon oral exposure but the phosphonate group remains intact.[175] It is not clear whether the ester hydrolysis is carried out by intestinal bacteria or by the animal itself.

Since the metabolism of only a few phosphonates has been examined in animals, it is difficult to generalize on the ability of higher animals to cleave the carbon-phosphorus bond or to metabolize the carbons of phosphonates. With increased use of phosphonates as pesticides,[12] antibiotics,[4,73-79] and chemotherapeutic agents[5,176] such studies will presumably be forthcoming.

REFERENCES

1. **Seaman, G. R.,** Inhibition of succinic dehydrogenase of *Tetrahymena geleii* S by a phosphono-substituted succinate analogue, *Arch. Biochem. Biophys.*, 39, 241, 1952.
2. **Seaman, G. R.,** Inhibition of the succinic dehydrogenase of parasitic protozoans by an arsono and a phosphono analog of succinic acid, *Exp. Parasitol.*, 2, 366, 1953.
3. **Horiguchi, M. and Kandatsu, M.,** Isolation of 2-Aminoethane phosphonic acid from rumen protozoa, *Nature (London)*, 184, 901, 1959.

4. **Hendlin, D., Stapley, E. O., Jackson, M., Wallick, H., Miller, A. K., Wolf, F. J., Miller, T. W., Chaiet, L., Kahan, F. M., Foltz, E. L., Woodruff, H. B., Mata, J. M., Hernandez, S., and Mochales, S.**, Phosphonomycin, a new antibiotic produced by strains of *Streptomyces*, *Science*, 166, 122, 1969.
5. **Boezi, J. A.**, The antiherpes action of phosphonoacetate, *Pharm. Ther.*, 4, 231, 1979.
6. **Clements, A. N. and May, T. E.**, Pharmacological studies on a locust neuromuscular preparation, *J. Exp. Biol.*, 61, 421, 1974.
7. **Cull-Candy, S. G., Donnellan, J. F., James, R. W., and Lunt, G. G.**, 2-Amino-4-phosphonobutyric acid as a glutamate antagonist on locust muscle, *Nature (London)*, 262, 408, 1976.
8. **Bioulac, B., de Tinguey-Moreaud, E., Vincent, J.-D., and Neuzil, E.**, Effets de quelques acides aminés phosphonique sur l'activité de neurones centreaux, *C. R. Acad. Sci. Paris*, 285D, 555, 1977.
9. **Bioulac, B., de Tinguey-Moreaud, E., Vincent, J.-D., and Neuzil, E.**, Neuroactive properties of phosphonic amino acids, *Gen. Pharmacol.*, 10, 121, 1979.
10. **Biziere, K., Thompson, H., and Coyle, J. T.**, Characterization of specific high affinity binding sites for L-[^{3}H]glutamic acid in rat brain membranes, *Brain Res.*, 183, 421, 1980.
11. **Slaughter, M. M. and Miller, R. F.**, 2-Amino-4-phosphonobutyric acid: a new pharmacological tool for retina research, *Science*, 211, 182, 1981.
12. **Eto, M.**, *Organophosphorus Pesticides: Organic and Biological Chemistry*, CRC Press, Boca Raton, Fla., 1974.
13. **Kittredge, J. S., Roberts, E., and Simonsen, D. G.**, The occurrence of free 2-aminoethylphosphonate in the sea anemone, *Anthopleura elegantissima*, *Biochemistry*, 1, 624, 1962.
14. **Rouser, G., Kritchevsky, G., Heller, D., and Lieber, E.**, Lipid composition of beef brain, beef liver and the sea anemone: two approaches to quantitative fractionation of complex lipid mixtures, *J. Am. Oil Chem. Soc.*, 40, 425, 1963.
15. **Kandatsu, M. and Horiguchi, M.**, The occurrence of ciliatine (2-aminoethylphosphonic acid) in the goat liver, *Agric. Biol. Chem. (Tokyo)*, 29, 781, 1965.
16. **Shimizu, H., Nakimoto, Y., Nakajima, T., Kanazawa, A., and Sano, I.**, Isolation and identification of 2-aminoethylphosphonic acid from bovine brain, *Nature (London)*, 207, 1197, 1965.
17. **Simon, G. and Rouser, G.**, Phospholipids of the sea anemone: quantitative distribution; absence of carbon-phosphorus linkages in glycerol phospholipids; structural elucidation of ceramide aminoethylphosphonate, *Lipids*, 2, 55, 1967.
18. **Kittredge, J. S., Isbell, A. F., and Hughes, R. R.**, Isolation and characterization of the *N*-methyl derivatives of 2-aminoethylphosphonate from the sea anemone, *Anthopleura xanthogrammica*, *Biochemistry*, 6, 289, 1967.
19. **Hori, T., Sugita, M., and Itasaka, O.**, Biochemistry of shellfish lipids. X. Isolation of a sphingolipid containing 2-monomethylaminoethylphosphonic acid from shellfish, *J. Biochem. (Tokyo)*, 65, 451, 1969.
20. **Hayashi, A., Matsuura, F., and Matsubara T.**, Isolation and characterization of a new sphingolipid containing 2-*N*-methylaminoethylphosphonic acid from the viscera of *Turbo cornutus*, *Biochim. Biophys. Acta*, 176, 208, 1969.
21. **Alhadeff, J. A. and Daves, G. D., Jr.**, Occurrence of 2-aminoethylphosphonic acid in human brain, *Biochemistry*, 9, 4866, 1970.
22. **Alhadeff, J. A. and Daves, G. D., Jr.**, 2-Aminoethylphosphonate: distribution in human tissues, *Biochim. Biophys. Acta*, 244, 211, 1971.
23. **Tan, S. A.**, The occurrence of aminoethylphosphonic acids in human tissues, *Tex. Rep. Biol. Med.*, 29, 408, (Abstr.), 1971.
24. **Tamari, M. and Kametaka, M.**, Isolation of ciliatine (2-aminoethylphosphonic acid) from the bile of the bovine, *Agric. Biol. Chem. (Tokyo)*, 37, 933, 1973.
25. **Tamari, M., Ogawa, M., Hasegawa, S., and Kametaka, M.**, Etudes sur les phosphonolipides de la bile de boeuf, *Agric. Biol. Chem. (Tokyo)*, 40, 2057, 1976.
26. **Hasegawa, S., Tamari, M., and Kametaka, M.**, The distribution of ciliatine (2-aminoethylphosphonic acid) in bovine liver, *Agric. Biol. Chem. (Tokyo)*, 40, 2097, 1976.
27. **Tamari, M., Ogawa, M., and Kametaka, M.**, A new bile acid conjugate, ciliatocholic acid, from bovine gall bladder bile, *J. Biochem. (Tokyo)*, 80, 371, 1976.
28. **Tamari, M. and Kametaka, M.**, Distribution de la phosphonates dans le sang et le sperme de boeuf, *Agric. Biol. Chem. (Tokyo)*, 44, 1957, 1980.
29. **Kandatsu, M. and Horiguchi, M.**, Occurrence of ciliatine (2-aminoethylphosphonic acid) in *Tetrahymena*, *Agric. Biol. Chem. (Tokyo)*, 26, 721, 1962.
30. **Sugita, M. and Hori, T.**, Isolation of diacylglycerol-2-aminoethylphosphonate from *Tetrahymena pyriformis*, *J. Biochem. (Tokyo)*, 60, 1149, 1971.
31. **Korn, E. D., Dearborn, D. G., Fales, H. M., and Sokoloski, E. A.**, Phosphonoglycan: a major polysaccharide constituent of the amoeba plasma membrane contains 2-aminoethylphosphonic acid and 1-hydroxy-2-aminoethylphosphonic acid, *J. Biol. Chem.*, 248, 2257, 1973.

32. **Glonek, T., Henderson, T. O., Hilderbrand, R. L., and Myers, T. C.,** Biological phosphonates: determination by phosphorus-31 nuclear magnetic resonance, *Science*, 169, 192, 1970.
33. **Henderson, T. O., Glonek, T., Hilderbrand, R. L., and Myers, T. C.,** Phosphorus-31 nuclear magnetic resonance studies of the phosphonate and phosphate composition of the sea anemone, *Bundasoma* sp., *Arch. Biochem. Biophys.*, 149, 484, 1972.
34. **Ugurbil, K., Rottenberg, H., Glynn, P., and Shulman, R. G.,** ^{31}P nuclear magnetic resonance studies of bioenergetics and glycolysis in anaerobic *Escherichia coli* cells, *Proc. Natl. Acad. Sci. U.S.A.*, 75, 2244, 1978.
35. **Kittredge, J. S. and Hughes, R. R.,** The occurrence of α-amino-β-phosphonopropionic acid in the zoanthid, *Zoanthus societus*, and the ciliate, *Tetrahymena pyriformis*, *Biochemistry*, 3, 991, 1964.
36. **Rosenberg, H.,** Distribution and fate of 2-aminoethylphosphonic acid in *Tetrahymena*, *Nature (London)*, 203, 299, 1964.
37. **Liang, C. R. and Rosenberg, H.,** On the distribution and biosynthesis of 2-aminoethylphosphonate in two terrestrial molluscs, *Comp. Biochem. Physiol.*, 25, 673, 1968.
38. **Kittredge, J. S., Horiguchi, M., and Williams, P. M.,** Aminophosphonic acids: biosynthesis by marine phytoplankton, *Comp. Biochem. Physiol.*, 29, 859, 1969.
39. **Marmouyet, J., Maget-Dana, R., and Douste-Blazy, L.,** Possibilité de biosynthèse "de novo" de phosphonates dans le foie de rat, *Biochimie*, 57, 261, 1975.
40. **Steiner, S., Conti, S. F., and Lester, R. L.,** Occurrence of phosphonosphingolipids in *Bdellovibrio bacteriovorus* strain UK$_i$2, *J. Bacteriol.*, 116, 1199, 1973.
41. **Sarma, G. R., Chandramouli, V., and Venkitasubramanian, T. A.,** Occurrence of phosphonolipids in mycobacteria, *Biochim. Biophys. Acta*, 218, 561, 1970.
42. **Snyder, W. R. and Law, J. H.,** A quantitative determination of phosphonate phosphorus in naturally occurring aminophosphonates, *Lipids*, 5, 800, 1970.
43. **LaBelle, Y.-S. L. and Walker, R. W.,** Phospholipids of *Mycobacterium smegmatis*, *Am. Rev. Respir. Dis.*, 105, 625, 1972.
44. **Horiguchi, M.,** Biochemical studies of ciliatine (in Japanese), *J. Agric. Chem. Soc. (Jpn.)*, 40, R25, 1966.
45. **Coleman, G. S., Kemp, P., and Dawson, R. M. C.,** The catabolism of phosphatidylethanolamine by the rumen protozoan *Entodinium caudatum* and its conversion into the *N*-(1-carboxyethyl) derivative, *Biochem. J.*, 123, 97, 1971.
46. **Stein, R. B. and Blum, J. J.,** Quantitative analysis of intermediary metabolism in *Tetrahymena*. Cells grown in proteose-peptone and resuspended in a defined nutrient-rich medium, *J. Biol. Chem.*, 254, 10385, 1979.
47. **Stein, R. B. and Blum, J. J.,** Quantitative analysis of intermediary metabolism in *Tetrahymena*. Cells grown in glucose-supplemented medium, *J. Biol. Chem.*, 255, 4198, 1980.
48. **Berger, H., Jones, P., and Hanahan, D. J.,** Structural studies on lipids of *Tetrahymena pyriformis* W., *Biochim. Biophys. Acta*, 260, 617, 1972.
49. **Liang, C. R. and Rosenberg, H.,** The biosynthesis of the carbon-phosphorus bond in *Tetrahymena*, *Biochim. Biophys. Acta*, 156, 437, 1968.
50. **Smith, J. D. and Law, J. H.,** Phosphonic acid metabolism in *Tetrahymena*, *Biochemistry*, 9, 2152, 1970.
51. **Horiguchi, M. and Rosenberg, H.,** Phosphonopyruvic acid: a probable precursor of phosphonic acids in cell-free preparations of *Tetrahymena*, *Biochim. Biophys. Acta*, 404, 333, 1975.
52. **Dennis, E. A. and Kennedy, E. P.,** Enzymatic synthesis and decarboxylation of phosphatidylserine in *Tetrahymena pyriformis*, *J. Lipid Res.*, 11, 394, 1970.
53. **Itasaka, O., Hori, T., and Sugita, M.,** Biochemistry of shellfish lipids. XI. Incorporation of [^{32}P]orthophosphate into ceramide ciliatine (2-aminoethylphosphonic acid) of the freshwater mussel, *Hyriopsis schlegelii*, *Biochim. Biophys. Acta*, 176, 783, 1969.
54. **Miceli, M. V., Myers, T. C., and Henderson, T. O.,** Alkylphosphonic acid metabolism in the planorbid snail *Helisoma* Sp., *Fed. Proc.*, 39, 2820 (Abstr.), 1980.
55. **Sampugna, J., Johnson, L., Bochman, K., and Keeney, M.,** Lipids of *Crossostrea virginica*. I. Preliminary investigations of aldehyde and phosphorus lipids in oyster tissue, *Lipids*, 7, 339, 1972.
56. **Alhadeff, J. A., Van Brugen, J. T., and Daves, G. D.,** Biosynthetic studies on 2-aminoethylphosphonic acid in a mammalian (rat) system, *Biochim. Biophys. Acta*, 286, 103, 1972.
57. **Trebst, A. and Geike, F.,** Zur Biosynthesis von Phosphonoaminosäuren. Die Verteilung der Radioactivität in Aminoäthylphosphonsäure nach Einbau von positionsmarkierter Glucose durch *Tetrahymena*, *Z. Naturforsch.*, 22b, 989, 1967.
58. **Warren, W. A.,** Biosynthesis of phosphonic acids in *Tetrahymena*, *Biochim. Biophys. Acta*, 156, 340, 1968.
59. **Horiguchi, M., Kittredge, J. S., and Roberts, E.,** Biosynthesis of 2-aminoethylphosphonic acid in *Tetrahymena*, *Biochim. Biophys. Acta*, 165, 164, 1968.
60. **Hill, D. L.,** *The Biochemistry and Physiology of Tetrahymena*, Academic Press, New York, 1972, 39.

61. **Segal, W.**, Biosynthesis of 2-aminoethylphosphonic acid: a phosphoramidic acid rearrangement?, *Nature (London)*, 208, 1284, 1965.
62. **Horiguchi, M.**, Biosynthesis of 2-aminoethylphosphonic acid in cell-free preparations from *Tetrahymena*, *Biochim. Biophys. Acta*, 261, 102, 1972.
63. **Horiguchi, M.**, Biosynthesis of phosphonic acids in *Tetrahymena pyriformis*, in *Progress in Protozoology*, 4th Int. Congr., de Puytorac, P. and Grain, J., Eds., UER Sciences, Clermont, 1973, 188.
64. **Horigane, A., Horiguchi, M., and Matsumoto, T.**, Metabolism of 2-amino-3-phosphono[3-^{14}C]propionic acid in cell-free preparations of *Tetrahymena*, *Biochim. Biophys. Acta*, 618, 383, 1980.
65. **Geike, F.**, Zur Biosynthese der Phosphonoaminosäuren II, *Naturwissenschaften*, 56, 462, 1969.
66. **Smith, J. D. and Law, J. H.**, Phosphatidylcholine biosynthesis in *Tetrahymena pyriformis*, *Biochim. Biophys. Acta*, 202, 141, 1970.
67. **Viswanathan, C. V. and Rosenberg, H.**, Isolation of ceramide-monomethylaminoethylphosphonate from the lipids of *Tetrahymena pyriformis* W, *J. Lipid Res.*, 14, 327, 1973.
68. **Smith, J. D. and O'Malley, M. A.**, Control of phosphonic acid and phosphonolipid synthesis in *Tetrahymena*, *Biochim. Biophys. Acta*, 528, 394, 1978.
69. **Smith, J. D. and Giegel, D. A.**, Replacement of ethanolamine phosphate by 3-aminopropylphosphonate in the phospholipids of *Tetrahymena*, *Arch. Biochem. Biophys.*, 206, 420, 1981.
70. **Rogers, T. O. and Birnbaum, J.**, Biosynthesis of fosfomycin by *Streptomyces fradiae*, *Antimicrob. Agents Chemother.*, 5, 121, 1974.
71. **Rogers, T. O., Birnbaum, J., and Demain, A.**, Fosfomycin biosynthesis: nutritional and radioisotope incorporation studies, *Bacteriol. Proc.*, A42 (Abstr.), 1971.
72. **White, R. F., Birnbaum, J., Meyer, R. T., ten Broeke, J., Chemerda, J. M., and Demain, A. L.**, Microbial epoxidation of cis-propenylphosphonic acid to (-)cis-1,2-epoxypropylphosphonic acid, *Appl. Microbiol.*, 22, 55, 1971.
73. **Park, B. K., Hirota, A., and Sakai, H.**, 2-Amino-5-phosphono-3-pentenoic acid, a new amino acid from N-1409 substance, an antagonist of threonine, *Agric. Biol. Chem. (Tokyo)*, 40, 1905, 1976.
74. **Kamiya, T., Hemmi, K., Takeno, H., and Hashimoto, M.**, Studies on phosphonic acid antibiotics. I. Structure and synthesis of 3-(*N*-Acetyl-*N*-hydroxyamino)propylphosphonic acid (FR-900098) and its *N*-formyl analogue (FR-31564), *Tetrahedron Lett.*, 21, 95, 1980.
75. **Hashimoto, M., Hemmi, K., Takeno, H., and Kamiya, T.**, Studies on phosphonic acid antibiotics. II. Synthesis of 3-(*N*-acetyl-*N*-hydroxyamino)-2(R)-hydroxypropylphosphonic acid (FR-33289) and 3-(*N*-formyl-*N*-hydroxyamino)-1-trans-propenylphosphonic acid (FR-32863), *Tetrahedron Lett.*, 21, 99, 1980.
76. **Okuhara, M., Kuroda, Y., Goto, T., Okamoto, M., Terano, H., Kohsaka, M., Aoki, H., and Imanaka, H.**, Studies on new phosphonic acid antibiotics. I. FR-900098, isolation and characterization, *J. Antibiot.*, 23, 13, 1980.
77. **Iguchi, E., Okuhara, M., Kohsaka, M., Aoki, H., and Imanaka, H.**, Studies on new phosphonic acid antibiotics. II. Taxonomic studies on producing organisms of the phosphonic acid and related compounds, *J. Antibiot.*, 23, 18, 1980.
78. **Okuhara, M., Kuroda, Y., Goto, T., Okamoto, M., Terano, H., Kohsaka, M., Aoki, H., and Imanaka, H.**, Studies on new phosphonic acid antibiotics. III. Isolation and characterization of FR-31564, FR-32863 and FR-33289, *J. Antibiot.*, 23, 24, 1980.
79. **Bayer, E., Gugel, K. H., Hägele, K., Hagenmaier, H., Jessipow, S., König, W. A., and Zähner, H.**, Phosphinothricin und phosphinothricyl-alanyl-alanin, *Helv. Chim. Acta*, 55, 224, 1972.
80. **Allen, J. G., Atherton, F. R., Hall, M. J., Hassall, C. H., Holmes, S. W., Lambert, R. W., Nisbet, L. J., and Ringrose, P. S.**, Phosphonopeptides, a new class of synthetic antibacterial agents, *Nature (London)*, 272, 56, 1978.
81. **Hori, T. and Arakawa, I.**, Isolation and characterization of new sphingolipids containing *N,N*-acylmethylaminoethylphosphonic acid and *N*-acyl aminoethylphosphonic acid from the mussel, *Corbicula sandai*, *Biochim. Biophys. Acta*, 176, 898, 1969.
82. **Ullman, M. D. and Radin, N. S.**, The enzymatic formation of sphingomyelin from ceramide and lecithin in mouse liver, *J. Biol. Chem.*, 249, 1506, 1974.
83. **Becker, G. W. and Lester, R. L.**, Biosynthesis of phosphoinositol-containing sphingolipids from phosphatidylinositol by a membrane preparation from *Saccharomyces cerevisiae*, *J. Bacteriol.*, 142, 747, 1980.
84. **Liang, C.-R., Segura, M., and Strickland, K. P.**, Phospholipid metabolism in the molluscs. II. Activities of choline kinase, ethanolamine kinase and CTP: phosphorylethanolamine cytidyltransferase in the mollusc *Helix lactea*, *Can. J. Biochem.*, 48, 580, 1970.
85. **Dearlove, G. E., Bishop, S. H., and Viles, J. M.**, Aminoethylphosphonic acid containing glycoproteins as acidic sea anemone mucus components, *Fed. Proc.*, 38, 2229 (Abstr.), 1979.
86. **Dearlove, G. E., Viles, J. M., and Bishop, S. H.**, Aminoethylphosphonic acid (AEP) containing glycoproteins in *Metridium senile*: ^{14}C-AEP incorporation and electron microprobe microanalysis, *J. Exp. Zool.*, 210, 169, 1979.

87. **Nandedkar, A. K.,** Report on the utilization of ethanolamine-1-^{14}C by *Mycobacterium* 607, *Biochem. Med.,* 11, 67, 1974.
88. **Nandedkar, A. K. N.,** Biosynthesis of phosphatidylethanolamine in *Mycobacterium* 607, *Biochem. Med.,* 12, 116, 1975.
89. **Tyhach, R. J., Engel, R., and Tropp, B. E.,** Metabolic fate of 3,4-dihydroxybutyl-1-phosphonate in *Escherichia coli, J. Biol. Chem.,* 251, 6717, 1976.
90. **Cheng, P.-J., Nunn, W. D., Tyhach, R. J., Goldstein, S. L., Engel, R., and Tropp, B. E.,** Investigations concerning the mode of action of 3,4-dihydroxybutyl-1-phosphonate on *Escherichia coli, J. Biol. Chem.,* 250, 1633, 1975.
91. **Klein, D. A., Engel, R., and Tropp, B. E.,** The incorporation of 3,4-dihydroxybutyl-1-phosphonate, a glycerol-3-phosphate analogue, into the cell wall of *Bacillus subtilis, J. Bacteriol.,* 129, 550, 1977.
92. **Viswanathan, C. V. and Nagabhushanam, A.,** Preparative isolation of phosphonolipids by ascending dry-column chromatography, *J. Chromatogr.,* 75, 227, 1973.
93. **Liang, C.-R. and Rosenberg, H.,** The biosynthesis of the phosphonic analogue of cephalin in *Tetrahymena, Biochim. Biophys. Acta,* 125, 548, 1966.
94. **Nozawa, Y., Iidi, H., Fukushima, H., Ohki, K., and Ohnishi, S.,** Studies on *Tetrahymena* membranes: temperature-induced alterations in fatty acid composition of various membrane fractions in *Tetrahymena pyriformis* and its effect on membrane fluidity as inferred by spin-label study, *Biochim. Biophys. Acta,* 367, 134, 1974.
95. **Fukushima, H., Martin, C. E., Iida, H., Kitajima, Y., Thompson, G. A., Jr., and Nozawa, Y.,** Changes in membrane lipid composition during temperature adaptation by a thermotolerant strain of *Tetrahymena pyriformis, Biochim. Biophys. Acta,* 431, 165, 1976.
96. **Nozawa, Y. and Kasai, R.,** Mechanism of thermal adaptation of membrane lipids in *Tetrahymena pyriformis* NT-1. Possible evidence for temperature-mediated induction of palmitoyl-CoA desaturase, *Biochim. Biophys. Acta,* 529, 54, 1978.
97. **Pieringer, J.-A. and Conner, R. L.,** Positional distribution of fatty acids in the glycerophospholipids of *Tetrahymena pyriformis, J. Lipid Res.,* 20, 363, 1979.
98. **Nandini-Kishore, S. G., Mattox, S. M., Martin, C. E., and Thompson, G. A., Jr.,** Membrane changes during growth of *Tetrahymena* in the presence of ethanol, *Biochim. Biophys. Acta,* 551, 315, 1979.
99. **Shug, A. L., Elson, C., and Shrago, E.,** Effect of iron on growth, cytochromes, glycogen and fatty acids of *Tetrahymena pyriformis, J. Nutr.,* 99, 379, 1969.
100. **Peng, Y.-M. and Elson, C.,** Effect of iron on lipid metabolism of *Tetrahymena pyriformis, J. Nutr.,* 101, 1177, 1971.
101. **Mattox, S. M. and Thompson, G. A., Jr.,** The effects of high concentrations of sodium or calcium ions on the lipid composition and properties of *Tetrahymena* membranes, *Biochim. Biophys. Acta,* 599, 24, 1980.
102. **Smith, J. D., Snyder, W. R., and Law, J. H.,** Phosphonolipids in *Tetrahymena* cilia, *Biochem. Biophys. Res. Commun.,* 39, 1163, 1970.
103. **Kennedy, K. E. and Thompson, G. A., Jr.,** Phosphonolipids: localization in surface membranes of *Tetrahymena, Science,* 168, 989, 1970.
104. **Nozawa, Y. and Thompson, G. A., Jr.,** Studies of membrane formation in *Tetrahymena pyriformis.* II. Isolation and lipid analysis of cell fractions, *J. Cell Biol.,* 49, 712, 1971.
105. **Jonah, M. and Erwin, J. A.,** The lipids of membranous cell organelles isolated from the ciliate, *Tetrahymena pyriformis, Biochim. Biophys. Acta,* 231, 80, 1971.
106. **Bridges, R. G. and Ricketts, J.,** Formation of a phosphonolipid by larvae of the housefly, *Musca domestica, Nature (London),* 211, 199, 1966.
107. **Bieber, L. L.,** Incorporation of trimethylaminoethylphosphonic acid and dimethylaminoethylphosphonic acid into lipids of housefly larvae, *Biochim. Biophys. Acta,* 152, 778, 1968.
108. **Bridges, R. G. and Ricketts, J.,** Incorporation of *N*-methylaminoethanol and *N*-dimethylaminoethanol into the phospholipids of the housefly, *Musca domestica, Biochem. J.,* 95, 41P, 1965.
109. **Bridges, R. G., Ricketts, J., and Cox, J. T.,** The replacement of lipid-bound choline by other bases in the phospholipids of the housefly, *Musca domestica, J. Insect Physiol.,* 11, 225, 1965.
110. **Swift, M. L.,** Phosphono-lipid content of the oyster, *Crassostrea virginica* in three physiological conditions, *Lipids,* 12, 449, 1977.
111. **Hasegawa, S., Tamari, M., and Kametaka, M.,** Isolation of diacylglyceryl-2-aminoethylphosphonate from bovine liver, *J. Biochem. (Tokyo),* 80, 531, 1976.
112. **Nair, C. R., Stith, I. E., Nair, R. R., and Das, S. K.,** The effect of aminoethylphosphonic acid on hepatic microsomal drug-metabolism and ultrastructure of normal and cholesterol fed rats, *J. Nutr.,* 108, 1234, 1978.
113. **Kandatsu, M., Horiguchi, M., and Tamari, M.,** The incorporation of ciliatine (2-aminoethylphosphonic acid) into lipids of the rat liver, *Agric. Biol. Chem. (Tokyo),* 29, 779, 1965.

114. **Curley, J. M. and Henderson, T. O.**, The incorporation of 2-aminoethylphosphonic acid into rat liver diacylglyceroaminoethylphosphonate, *Lipids*, 7, 676, 1972.
115. **Tamari, M., Maget-Dana, R., Marmouyet, J., and Douste-Blazy, L.**, CMP-aminoéthylphosphonate: intérmediare de la biosynthèse des phosphonolipides dans le foie de rat, *Biochimie*, 55, 1311, 1973.
116. **Maget-Dana, R., Tamari, M., Marmouyet, J., and Douste-Blazy, L.**, Incorporation de l'acide 2-aminoéthylphosphonique dans les lipides tissulaires de rat, *Eur. J. Biochem.*, 42, 129, 1974.
117. **Tamari, M., Cassaigne, A., Lacoste, A.-M., and Neuzil, E.**, *In vivo* incorporation of CMP-ciliatine into rat liver lipids, *Biochimie*, 57, 97, 1975.
118. **Curley-Joseph, J. and Henderson, T. O.**, 2-Aminoethylphosphonic acid metabolism in the rat, *Lipids*, 12, 75, 1977.
119. **Bjerve, K. S.**, Lecithin biosynthesis in the rat studied with phosphonate analogues of phosphorylcholine, *Biochim. Biophys. Acta*, 270, 348, 1972.
120. **Krause, R. F., Beamer, K. C., and Lotspeich, H. J.**, Phosphonolipid metabolism in control and Vitamin A deficient rats, *Proc. Soc. Exp. Biol. Med.*, 140, 544, 1972.
121. **Plantavid, M., Maget-Dana, R., and Douste-Blazy, L.**, Interactions d'analogues de la phosphoryléthanolamine avec la phosphoéthanolaminecytidyltransférase, *Biochimie*, 57, 951, 1975.
122. **Dana, R. and Douste-Blazy, L.**, Influence de l'acide amino-2-éthylphosphonique sur l'incorporation *in vitro* de ^{32}P dans les phospholipides cérébraux, *C. R. Acad. Sci. Paris*, 268D, 185, 1969.
123. **Dana, R. and Douste-Blazy, L.**, Effet de l'acide amino-2-éthylphosphonique sur la biosynthèse *in vitro* de phospholipides cérébraux, *Bull. Soc. Chim. Biol.*, 52, 405, 1970.
124. **Allen, A. K. and Rosenberg, H.**, The mechanism of action and some properties of serine ethanolamine phosphate synthetase, *Biochim. Biophys. Acta*, 151, 504, 1968.
125. **Quin, L. D.**, 2-aminoethylphosphonic acid in insoluble protein of the sea anemone, *Metridium dianthus*, *Science*, 144, 1133, 1964.
126. **Quin, L. D.**, The presence of compounds with a carbon-phosphorus bond in some marine invertebrates, *Biochemistry*, 4, 324, 1965.
127. **Shelburne, F. A. and Quin, L. D.**, Isolation of 2-(methylamino)-ethylphosphonic acid from the proteinaceous residue of a sea anemone, *Biochim. Biophys. Acta*, 148, 595, 1967.
128. **Dearborn, D. G., Smith, S., and Korn, E. D.**, Lipophosphonoglycan of the plasma membrane of *Acanthamoeba castellanii*. Inositol and phytosphingosine content and general structural features, *J. Biol. Chem.*, 251, 2976, 1976.
129. **Lepak, N. M. and Smith, J. D.**, Characterization of phosphonoglycoproteins from *Tetrahymena*, *Fed. Proc.*, 40, 1845 (Abstr.), 1981.
130. **Lubansky, H. J. and Henderson, T. O.**, *In vivo* incorporation of 2-aminoethylphosphonic acid into the phosphoglycoproteins of *Metridium dianthus*, *Fed. Proc.*, 37, 1729 (Abstr.), 1978.
131. **Stevenson, K. J., Gibson, D., and Dixon, G. H.**, Amino acid analyses of chymotrypsin-like proteases from the sea anemone *(Metridium senile)*, *Can. J. Biochem.*, 52, 93, 1974.
132. **Zeleznick, L. D., Myers, T. C., and Titchener, E. B.**, Growth of *Escherichia coli* on methyl- and ethylphosphonic acids, *Biochim. Biophys. Acta*, 78, 546, 1963.
133. **James, E. A., Jr., Myers, T. C., and Titchener, E. B.**, Bacterial cleavage of methylphosphonic acid, *Fed. Proc.*, 24, 440 (Abstr.), 1965.
134. **Mastalerz, P., Wieczorek, Z., and Kochman, M.**, Utilization of carbon-bound phosphorus by microorganisms, *Acta Biochim. Pol.*, 12, 151, 1965.
135. **Harkness, D. R.**, Bacterial growth on aminoalkylphosphonic acids, *J. Bacteriol.*, 92, 623, 1966.
136. **Rosenberg, H. and La Nauze, J. M.**, The metabolism of phosphonates by microorganisms. The transport of aminoethylphosphonate in *Bacillus cereus*, *Biochim. Biophys. Acta*, 141, 79, 1967.
137. **Labeyrie, S., Valette, J.-P., Lacoste, A.-M., and Neuzil, E.**, Chromogenèse des souches érythrogènes de *Pseudomonas aeruginosa* en mileau synthétique, *C. R. Acad. Sci. Paris*, 266D, 1898, 1968.
138. **Alam, A. U. and Bishop, S. H.**, Growth of *Escherichia coli* on some organophosphonic acids, *Can. J. Microbiol.*, 15, 1043, 1969.
139. **Lacoste, A.-M. and Neuzil, E.**, Transamination de l'acide amino-2-éthylphosphonique par *Pseudomonas aeruginosa*, *C. R. Acad. Sci. Paris*, 269D, 254, 1969
140. **Tamari, M., Horiguchi, M., and Kandatsu, M.**, Isolation and characterization of ciliatine decomposing bacteria from the feces of sheep (studies on the metabolism of ciliatine (2-aminoethylphosphonic acid) Part III) (in Japanese), *J. Agric. Chem. Soc. (Jpn.)*, 49, 653, 1975.
141. **Cassaigne, A., Lacoste, A.-M., and Neuzil, E.**, Recherches sur le catabolism des acides phosphoniques: biodégradation de la liason C–P par *Pseudomonas aeruginosa*, *C. R. Acad. Sci. Paris*, 282D, 1637, 1976.
142. **Lacoste, A. M., Cassaigne, A., Tamari, M., and Neuzil, E.**, Transport de l'acide amino-2-éthylphosphonique chez *Pseudomonas aeruginosa*, *Biochimie*, 58, 703, 1976.
143. **Tamari, M., Horiguchi, M., and Kandatsu, M.**, Growth of *Escherichia coli* TM-1 on natural phosphonic acids and their related compounds, *J. Gen. Appl. Microbiol.*, 23, 49, 1977.

144. **Cook, A. M., Daughton, C. G., and Alexander, M.,** Phosphonate utilization by bacteria, *J. Bacteriol.*, 133, 85, 1978.
145. **Cook, A. M., Daughton, C. G., and Alexander, M.,** Benzene from bacterial cleavage of the carbon-phosphorus bond of phenylphosphonates, *Biochem. J.*, 184, 453, 1979.
146. **Daughton, C. G., Cook, A. M., and Alexander, M.,** Biodegradation of phosphonate toxicants yields methane or ethane on cleavage of the C–P bond, *FEMS Microbiol. Lett.*, 5, 91, 1979.
147. **Daughton, C. G., Cook, A. M., and Alexander, M.,** Bacterial conversion of alkylphosphonates to natural products via carbon-phosphorus bond cleavage, *J. Agric. Food Chem.*, 27, 1375, 1979.
148. **Holden, J. T., Van Balgooy, J. N. A., and Kittredge, J. S.,** Transport of aminophosphonic acids in *Lactobacillus plantarum* and *Streptococcus faecalis*, *J. Bacteriol.*, 96, 950, 1968.
149. **Shopsis, C. S., Engel, R., and Tropp, B. E.,** Effects of phosphonic acid analogues of glycerol-3-phosphate on the growth of *Escherichia coli*, *J. Bacteriol.*, 112, 408, 1972.
150. **Leifer, Z., Engel, R., and Tropp, B. E.,** Transport of 3,4-dihydroxybutyl-1-phosphonate, an analogue of *sn*-glycerol-3-phosphate, *J. Bacteriol.*, 130, 968, 1977.
151. **Ames, G. F. and Roth, J. P.,** Histidine and aromatic permeases of *Salmonella typhimurium*, *J. Bacteriol.*, 96, 1742, 1968.
152. **Atherton, F. R., Hall, M. J., Hassall, C. H., Lambert, R. W., and Ringrose, P. S.,** Phosphonopeptides as antibacterial agents: rationale, chemistry, and structure activity relationships, *Antimicrob. Agents Chemother.*, 15, 677, 1979.
153. **Atherton, F. R., Hall, M. J., Hassall, C. H., Lambert, R. W., Lloyd, W. J., and Ringrose, P. S.,** Phosphonopeptides as antibacterial agents: mechanism of action of alaphosphin, *Antimicrob. Agents Chemother.*, 15, 696, 1979.
154. **Allen, J. G., Havas, L., Leicht, E., Lenox-Smith, I., and Nisbet, L. J.,** Phosphonopeptides as antibacterial agents: metabolism and pharmacokinetics of alafosfalin in animals and humans, *Antimicrob. Agents Chemother.*, 16, 306, 1979.
155. **Dulaney, E. L.,** 1-aminoethylphosphonic acid, an inhibitor of bacterial cell wall synthesis, *J. Antibiot.*, 23, 567, 1970.
156. **Lacoste, A. M., Cassaigne, A., and Neuzil, E.,** Acides aminés phosphoniques et croissance de *Pseudomonas aeruginosa*, *C. R. Acad. Sci. Paris*, 280D, 1173, 1975.
157. **La Nauze, J. M. and Rosenberg, H.,** The breakdown of aminoethylphosphonate by cell-free extracts of *Bacillus cereus*, *Biochim. Biophys. Acta*, 148, 811, 1967.
158. **La Nauze, J. M. and Rosenberg, H.,** The identification of 2-phosphonoacetaldehyde as an intermediate in the degradation of 2-aminoethylphosphonate by *Bacillus cereus*, *Biochim. Biophys. Acta*, 165, 438, 1968.
159. **La Nauze, J. M., Rosenberg, H., and Shaw, D. C.,** The enzymatic cleavage of the carbon-phosphorus bond; purification and properties of phosphonatase, *Biochim. Biophys. Acta*, 212, 332, 1970.
160. **La Nauze, J. M., Coggins, J. R., and Dixon, H. B. F.,** Aldolase-like imine formation in the mechanism of action of phosphonoacetaldehyde hydrolase, *Biochem. J.*, 165, 409, 1977.
161. **Roberts, E., Simonsen, D. G., Horiguchi, M., and Kittredge, J. S.,** Transamination of aminoalkylphosphonic acids with alpha ketoglutarate, *Science*, 159, 886, 1968.
162. **Cassaigne, A., Lacoste, A. M., and Neuzil, E.,** Transamination non enzymatique des acides aminés phosphoniques, *Biochim. Biophys. Acta*, 252, 506, 1971.
163. **Martel, A. E. and Langlohr, M. F.,** Metal ion- and pyridoxal-catalyzed transamination and dephosphonylation of 2-amino-3-phosphonopropionic acid. A new phosphonatase model, *J. Chem. Soc. Chem. Commun.*, 10, 342, 1977.
164. **Neuzil, E., Cassaigne, A., and Lacoste, A.-M.,** Transamination non enzymatique des acides aminoalkylphosphoniques par l'acide glyoxylique, *C. R. Acad. Sci. Paris*, 270D, 2724, 1970.
165. **Smith, J. D., King, J. C., and Holland, E. C.,** *J. Protozool.*, 29, 103, 1982.
166. **Henderson, T. O., Miceli, M. V., and Myers, T. C.,** Occurrence of 2-aminoethylphosphonic acid in the eggs of the schistosomal vector *Biomphalara glabrata* and its metabolism during embryonic development, *Fed. Proc.*, 39, 2821 (Abstr.), 1980.
167. **Miceli, M. V., Henderson, T. O., and Myers, T. C.,** 2-Aminoethylphosphonic acid metabolism during embryonic development of the planorbid snail *Helisoma*, *Science*, 209, 1245, 1980.
168. **Cregar, C. R. and Kubena, L. F.,** The metabolism of [^{14}C]-2-aminoethylphosphonic acid by the laying hen, *Poult. Sci.*, 48, 1799 (Abstr.), 1969.
169. **Tamari, M., Hasegawa, S., and Kametaka, M.,** Metabolisme de l'acide 2-aminoéthylphosphonique (ciliatine) chez le poulet, *Agric. Biol. Chem. (Tokyo)*, 40, 2117, 1976.
170. **Tamari, M., Horiguchi, M., and Kandatsu, M.,** Studies on metabolism of ciliatine (2-aminoethylphosphonic acid) part I. Digestion, absorption and excretion of free-form-ciliatine in rats (in Japanese), *J. Agric. Chem. Soc. (Jpn.)*, 45, 433, 1971.
171. **Horigane, A., Horiguchi, M., and Matsumoto, T.,** Metabolism of 2-amino-3-phosphonopropionic acid in rats, *Biochim. Biophys. Acta*, 572, 385, 1979.

172. **Horigane, A., Horiguchi, M., and Matsumoto, T.,** Metabolism of 2-amino-3-phosphono[3-^{14}C]propionic acid in cell-free preparations of rat liver, *Biochim. Biophys. Acta*, 584, 254, 1979.
173. **Bopp, B. A., Estep, C. B., and Anderson, D. J.,** Disposition of disodium phosphonoacetate-^{14}C in rat, rabbit, dog and monkey, *Fed. Proc.*, 36, 3524 (Abstr.), 1977.
174. **Hoskin, F. C. G.,** Some observations concerning the biochemical inertness of methylphosphonic acid and isopropyl methylphosphonic acid, *Can. J. Biochem. Physiol.*, 34, 743, 1956.
175. **Chrzanowski, R. L., Han, J. C.-Y., and McIntosh, C. L.,** Metabolism of carbon-14 fosamine ammonium in the rat, *J. Agric. Food Chem.*, 27, 550, 1979.
176. **Helgstrand, E., Eriksson, B., Johansson, N. G., Lannerö, B., Larsson, A., Misiorny, A., Norén, J. O., Sjöberg, B., Stenberg, K., Stenning, G., Stridh, S., Öberg, B., Alenius, S., and Philipson, L.,** Trisodium phosphonoformate, a new antiviral compound, *Science*, 201, 819, 1978.
177. **Thompson, G. A., Jr.,** Studies of membrane formation in *Tetrahymena pyriformis*. I. Rates of phospholipid biosynthesis, *Biochemistry*, 6, 2015, 1967.
178. **Smith, J. D. and Giegel, D. A.,** Effect of a phosphonic acid analog of choline phosphate on phospholipid metabolism in *Tetrahymena*, *Arch. Biochem. Biophys.*, 213, 595, 1982.
179. **Smith, J. D, and Lepak, N. M.,** Purification and characterization of a phosphonic acid-containing glycoprotein from the cell membranes of *Tetrahymena*, *Arch. Biochem. Biophys.*, 213, 565, 1982.

Chapter 4

CHEMICAL, BIOCHEMICAL, AND MEDICINAL PROPERTIES OF THE DIPHOSPHONATES

Marion D. Francis and Raymond R. Martodam

TABLE OF CONTENTS

I. INTRODUCTION

The diphosphonates and polyphosphonates are characterized in general by broad structural variations around the geminal grouping

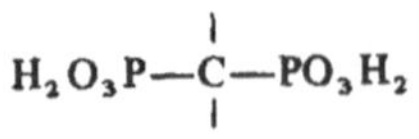

or the vicinal

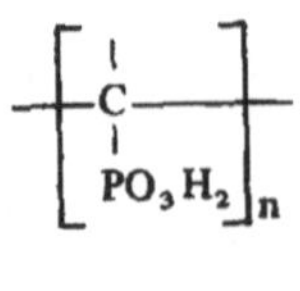

phosphonate grouping.

These structural arrangements of the phosphonate group ($-PO_3H_2$) impart unique physical chemical properties relative to the alkaline earth cations (e.g., calcium, magnesium) or the heavy metal cations (e.g., tin, iron) that have led to their applications in industry and medicine. Industrially, their chief use is in the prevention of calcium and magnesium scale formation in boilers and pipes. Medically, these compounds are used as drugs for prophylaxis and therapy of abnormal calcium phosphate metabolism and for diagnosis of bone pathologies. In this chapter, the physical chemical properties of some of the diphosphonates which have led to their application in industry and medicine will be discussed. We will elaborate primarily on their medical and dental uses and the physical chemical and biological mechanisms involved.

II. SYNTHESIS OF THE DIPHOSPHONATES

The diphosphonates are synthesized by two major methods — the Arbuzov reaction and the condensation of phosphorus trichloride and an acid.

A. Arbuzov Reaction

An example of the Arbuzov[1] reaction applied to the synthesis of a geminal diphosphonate is shown in Equation 1.

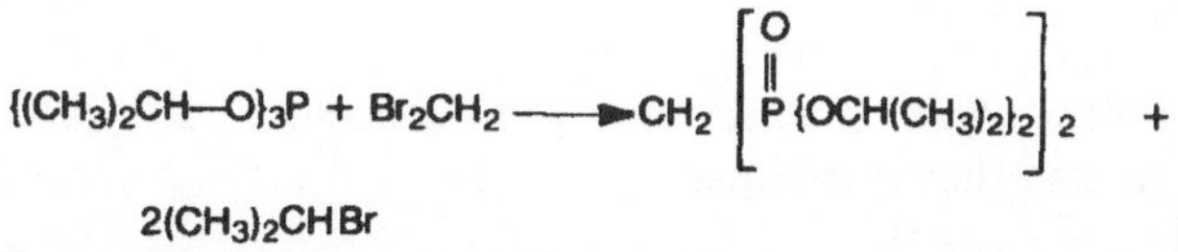

(1)

The diisopropylester of the geminal disphosphonate is converted to the phosphonic acid[2,3] by acid catalyzed hydrolysis or by pyrolysis as shown in Equation 2.

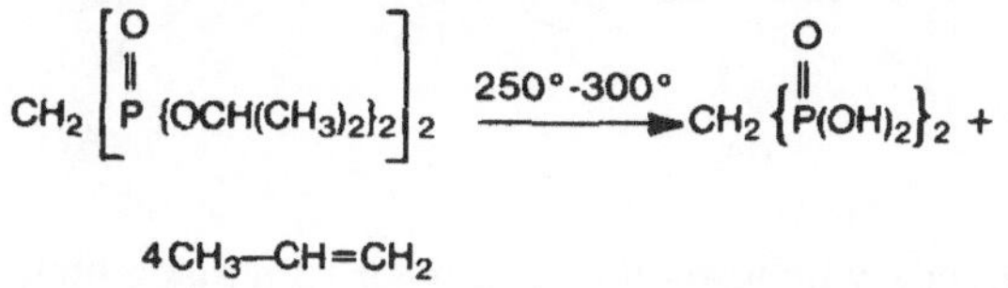

(2)

B. Condensation Reaction

The condensation reaction of an acid and phosphorus trichloride[4] is one of the simplest methods of forming a hydroxydiphosphonate such as 1-hydroxyethlidene diphosphonate (HEDP) and is shown in Equation 3.

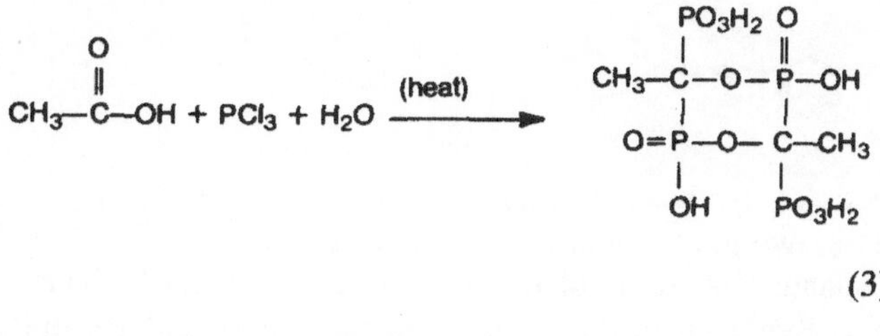

(3)

The cyclic condensation product is hydrolyzed to the geminal hydroxydiphosphonic acid as in Equation 4.

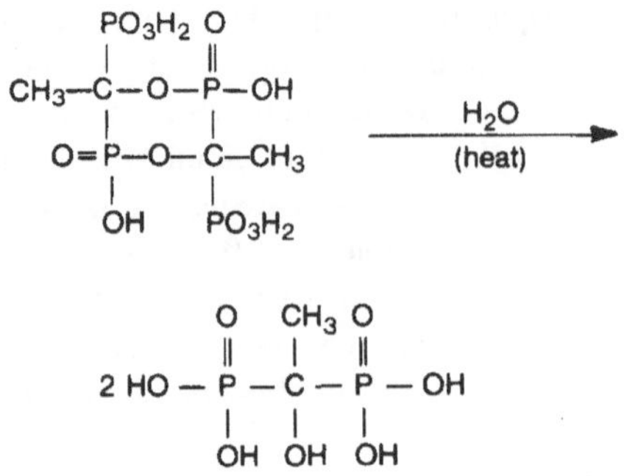

(4)

Because the 1-hydroxy group helps to impart a special geometric relation with the 1,1-diphosphonate groups and calcium (see Section III. A and B), a useful variant on the above methods to form hydroxydiphosphonates is the recent synthesis involving tris (trimethylsilyl) phosphate and acylchlorides[5] as in Equation 5.

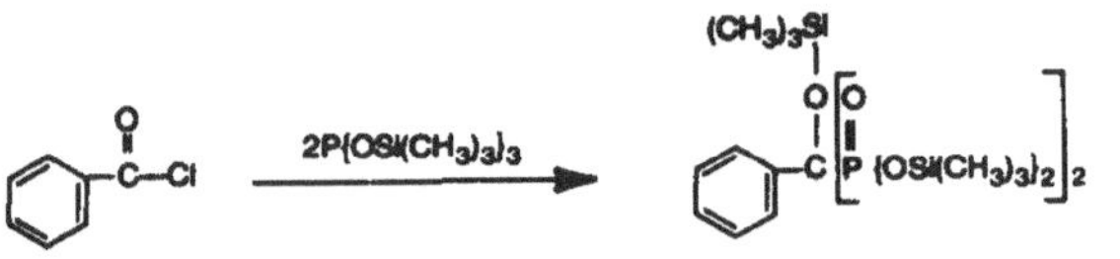

(5)

The bisphosphonic acid is obtained by reaction with methanol in the presence of analine to give the crystalline salt as shown in Equation 6.

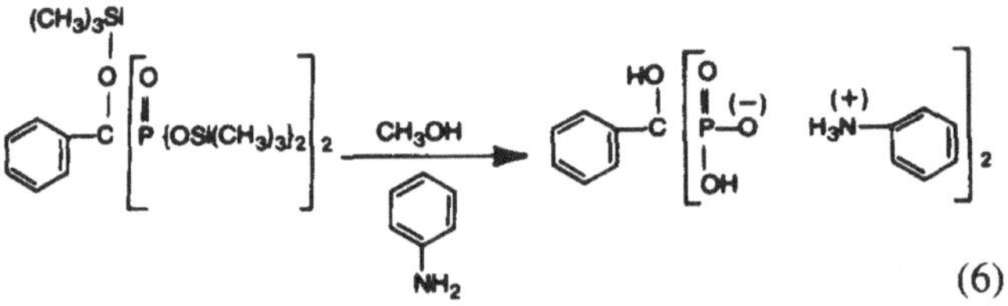

(6)

A method of forming the hydroxydiphosphonates through an aminodiphosphonate intermediate has recently been demonstrated by Worms et al.[6] as in Equation 7.

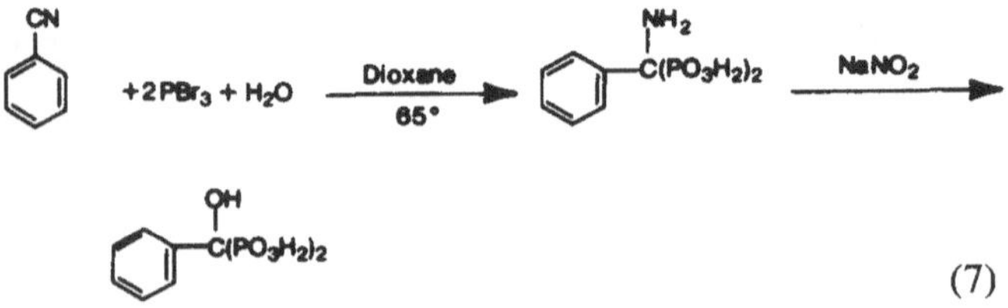

(7)

Many other methods of synthesizing diphosphonates are recorded in the literature[7] but reflect variations on the two major methods of synthesis. Structural formulas for the natural metabolite, pyrophosphate, and some of the most investigated short chain diphosphonates are shown in Figure 1. Synthesis of the polyphosphonates is considered in the next section (III.A).

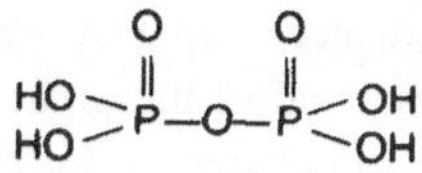

pyrophosphoric acid (PPi)

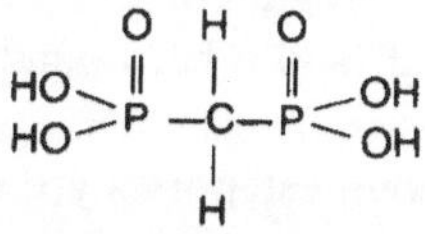

methylene diphosphonic acid (MDP)

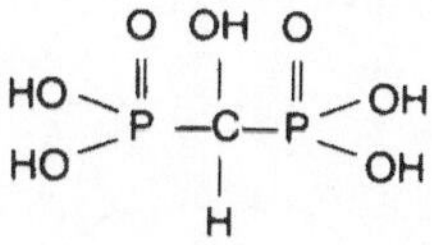

hydroxymethylene diphosphonic acid (HMDP)

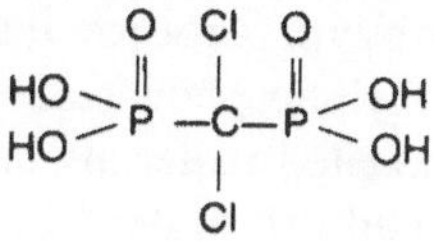

dichloromethylene diphosphonic acid (Cl_2MDP)

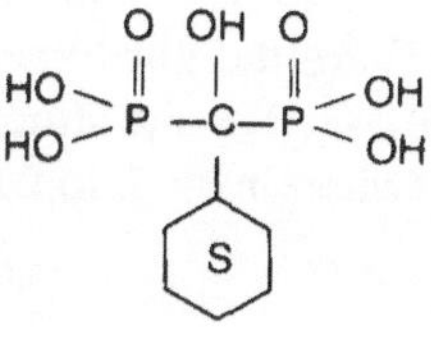

hydroxycyclohexylmethylene diphosphonic acid (HCMDP)

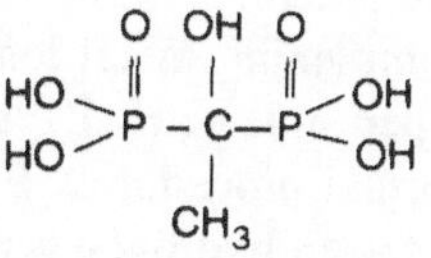

1-hydroxyethylidene diphosphonic acid (HEDP)

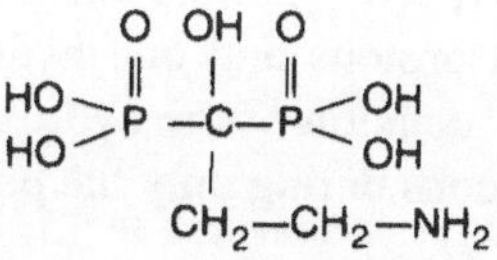

1-hydroxy-3-aminopropane-1,1-diphosphonic acid (APD)

FIGURE 1. Structural formulae in the acid form for the condensed orthophosphate, pyrophosphoric acid, and some of the more frequently investigated short chain geminal diphosphonates.

III. PHYSICAL CHEMICAL PROPERTIES OF THE DIPHOSPHONATES LEADING TO BIOLOGICAL ACTIVITY

A. Geometric Requirements for Activity

The two broadest chemical configurations of the di- and polyphosphonates which seem to be required for biological activity in mammalian systems are the geminal arrangements of the phosphonate groups as exemplified by the simplest structure methylenediphosphonic acid, $CH_2(PO_3H_2)_2$, and the vicinal configuration as exemplified by propane-1,2,3-triphosphonic acid, $CH_2(PO_3H_2)$–$CH(PO_3H_2)$–$CH_2(PO_3H_2)$.

It is interesting that the simplest vicinal compound, $CH_2(PO_3H_2)$–$CH_2(PO_3H_2)$ has little or no biological activity, probably because it has little adsorptive capacity for the hydroxyapatite of mammalian hard tissue (see Section III.C). Additionally, the alpha, omega type diphosphonates, $H_2PO_3(CH_2)_nPO_3H_2$, where $n > 3$, have no currently known biological effects. They do, however, have the potential to form complexes with calcium ions to varying degrees, in a manner analogous to nonphosphonate calcium chelating molecules such as ethylenediaminetetraacetic acid (EDTA), lactic acid, and other nonadsorptive molecules for insoluble calcium salts.

Polyphosphonated molecules that do not fall into the above categories are the polymeric phosphonates such as phosphonated polyethylene, synthesized using PCl_3, oxygen and polyethylene[8] to yield on hydrolysis the phosphonated polymer

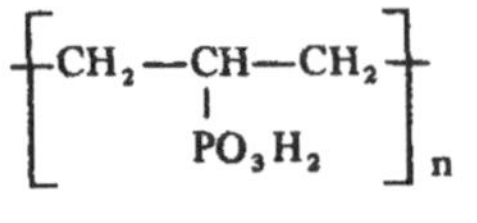

In addition, synthesis of vinylphosphonates using PCl_3[9] or by the Arbuzov reaction[10] allow the formation of polyvinylphosphonates or copolymers of vinylphosphonates with monomers such as methylmethacrylate. Other polymeric polyphosphonated materials have also been synthesized.[11,12] These polymeric materials are reported to adsorb to dental hard tissue[10,12] which may make them useful in dental caries treatment.[13] A group of phosphonates that also bind to hydroxyapatite and have biological activity[14,15] are the nitrilomethylene phosphonates such as nitrilotrimethylene phosphonic acid, $N(CH_2PO_3H_2)_3$ and structurally related to compounds[16] like $(H_2O_3PCH_2)_2N(CH_2)_nN(CH_2PO_3H_2)_2$ (where n = 2 to 6).

B. Physical Chemistry of the Diphosphonates

The diphosphonates such as $CH_3C(OH)(PO_3H_2)_2$, HEDP are polyprotic acids. The titration of HEDP with base has been carried out by a number of workers.[17-19] The variation in pH with addition of NaOH in the absence and presence of complexing metal ions is shown in Figure 2. In the absence of metal ions, only 3 mol of base per mole of HEDP are consumed. The fourth hydrogen is too firmly held to titrate using normal procedures. In the presence of metal ions such as Ca(II), Mg(II), Fe(II), or Fe(III), the fourth hydrogen is either partially neutralized [Ca(II) or Mg(II)] or completely neutralized [Fe(III)]. Some interpret the titration data in terms of dissociation of four phosphonic acid protons and one hydroxyl proton[17] while others indicated it is unnecessary to consider dissociation of the hydroxyl proton.[18,19] The ionization constants of the tetraprotic acid HEDP, considering only the phosphonic acid groups, are $pK_1 = 1.4$, $pK_2 = 2.8$, $pK_3 = 7.0$, $pK_4 = 11.1$.[20]

In addition to mononuclear complexes of HEDP with calcium, it has been found that under certain conditions polynuclear complexes of calcium and HEDP of the "core + links" type[21] are present.[22] The complexes are thought to be of the type $[Ca_7(HEDP)_4]_n^{-2n}$ where (n) may be as high as 10 to 20. The presence of these high molecular weight aggregates has been confirmed at high pH by light scattering, sedimentation, electrophoretic mobility, and dialysis measurements.[23] The existence, importance, or function of these polynuclear complexes of calcium and HEDP in mammalian biological systems is not well-established although their presence may partially explain the hypercalcemia seen in rats receiving high doses of HEDP. In general, the 1-hydroxy type geminal diphosphonates such as HEDP, hydroxycyclohexylmethylene diphosphonate, and hydroxymethylene diphosphonate (HMDP) are active in forming polynuclear complexes with calcium and with other heavy metal type cations such as Sn(II),[24] Pb(II), Tc(IV), and Fe(III). The geminal diphosphonic acids $Cl_2C(PO_3H_2)_2$, dichloromethylene diphosphonic acid (Cl_2MDP), and $CH_2(PO_3H_2)_2$, methylene disphosphonic acid (MDP) do not form similar polynuclear complexes[24] presumably

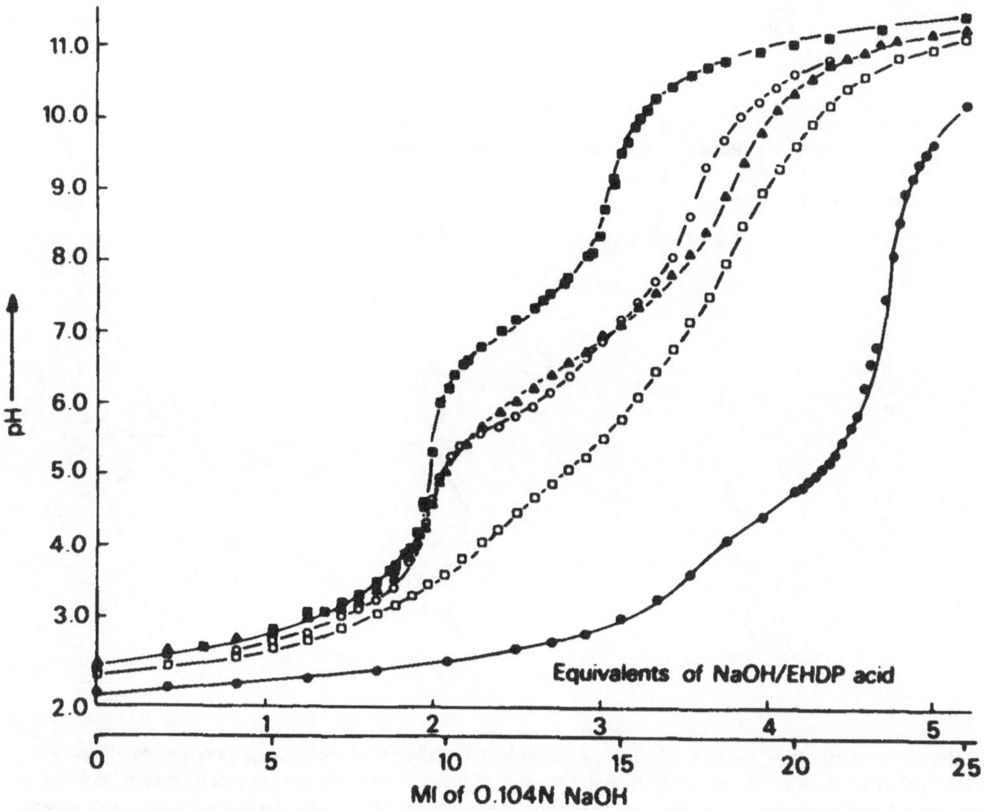

FIGURE 2. Variation in pH as a function of the titration of 100 mℓ of 0.005*M* 1-hydroxyethylidene diphosphonic acid (HEDP) with 0.104 *M* sodium hydroxide in the presence and absence of 0.005 *M* calcium, magnesium, or iron. (■) HEDP acid alone; (○) HEDP acid plus Ca^{2+}; (▲) HEDP acid plus Mg^{2+}; (□) HEDP acid plus Fe^{2+}; (●)HEDP acid plus Fe^{3+}. (From Francis, M. D. and Centner, R. L., *J. Chem. Ed.*, 55, 762, 1978. With permission.)

because of the absence of the hydroxyl group. Few other phosphonates have been investigated thoroughly for these properties either in vitro or in vivo. In the solid state, however, the medically important technetium phosphonate complexes such as technetium-MDP exist as polymeric chains having an MDP/Tc ratio of 1:1 within the polymer[25] a shown in Figure 3.

A significant structural factor of the diphosphonates is the "W" configuration made by the PO_3 groups with respect to the P–C–P plane[26] and can be seen in Figure 3(a) and 3(b) inverted "W" - as O2–P1–C–P2–O4. This configuration in the case of MDP allows bidentate-bidentate coordination to metal atom centers on either side of the "W" plane and bidentate-tridentate coordination to metal atom centers in the case of the hydroxydiphosphonates such as HEDP, HMDP, etc.[25] This latter specialized coordination potential could be part of the reason that the 1-hydroxydiphosphonates are so effective in blocking the crystal growth of hydroxypatite by adsorption since the tridentate ligation fits the trigonal antiprismatic coordination of calcium in the most rapidly growing 001 face or (c) axis face of hydroxyapatite[25-28] (Figure 4).

C. Adsorptive Properties of Disphosphonates for Calcium Phosphate

The most unique chemical property of the geminal and vicinal diphosphonates that relate directly to their biological activity is their strong chemisorptive reaction with calcium salts and in particular with calcium hydroxyapatite, $Ca_5(PO_4)_3OH$, the main inorganic constituent

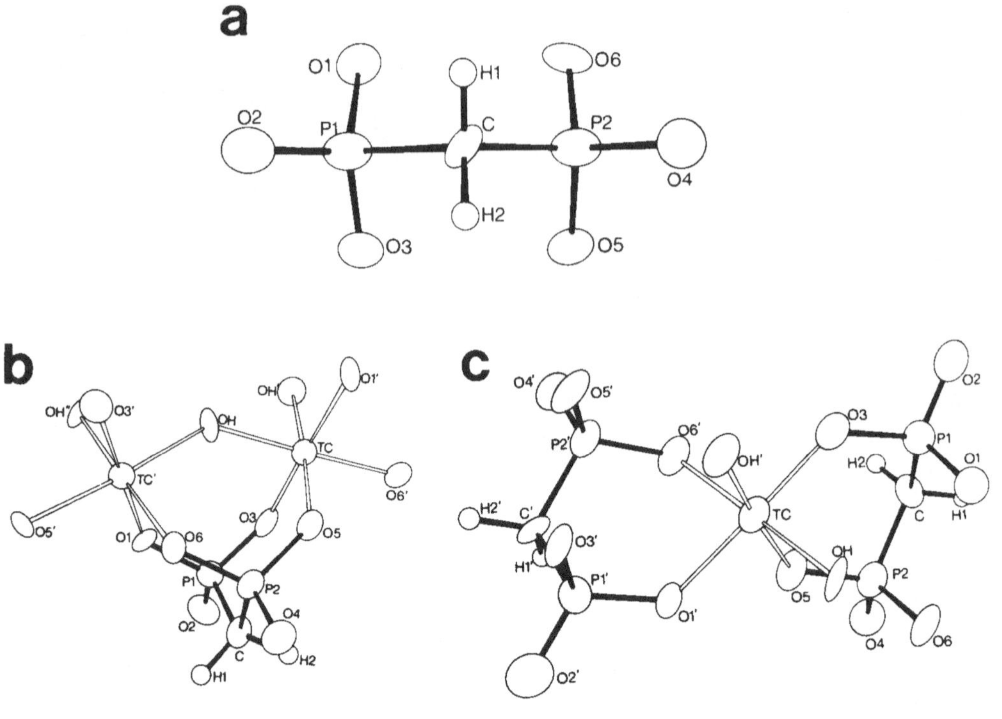

FIGURE 3. A perspective view of portions of the technetium (Tc) methylene diphosphonate (MDP) polymer [Tc(MDP)(OH)$^-$]∞ showing (a) the MDP ligand, (b) one MDP ligand bridging two symmetry related technetium atoms, and (c) each technetium atom bound to two symmetry related MDP ligands. (From Deutsch, E. and Barnett, B. L., *Inorganic Chemistry in Biology and Medicine*, ACS Symposium Series, #140, American Chemical Society, Washington, D.C., 1980, 114. With permission.)

of the mammalian skeleton.[20,29] The physical chemical effect of adsorption on hydroxyapatite is to block crystal growth of certain calcium salts such as calcium hydroxyapatite, calcium oxalate, and calcium carbonate through adsorption of the phosphonate on the growing crystallites. The mechanism of inhibition of crystal growth of apatite is one of strong chemisorption (reaction) to form a chelate on the surface of the microcrystallites particularly at screw dislocation or kink sites of growth.[16,29] A schematic representation of this adsorption at a growth site is shown in Figure 5. This deposition of phosphonate blocks or slows the addition of calcium, orthophosphate, and hydroxide ions into the lattice of the nuclei of calcium phosphate and so blocks crystal growth if all such sites are covered. At this writing, the elucidation of the exact relationship of effectiveness of crystal growth blocking to such factors as molecular geometry of the phosphonates, strength of the chelate bond, solubility of the calcium salts of the phosphonates, or numbers of chelating groups has not been clearly defined.

The net effect of crystal growth inhibition in the nascent calcium phosphate solid is to produce an apparent increase in solubility of hydroxapatite. For example, in calcium phosphate precipitations at physiological pH (7.4) and similar initial calcium and orthophosphate concentrations (Table 1), the differences in filtrate concentrations of calcium and orthophosphate are shown in the presence and absence of effective crystal growth inhibitors for 16 systems. The progressive decrease in calcium concentration and rise in phosphate of the control systems (1 to 3) reflects the change in Ca–P phase of the precipitate with time[30] from the more acidic $CaHPO_4{\cdot}2H_2O$ to the more basic phase $Ca_5(PO_4)_3OH$. The alpha, omega types of diphosphonate (systems 4 to 6) are not effective crystal growth inhibitors

FIGURE 4. A molecular model showing tridentate binding of hydroxymethylene diphosphonate to the trigonal face of a calcium center in the 001 or most rapidly growing face of hydroxyapatite (Adapted from Deutsch, E. and Barnett, B. L. *Inorganic Chemistry in Biology and Medicine*, ACS Symposium, Series #140, American Chemical Society, Washington, D.C., 1980, 116. With permission.)

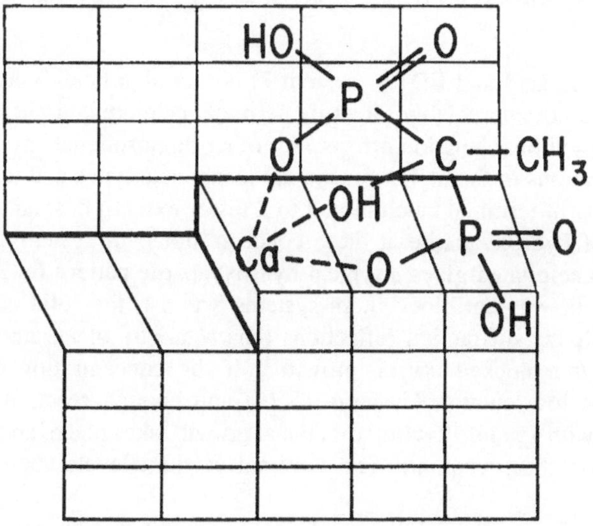

FIGURE 5. Possible mode of bonding of 1-hydroxyethylidene diphosphonate to a calcium ion at an active growth site "kink site" on the surface of a growing hydroxyapatite crystal. (From Meyer, J. L. and Nancollas, G. H., *Calcif. Tissue Res.*, 13, 295, 1973. With permission.)

and do not markedly affect the terminal calcium and phosphate concentration except through a soluble calcium chelation effect at higher concentrations (system 5). The known calcium

Table 1
TERMINAL CALCIUM (Ca) AND ORTHOPHOSPHATE (P) CONCENTRATIONS AS A FUNCTION OF INHIBITOR EFFECT ON CRYSTAL GROWTH

System number	Initial ($m/l \times 10^3$)		Inhibitor[a]	Inhibitor concentration ($m/l \times 10^3$)	Time of reaction (min)	Terminal ($m/l \times 10^3$)	
	Ca	P				Ca	P
1	3.87	4.00	—	—	40	0.2244	1.30
2	4.20	4.00	—	—	120	0.19	1.31
3	4.20	4.00	—	—	1176	0.0468	1.69
4	4.20	4.00	E-1,2-DP	0.20	116	0.210	1.41
5	3.87	4.00	E-1,2-DP	1.00	220	0.574	1.91
6	4.20	4.20	H-1,6-DP	0.21	168	0.226	1.44
7	3.87	4.00	EDTA	0.20	260	0.213	1.46
8	4.20	4.00	HMDP	0.20	424	1.11	2.54
9	3.87	4.00	HMDP	0.06	100	1.19	2.30
10	3.87	4.00	HEDP	0.20	1080	1.55	2.81
11	3.88	4.00	HCMDP	0.20	1056	0.936	1.36
12	4.20	4.00	Cl_2MDP	0.20	164	1.60	2.56
13	3.87	4.00	E-1-HTP	0.20	112	0.760	2.05
14	3.87	4.00	E-2-HTP	0.21	200	0.424	1.99
15	3.92	4.00	HEDP	0.05	1364	0.203	1.40
16	3.87	4.00	Cl_2MDP	0.05	220	0.280	1.56

[a] E-1,2-DP = ethane-1,2-diphosphonate; H-1,6-DP = hexane-1,6-diphosphonate; EDTA = ethylenediaminetetraacetic acid; HMDP = methylenehydroxydiphosphonate; HEDP = 1-hydroxyethylidene diphosphonate; HCMDP = hydroxycyclohexylmethylene diphosphonate; Cl_2MDP = dichloromethylenediphosphonate; E-1-HTP = ethane-1-hydroxy-1,1,2-triphosphonate; E-2-HTP = ethane-2-hydroxy-1,1,2-triphosphonate.

chelators nitrilotriacetic acid and EDTA (system 7) do not significantly adsorb on hydroxyapatite and hence do not appreciably affect the terminal calcium and phosphate concentrations except through soluble chelation effects. All other diphosphonate systems (8 to 14) at comparable concentrations to the alpha, omega diphosphonate types and to the EDTA show a significant elevation in terminal calcium and to a lesser extent phosphate due to blocking the crystal growth of hydroxyapatite at these concentrations. In systems 1 to 7, the solid calcium phosphate precipitated gives a typical hydroxyapatite pattern by X-ray diffraction. That is, crystal growth was not blocked. In systems 8 to 14, the solid calcium phosphate precipitate gave no X-ray diffraction reflections (characteristic of an amorphous material) and this confirmed the blocked crystal growth.[31] If the concentration of very effective inhibitors is dropped low enough (systems 15,16) and enough reaction time is allowed (compare system 10 with 15 and 12 with 16), crystal growth takes place, an apatite diffraction pattern is obtained on the precipitate, and terminal calcium levels approach the controls (systems 1 to 3).

The effectiveness of potential inhibitors of apatite crystal growth can also be followed titrimetically using a pH stat instrument.[32] When calcium ion is added to orthophosphate ion at physiological pH at levels to produce precipitation and crystallization, sodium hydroxide or base is consumed to maintain constant pH according to the overall Equation 8.

$$5Ca^{+2} + 3HPO_4^{-2} + 4OH^{-1} \longrightarrow Ca_5(PO_4)_3OH\downarrow + 3H_2O \quad (8)$$

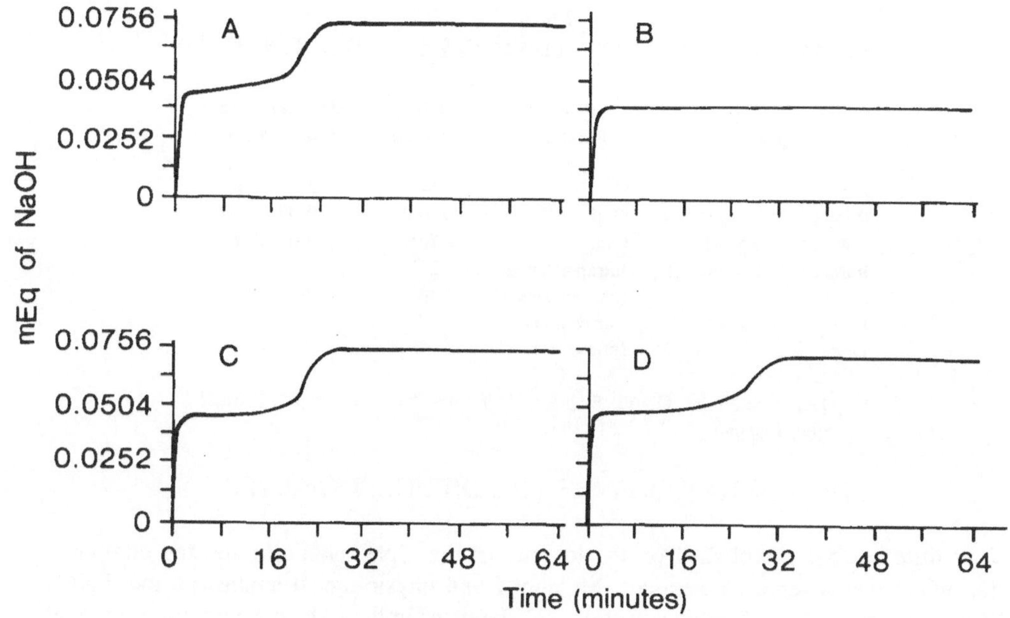

FIGURE 6. Rate of formation of hydroxyapatite followed titrimetrically at pH 7.4 using a pH-stat instrument (a) control, no inhibitor and calcium chloride and sodium phosphate initially each at 4×10^{-3} *M*. (b, c, d) Initial calcium and phosphate as in (a) but 2×10^{-4} *M* inhibitor present and inhibitors were (b) 1-hydroxyethylidene diphosphonate, (c) ethylenediaminetetraacetate, and (d) *n*-pentane-1-phosphonate.

The rate of base consumption in Equation 8 in the presence and absence of inhibitors is a measure of the rate of formation of crystalline apatite. Typical titrimetric curves for an effective crystal growth inhibitor, HEDP (Curve B), and two ineffective crystal growth inhibitors, EDTA (Curve C), and *n*-pentane-1-phosphonate, PMP (Curve D), are shown in Figure 6 compared to a control system with no inhibitor (Curve A). The first rapid rise in base (OH^-) consumption is due to reaction with protons released by hydrogen phosphate upon complexation by calcium and to formation and aggregation of nuclei of calcium phosphates. This initially formed precipitate is amorphous by X-ray techniques. The second rapid rise in base consumed at about 24 to 35 min in Figure 6 (Curves A, C, and D) is due to the base consumed on rapid crystal growth of the apatite,[33] primarily along the c axis or at the 001 face of apatite.[27] Thus, there is essentially no crystal growth inhibition by the two materials EDTA and PMP which lack the structural requirements indicated in Section III.A. The only difference among these curves is a slightly lower base consumption by 80 to 100 min for EDTA and PMP than the control system because of some soluble calcium complexation formation. In addition, all the precipitates formed in the systems (Curves A, C, and D) gave an apatite pattern by X-ray. The effective inhibitor HEDP, however, has totally blocked crystal growth (no sigmoid shaped curve) and overall base consumption is markedly depressed (Curve B). Inhibition of crystalline apatite formation[32] was confirmed by an amorphous X-ray diffraction pattern on the solid and elevated terminal calcium concentrations (Table 1). This decreased apatite formation or apparent increased solubility of apatite forms the physical chemical basis for the effectiveness of the diphosphonates in blocking calculus formation in both animals[14,15] and in man[34-36] and for their use in heterotopic ossification such as in para- and quadraplegia[37] and in hip replacements in man.[38,39] Furthermore, the strong adsorption or ligand chelation effects with metal ions seen with the geminal diphosphonates (see III.B), form the basis for their use with the 140 keV gamma emitting radionuclide 99m-technetium utilized for bone scanning[27,28] (see Section VII.F).

Table 2
SUMMARY OF ACUTE TOXICITY STUDIES ON HEDP

Species	Route of administration	LD_{50}, mg/kg	95% Confidence limits, mg/kg
Rat	Oral	1340[a]	1254—1389
Rabbit (2250—2800 g)	Oral	1100[a]	961—1250
Rabbit (> 3300 g)	Oral	701[a]	591—831
Rabbit (2250—2800 g)	Intraperitoneal	212[a]	—
Rat	Intraperitoneal	400[a]	290—540
Rat	Intravenous	79	
Dog	Intravenous	43.8	—

[a] Data from Nixon, G. A., Buehler, E. V., and Newman, E. A., *Toxicol. Appl. Pharmacol.*, 22, 661, 1972.

IV. TOXICOLOGY OF THE DIPHOSPHONATES

Very little has been published on the toxicity of the diphosphonates and the compound HEDP will have to serve as a model. Metabolic and physiological studies on the diphosphonates are covered in subsequent sections. Differences in the LD_{50} dose required for death of 50% of animals in 14 days as a function of the administration route (Table 2)[40] are a result of a very low absorption (highest LD_{50}) by the oral route. In normal human subjects, 6.4 (± 2.4)% of an oral dose of 20 mg/kg body weight per day is absorbed and only 0.8 (± 0.3)% of a 2.5 mg/kg/day dose is absorbed.[41]

A greater systemic load is obtained with both s.c. and i.p. injection of HEDP while direct i.v. infusion gives the lowest LD_{50} depending upon the rate of infusion.[42] Regardless of the route of administration (oral or parenteral), the mechanism of death in LD_{50} studies appears to be muscular tetany with cessation of respiration resulting from the complexing of calcium (decreased ionized calcium) as very high blood levels of HEDP and/or Cl_2MDP are reached.[42] This tetany can be prevented by the administration of parenteral Ca^{++} ion as calcium gluconate solution.[42] In rats, we have shown[43] that the rate of administration of HEDP into the blood system (systemic concentration) is the key to toxicity. For example, a lethal dose of HEDP as the disodium salt administered as a bolus (0.7 to 1.0 cc) in 2 to 5 sec into the jugular vein of rats is about 17.5 mg/kg. This bolus reaches the most sensitive pacemaker (the sinoatrial node at the junction of the superior vena cava and right atrium) almost instantly with little or no mixing and dilution. If the same route and 20 mg/kg is given over a 2 min interval, the dose is tolerated. If the same dose is given via the tail vein in only 30 sec, the much greater distance from the sinoatrial node results in even greater dilution and, in spite of the shorter time of injection, a marked decrease in toxicity occurs such that the LD_{50} is now 79 mg/kg.

The diphosphonate HEDP as the disodium salt has been administered subcutaneously to dogs at levels of 2, 4, 6, 8, 10, 15, 20, and 60 mg/kg/day over at least a 16-day period.[44] No statistically different hematological results with these animals were obtained compared to the control animals. Changes in bone morphology, growth, food consumption, and alkaline phosphatase levels were seen at the higher doses of HEDP (greater than 6 mg/kg for 28 days, or 20 mg/kg or greater for 16 days) but these values returned to normal by the end of 5 months after withdrawal. Parenteral injection studies were also performed in weanling (50 to 60 g) and young adult (200 g) rats at doses of 2.5 to 50 mg/kg.[44] At dose levels above 20 mg/kg HEDP in weanlings and 15 mg/kg in 200 g rats for 14 days, significant increases were seen in epiphyseal plate width and in osteoid on the periosteal and endosteal bone surfaces. At the highest level (50 mg/kg for 28 days) changes in bone and renal tubular morphology and in RBC, hemoglobin and hematocrit were seen; however, these had returned to normal by the end of 21 weeks after treatment was stopped.

Teratogenic studies have been carried out on both rats and rabbits administering HEDP both in the diet and by gavage.[45] No evidence of teratogenicity was found in either the rats or rabbits.

Two year feeding studies of HEDP at 0.0, 0.21, 0.6, and 1.0% in the diet[46] indicated there was no gross toxicity regardless of the level of the disodium salt added to the diet. The only other significant finding in this study was scattered increases in urinary phosphorus at the highest feeding level (1%). This phosphaturic effect will be discussed later in Section VI.D.2.

In a long-term study on bone in dogs, HEDP was administered subcutaneously (s.c.) at 0.1, 0.5, 2.0, 5.0, and 10 mg/kg and Cl_2MDP s.c. at 0.6, 2.5, 10, and 25 mg/kg[47] (both as disodium salts). HEDP was also given at 100 and 400 mg/kg by gavage.[48] Both HEDP and Cl_2MDP have effects on the skeleton of the dog. At one year in the oral HEDP study and at greater than or equal to 2 mg/kg (s.c.), there was suppression of osteoblastic activity in the bone, reduced resorbing surfaces, accumulation of osteoid tissue, and increased fractures of the bone. At the 0.1 and 0.5 mg/kg s.c. HEDP doses at 2 years, there was only slight suppression of osteoblastic activity and decrease in bone surfaces undergoing resorption. There were no fractures at the 0.1 mg/kg HEDP level but a possible slight increase in fractures at 0.5 mg/kg. Cl_2MDP caused a significant increase in osteoid seam width and numbers of osteoid seams only at the highest dose level (25 mg/kg) but at 11 months of treatment the difference in seam width from control was not significant. Cl_2MDP at 10 and 25 mg/kg slowed but did not completely block the mineralization process as occurred for >2 mg/kg HEDP. Fracturing of bone with doses of Cl_2MDP was seen at 2.5 mg/kg or greater or at doses about 5 times that producing fractures with HEDP.

In studies of 3-amino-1-hydroxypropylidene diphosphonate (APD) and Cl_2MDP[49] as disodium salts administered subcutaneously at 0.05, 0.5, and 5.0 and 0.5, 5.0, and 50 mg/kg respectively, both high doses of APD but only the highest dose of Cl_2MDP induced interference in bone formation and caused excess accumulation of osteoid in alveolar bone. None of the above doses, however, had the above effect on the slower mineralizing vertebrae but all doses induced inhibition of bone resorption.

V. PHARMACOLOGY OF THE DIPHOSPHONATES

The pharmacodynamics of HEDP and Cl_2MDP have been studied in the rat, rabbit, cat, dog, monkey, and man.[50-64] In general the gastrointestinal absorption of these drugs is limited to less than 11% of the oral dose in the rat, rabbit, monkey, and man, but is higher in the cat and dog.[50,51] This variability in absorption is both an intra- and interspecies phenomenon. Nothing has been found to enhance the absorption of diphosphonates; however, a number of factors have an influence on the absorptive process. Young rats and dogs absorb more HEDP than do older members of the species.[51] In dogs and man, but not rats, the percent of the dose of HEDP which is absorbed is directly related to the size of the dose. Prior exposure to diphosphonates does not seem to alter either their absorption or distribution.[50] If HEDP is given as the calcium salt[52] the absorption of the compound is decreased considerably. Antacids, milk, and EDTA also interfere with the absorption of HEDP.[53-55] The percent of the dose absorbed by rats receiving HEDP in the diet is lower than the amount absorbed in fasting animals.[51] Diphosphonates can be absorbed from the stomach[56] but are preferentially absorbed from the duodenum in the dog[55] and the ileum in the chick.[57] Unlike pyrophosphate which is hydrolyzed, diphosphonates are completely stable to the environment of the stomach and intestines and are either absorbed or excreted intact.[50]

In rats, dogs, and man, 20 to 50% of the systemic load of HEDP is distributed in the bone with the remaining drug being excreted in the urine within 24 hr.[50,51,58-60] The distribution of HEDP can be significantly altered by age, species, and disease state but is not believed to be altered by the route of administration within fairly wide limits. The renal clearance of HEDP and Cl_2MDP exceeds that of inulin in rats, indicating the existence of a secretory pathway.[61] At the present time, further studies are being undertaken to evaluate this unexpected finding. There is virtually no soft tissue uptake, retention, or metabolism of HEDP,[50] although phosphonatases capable of catalyzing the cleavage of certain types of C–P bonds are present in some micro-organisms (see Chapter 3).[62] Following parenteral administration to rats, HEDP is either found in the bone or excreted in the urine suggesting that there is no enterohepatic circulation of the drug.[50]

The half life of HEDP in bone has been found to be about 12 days following s.c. dosing in young (125 to 175 g) rats.[50] Other studies conducted in our laboratory with a similar aged rat suggest that the half-life of HEDP in bone is considerably longer than 4 months.[63] The major experimental difference between the two studies was the 1 year follow up in the second study (4 month half-life) vs. the 28 day follow up in the first study (12 day half-life). This indicates that the bone has to be treated as more than a single compartment in terms of a reservoir for the diphosphonate. In man, the half-life of HEDP in bone from normal adult males appears to be about 17 days.[64] This observation also needs to be considered in light of the extremely short study time (96 hr). More work is necessary before the half-life(s) of HEDP in human bone can be known with certainty. Reverse isotope dilution analysis was performed on different biological samples obtained from rats (urine) and from dogs (urine, feces, serum, and bone) given ^{14}C-HEDP. Within the limits of detection, HEDP was not metabolized.[50] Although it has not been studied as exhaustively, the absorption, distribution, metabolism, and excretion of Cl_2MDP in the rat appears to mimic that of HEDP.[61]

VI. CELLULAR AND BIOLOGICAL EFFECTS OF THE DIPHOSPHONATES

The unique physical chemical properties of the diphosphonates with respect to the main inorganic constituent of mammalian skeleton, hydroxyapatite, suggested their potential use in metabolic aberrations of calcium phosphate metabolism.[29,31] The following sections of this paper deal with the many biological studies in vitro and in vivo which have been carried out on these interesting and biologically active materials.

A. The Effects of the Diphosphonates Related to Bone Resorption in Tissue and Cell Culture

1. Bone Resorption Stimulated by Metabolic Factors

Cell and organ culture experiments have made valuable contributions to our knowledge of the biochemical events related to the interaction of diphosphonates with bone tissue. Cl_2MDP and HEDP added to culture media inhibit the endogenous bone resorption associated with cultured calvaria bone but do not influence the movement of calcium out of dead bones.[65] In vitro endogenous bone resorption is stimulated by parathyroid hormone (PTH),[66,67] lipopolysaccharide,[68] osteoclast activating factor (OAF),[69] prostaglandin E_1, or 1,25-dihydroxy vitamin D_3 [$1,25(OH)_2D_3$].[70] This stimulated in vitro resorption is inhibited by diphosphonates. Inhibition of bone resorption in culture is also seen if the diphosphonates are administered in vivo before explantation of the tissue. In culture, diphosphonates appear to inhibit bone resorption independent of the resorption stimulating agent. Cl_2MDP is the most potent diphosphonate[65,66,68,70] with other diphosphonates being weaker inhibitors of bone resorption in vitro.

2. Morphological Changes in Osteoclasts

In hydroxyapatite dissolution experiments, Cl_2MDP has been shown to be among the least effective of the diphosphonates in imparting acid antisolubility properties to the apatite.[67] In bone culture and in vivo, however, Cl_2MDP has been among the most effective of the diphosphonates in blocking dissolution.[65,66] Efforts have been made, therefore, to identify biochemical changes following diphosphonate treatment. Both in vitro[71] and in vivo[72-78] experiments have shown that Cl_2MDP and HEDP treatment results in morphologic alterations in osteoblasts and osteoclasts while other cell types are relatively unaffected. Under the influence of HEDP, active pump osteoblastic cells change to inactive flat cells. Osteoclastic changes include excessive storage of calcium by mitochondria, increased clumping of chromatin, decreased ruffled border area, and in rats, increased numbers of osteoclasts and increased numbers of nuclei per osteoclast. It is not known if these morphological changes are a result of the osteoclast's inability to dissolve diphosphonate-coated hydroxyapatite *in situ* or whether they are a result of a more direct cellular effect of the phosphonates.

3. Intermediary Metabolism

It is generally held that acid production by bone cells plays a role in bone resorption. Inhibition of acid production may be a partial explanation for the function of certain diphosphonates. Cl_2MDP and HEDP stimulate glycogen accumulation and inhibit glucose consumption and lactic acid formation in rabbit ear cartilage and rat calvarial cells.[79-81] These effects are seen after incubation for 7 days with 0.025 m*M* Cl_2MDP and with 0.25 m*M* HEDP. Similar results are seen with whole calvaria in culture.[80,82] This interference with glycolysis is apparently compensated for by increased oxidation of amino acids and fatty acids. The citric acid cycle is also stimulated by Cl_2MDP.[83] Incubation of cartilage cells with diphosphonates in an anaerobic atmosphere results in a return to glycolysis and lactic acid production indicating that there had been a regulatory shift rather than a change in the enzymatic profile. The above interference of these phosphonates in metabolism, however, is not of fundamental importance since some long-chain diphosphonates and APD which are effective inhibitors of bone resorption have little or no effect on lactate production, or in some cases, increase it.[84,85] The results of these experiments are also somewhat difficult to interpret because the in vitro drug concentrations are about an order of magnitude higher than pertain in vivo and long incubation times are required. However, these experiments do serve to point out possible diphosphonate effects on metabolism.

4. Adenylate Cyclase Activity

Cl_2MDP and HEDP inhibit rat kidney cortex and liver adenylate cyclase activity in vitro,[86,87] and high doses of HEDP (30 mg/kg/day, s.c.) in vivo reduce rat urinary cyclic adenosine monophosphate (cAMP) excretion.[88] In culture, levels of Cl_2MDP which inhibit PTH-induced bone resorption do not affect the associated rise in bone cAMP indicating that diphosphonates function at a different step in the resorptive process.

5. Lysosomal Enzymes

Diphosphonates, particularly Cl_2MDP, have effects on lysosomal enzymes which are thought to be important in the resorptive processes. Cl_2MDP inhibits the PTH-induced rise in acid phosphatase and acid pyrophosphatase in cultured calvaria.[82] In vivo, HEDP reduces acid phosphatase in rat long bones.[89] Cl_2MDP and HEDP inhibit rat liver acid pyrophosphatase at 10^{-4} *M* and Cl_2MDP (but not HEDP) inhibits acid phosphatase and acid glycerophosphatase at 10^{-3} *M*.[90] Arylsulfatase and DNAase II are only weakly inhibited, and cathepsin D and β-glucuronidase are not inhibited.[90] The direct inhibition of any of the lysosomal enzymes occurs at diphosphonate concentrations which are from 5 to 50 times the concentration needed to inhibit PTH-stimulated bone resorption in culture.

6. Calcium Transport

The diphosphonates, Cl_2MDP and HEDP, do not affect the uptake but do retard the release of calcium by rat kidney mitochondria in vitro.[91] High in vivo doses of HEDP also retard the release of calcium from chick mitochondria in culture.[92] Neither drug has a significant effect on calcium transport in red cells.[93] While it is obvious that diphosphonates affect the transport of calcium out of the bone by virtue of their ability to interfere with osteoclastic function, there is no evidence to show that this is due to a direct effect on a calcium transport mechanism.

7. Alkaline Phosphatase

Cl_2MDP does not have any effect on alkaline phosphatase in PTH-stimulated bone culture.[82] When calvaria cells are isolated free of matrix and cultured, incubation with Cl_2MDP results in approximately a 30-fold increase in the production of alkaline phosphatase.[94] This same rise in enzyme production is not seen with cartilage cells. In vivo, Cl_2MDP increases the bone alkaline phosphatase activity of thyroparathyroidectomized rats.[78] This effect is not seen if the rats receive vitamin D.[95] Alkaline phosphatase activity is elevated in cartilage or matrix vesicles isolated from the growth plate of chickens treated with HEDP or Cl_2MDP.[96] HEDP was more potent than Cl_2MDP, a finding opposite to the cell culture work where HEDP had little effect on the enzyme. Low millimolar concentrations of Cl_2MDP, HEDP, and MDP inhibit alkaline pyrophosphatase from *Streptococcus mutans* SI67101.[97] This effect is cation dependent with calcium and magnesium enhancing the activity of HEDP and zinc abolishing it. Diphosphonates can act on pyrophosphatase intracellularly by entering *S. mutans* or extracellularly on enzyme release by cell death and rupture.[98] Alkaline pyrophosphatase containing streptococci are present in large numbers in calcifying dental plaque[99,100] and inhibition of this enzyme may result in a reduction of calcification. This would be a mechanism in addition to or separate from the physical chemical mechanism of crystal growth inhibition.[15]

8. Diphosphonate Effects on Bone Matrix Components

Cl_2MDP but not HEDP increases the biosynthesis of extracellular macromolecules containing SO_4^{-2} by isolated rabbit articular chondrocytes.[101] Pulse chase experiments (a radioactive dose of high specific activity substance followed *in specific time* by a dose of the same unlabeled substance) reveal that the increased proteoglycan is a result of stimulated synthesis and not a reduction in degradation rate. Incubation of diphosphonates with cartilage slices results in a decrease in proteoglycan synthesis.[102] In vivo doses of HEDP which do not inhibit mineralization (5 mg/kg for 10 days) do inhibit glycosaminoglycan formation at the epiphyseal growth plate of the rabbit. In vivo administration of Cl_2MDP increases the mechanical strength of epiphyseal plate cartilage.[103] The synthesis of cartilage and bone collagen are increased by Cl_2MDP when the drug is administered to the whole animal or incubated with the cells in culture.[104] If 180 g rats are used rather than neonates, Cl_2MDP inhibits collagen synthesis but is far less potent than APD with HEDP having activity between these two.[85] Thus diphosphonates appear to have mixed effects on the biosynthesis of collagen and glycosaminoglycan, the two major macromolecules of bone matrix.

9. Internalization of Diphosphonates by Bone Cells

Many of the effects that have been seen on bone cell metabolism suggest that diphosphonates enter the cell. Calvarial cells and fibroblasts have been shown to take up diphosphonates in a dose-dependent manner.[80] After 48 hr, the concentration inside the cell was several times the drug concentration in the culture media (e.g., a 48-hr incubation of cells in 0.25 m*M* Cl_2MDP resulted in an intracellular "concentration" of about 0.7 m*M* Cl_2MDP). In order to relate the inhibition of biochemical events with an inhibition of bone resorption, it would be important to know what extracellular and intracellular concentrations of diphosphonate are achieved when the minimum amount of compound necessary to inhibit bone resorption is used. If isolated cells are used, the extracellular concentration can be defined, but the answer has been altered because the drug binding property of the bone which would affect the extracellular concentration of diphosphonate has been omitted. If organ culture or whole animal experiments are conducted, not only does the extracellular concentration become undefined, but the intracellular concentration may be altered during the isolation process. This problem, along with the fact that diphosphonates affect the various biochemical parameters in different ways, has made it impossible to develop a unifying hypothesis to totally explain their mechanism of action.

B. The Effect of Diphosphonates Related to Bone Resorption — Animal Studies

1. Normal Modeling and Remodeling

The potent effect that diphosphonates have on resorption in vitro is also seen when these compounds are administered in vivo. The primary effect seen in growing rats during modeling is to inhibit resorption in the metaphysis so that it becomes club-shaped and radiographically more dense than normal.[72,78,105,106] With Cl_2MDP at 10 mg P/kg/day, the gain in bone density in the metaphysis is probably due mainly to an increase in the thickness of individual trabeculae rather than the number. In the zones of cartilage calcification and vascular permeation, the mineralized longitudinal intercolumnar septa are retained rather than being removed by the action of chondroclasts as in the control animals. In contrast, similar doses of HEDP inhibit resorption but also inhibited mineralization of both bone and cartilage in trabeculae so that a widening of the epiphyseal plate is evident.[72] The vascular invasion of the growth cartilage is temporarily interrupted and then continues in an irregular manner suggesting a relationship between mineralization and angeogenesis.[72,107] Bone resorption is also inhibited by Cl_2MDP in experimental renal osteodystrophy.[108] For Cl_2MDP but not HEDP, the decrease in the rate of bone resorption is related in a linear fashion to the log dose of diphosphonate.[107] The range of potency of the various diphosphonates in vivo for blocking resorption goes from a maximum for APD, some long-chain diphosphonates, and Cl_2MDP to a minimum for HEDP.[84] This leads to an apparent discrepancy since Cl_2MDP is the most potent diphosphonate in tissue culture. This difference may be related to the method of delivery. In culture the drugs reach their site of action by diffusion while in vivo they are originally distributed to mineralized tissue by the capillary system. The effects of Cl_2MDP are probably related to host age as well as dose, at least this is the case with newborn mice.[109] The administration of Cl_2MDP at a dose rate of 10 mg P/kg/day leads to skeletal changes that are similar to those observed in gray-lethal osteopetrotic mice. It is probable that both in the gray-lethal and diphosphonate-treated mice, many of the abnormalities are secondary to decreased bone resorption.

The diphosphonates, Cl_2MDP and HEDP, prevent the hypercalcemic rise induced by PTH in thyroparathyroidectomized rats.[67,68] In contrast to its effects in hypercalcemia, HEDP can stabilize plasma calcium during the hypocalcemic challenges produced in cows during lactation immediately after parturition.[110-113] This effect is possibly related to the hypercalcemia produced with large doses of certain hydroxy-related diphosphonates.[66]

Inhibition of bone resorption and an accumulation of mineral has also been shown kinetically following diphosphonate treatment.[78,114,115] The increase in mineral content is probably limited to the amount of unmineralized osteoid available at the time bone resorption is stopped. If less than total inhibition of resorption is achieved, both resorption and formation slow down with bone turnover reduced as the net result. This effect on turnover is probably a manifestation of the coupling between formation and resorption.[47] Whether the accumulation of bone is transient or will be of long-term significance is an unanswered question. Diphosphonates also reduce urinary hydroxyproline excretion which is taken as a sign that bone turnover has been inhibited.[78,116]

2. Models of Osteoporosis

An animal model of immobilization osteoporosis can be generated by nerve section, which results in bone loss in the immobilized limb. Cl_2MDP and, at five- to tenfold higher doses, HEDP are very effective at preventing bone loss due to limb immobilization in the rat.[114,115,117-120] HEDP, at a dose which is effective in the rat (0.5 mg/kg), is not effective on the bone loss induced by denervation of the cat tail.[121]

In general, the effects of the diphosphonates are more striking on immobilization osteoporosis than in systemic or hormonal models of the disease. Cl_2MDP reduces the bone loss induced by cortisol in rabbits[122] and HEDP is partially effective at preventing heparin[123] or coumarin[124] mediated osteoporosis. However, in other experiments, HEDP was not found to be effective at preventing bone loss associated with steroid treatment[119] or uremia.[125] At the doses used, neither Cl_2MDP nor HEDP prevented the bone loss associated with a low calcium high phosphorus diet in the cat[126] or rat.[127] It is unlikely that higher doses of diphosphonates would be markedly more effective since kinetic measurements showed that doses of Cl_2MDP as high as 35 mg/kg administered s.c. do not decrease bone loss in this rat model of dietary osteoporosis.[127] The mechanisms of bone loss in the various models of osteoporosis are not well-understood nor is it understood why the diphosphonates are effective in some models and not in others.

Both Cl_2MDP and HEDP cause an increase in bone weight and bone calcium in the normal rat.[114,115,128-130] Cl_2MDP is more potent, suggesting that an inhibition of resorption without impairment of mineralization can lead to at least a transient accumulation of bone. Whether this will translate to the remodeling human skeleton remains to be seen. Notwithstanding the lack of evidence that diphosphonates have any positive effects on bone formation, inhibition of bone resorption would by itself suggest that these compounds should impede the bone loss in some forms of human osteoporosis.

3. Freund's Adjuvant Arthritis and Crystal Induced Inflammation

Adjuvant-induced polyarthritis, produced in rats by a s.c. injection of heat-killed mycobacteria in oil has some similarity to human rheumatoid arthritis. Both the human disease and the experimental rat model are characterized by synovial membrane cell proliferation, bone destruction, and heterotopic ossification,[131,132] and elevated levels of hydrolytic enzymes in the inflamed tissues. Cl_2MDP and HEDP are effective at inhibiting the bone resorption and mineralization in the adjuvant arthritis model and both drugs also reduce the paw edema and protect against the adjuvant-induced weight loss.[131,132] In another study, HEDP was effective at inhibiting pannus formation and the inflammatory erosion of the cartilage. The splenomegaly and increase in serum levels of β-glucuronidase, both of which reflect systemic inflammation, are inhibited by Cl_2MDP.[133] Circulating levels of polymorphonuclear leukocytes and plasma lysozyme are not reduced. Subplantar injection of hydroxyapatite or monosodium urate crystals also produces inflammation; however it differs from adjuvant-induced arthritis in that only soft tissue changes occur, and bone resorption need not be considered. Cl_2MDP inhibits hydroxyapatite but not monosodium urate induced inflammation. This is consistent with the in vitro observation that Cl_2MDP will bind to crystals of hydroxyapatite but not to crystals of monosodium urate.[133] If the animals rather than the hydroxyapatite crystals were pretreated with the diphosphonate, then at very high doses Cl_2MDP but not HEDP partially inhibited the paw edema. These studies suggest that some of the properties that diphosphonates have in the adjuvant model of arthritis relate to their ability to bind to bone. Certainly this is true for the effects on bone resorption and periarticular bone formation. It is unknown whether the effects of Cl_2MDP or HEDP on other aspects of inflammation such as inhibition of paw edema, serum β-glucuronidase levels, and splenomegaly are a consequence of their binding to hydroxyapatite.

In the guinea pig, s.c. injections of large doses (30 mg/kg) of MDP, to previously sensitized animals, inhibits the delayed and to a lesser extent the immediate hypersensitivity reactions associated with a challenge of sheep's red cells in Freund's adjuvant.[134] MDP was also effective at inhibiting the formation of antibody-secreting cells (B Cells).

4. Tumor-Induced Bone Resorption

Many human tumors, primarily of lung, breast, or prostatic origin, and multiple myeloma metastasize to the bone resulting in erosion and hypercalcemia. Generally the bone resorption is a consequence of osteoclast activation; however, in one animal model, the VX_2 virally induced carcinoma of the rabbit, there is evidence to suggest that a late stage of bone erosion takes place in the absence of osteoclasts.[135] Occasionally hypercalcemia exists without any evidence of bone metastasis and humoral factors which stimulate a generalized bone resorption have been implicated.[136] Efforts to identify the humoral factors responsible for malignant hypercalcemia have focused on immunoreactive parathyroid hormone (iPTH) and prostaglandins, particularly of the E series (see Besarab and Caro[137] for a review on the subject). Other osteolytic tumor products include phytosterols in breast cancer[138] and OAF in lymphoma and myeloma.[139] As was discussed in Section VI.A.1, in culture diphosphonates inhibit bone resorption stimulated by PTH, prostaglandins, and OAF. Diphosphonates have not been tested for their ability to inhibit phytosterol induced bone resorption but it is likely that they would, since they interfere with the action of osteoclasts on the bone.

Two animal models of malignant hypercalcemia have been used to study the effects of diphosphonates in vivo. The VX_2 carcinoma of the rabbit secretes PGE_2[140] with increased osteoclast activity in bones adjacent to the tumor.[135,141] Cl_2MDP inhibits the onset and slows the rise in hypercalcemia in this model as measured by both ionized and total calcium.[142] Hypercalcemia in the Walker 256 rat mammary carcinosarcoma model is probably mediated by tumor-secreted iPTH.[143] The diphosphonate APD is effective in retarding and suppressing the hypercalcemia which follows an intramuscular implantation of the tumor.[144] Aortic injection of the ascites form of Walker 256 cells results in lytic deposits in the bone with up to a 30% loss in calcium. Cl_2MDP and to a lesser extent HEDP were effective at inhibiting osteolysis as evidenced by a more normal calcium content of the bone.[145]

The coculturing of pieces of human or animal tumors, or media in which tumor cells have been cultured, with fetal or neonatal bones has been used as a model to study tumor-mediated bone resorption. Cl_2MDP or HEDP, whether given to the neonates before the bones were explanted[146] or added to the culture media,[147] inhibits the bone resorption associated with either animal (VX_2-carcinoma of the rabbit) or human mammary tumors. In one study, of 15 mammary tumors the osteolysis was partially or totally inhibited in 15/15 cases by Cl_2MDP and 14/15 with HEDP. Inhibition by indomethacin was significant in 10/15 cases indicating that prostaglandins were involved in the osteolysis. If devitalized bones (bones in which the cells have been killed by freezing and thawing three times) are cultured with tumors or tumor-conditioned media, they will undergo resorption.[148] In this model, which may mimic the direct effects that tumors have on bone in the absence of osteoclasts, both Cl_2MDP and HEDP inhibited the bone resorption in a dose dependent manner.[149]

C. The Effect of Diphosphonates Related to Bone Mineralization

1. General Background

Failure of the bone to accumulate mineral (widening of the epiphyseal plate and larger anulus ring of osteoid in osteones) following the administration of a diphosphonate (see Sections VI.B.1 and VI.C.2 following) could be due to the inhibition of one or more of three separate processes: (1) the direct chemisorption of the diphosphonate to the calcium phosphate nuclei leading to an inhibition of crystal growth; (2) an interference with the cellular production of matrix components resulting in a decreased or compositionally altered matrix with less area for mineralization; and (3) as a result of the homeostatic coupling in bone, an inhibition of resorption is followed by an inhibition of mineralization; i.e., bone turnover is slowed down. The above three processes will be discussed below.

a. Effect on Biological Mineralization

The chemisorption of the diphosphonates to calcium phosphate nuclei during their formation leads to a physical chemical inhibition of crystal growth of apatite. This effect produces an apparent increase in solubility of hydroxyapatite and can block apatite deposition as in calculus formation (see Section III.C). The possible role of chemisorption as it relates to biological mineralization of mammalian hard tissue is discussed below.

Matrix vesicles are extracellular membrane-containing vesicles which appear to play an initial role in the mineralization of hard tissue. The incubation of slices of rachitic cartilage, from the growth plate of rats, in the appropriate media leads to a matrix vesicle associated line of mineralization. Preincubation with HEDP but not Cl_2MDP abolishes the mineralization.[150] Similar results are obtained if the diphosphonates are given in vivo to chicks and the matrix vesicles isolated and used to study calcification in vitro.[151] The vesicles take up the diphosphonates from solution. HEDP is taken up to a greater extent than Cl_2MDP which parallels their respective avidity for hydroxyapatite.[150] Both diphosphonates stimulate alkaline phosphatase activity in matrix vesicles with HEDP showing the greatest effect. This rise must not be associated with the mineralization process since vesicles, from chicks receiving Cl_2MDP, mineralized at the same rate as the untreated controls. When matrix vesicles and HEDP or Cl_2MDP are simultaneously incubated in solutions of calcium and phosphate, the inhibitory concentration of diphosphonate is the same as in the absence of matrix vesicle.[150] The difference in the in vivo effect of these two diphosphonates on matrix vesicles is unexplained but may be related to both the greater uptake of HEDP and its greater potency at inhibiting calcium phosphate crystal growth.[151] There may also be some unappreciated direct effect of HEDP on the matrix vesicle.

Calcium — phospholipid — phosphate complexes which are obtained only from mineralized tissues have been associated with the initiation of mineral formation in bone.[152] These complexes, which contain the acidic phospholipids phosphatidyl serine and phosphatidyl inositol, induce hydroxyapatite formation in vitro. All biological membranes contain these lipids and it has been suggested that their calcium phosphate complexes may be components of the membranes of calcifying cells or matrix vesicles. Cl_2MDP and HEDP inhibit the hydroxyapatite which forms when calcium — phospholipid — phosphate complexes are added to solutions of metastable calcium phosphate.[152] Cl_2MDP and HEDP also prevent the formation of calcium — phospholipid — phosphate complexes and when added to preformed complexes reduce them to their individual components. The significance of lipid complexes to bone mineralization and consequently the importance of the effects of diphosphonates on these complexes remains to be determined.

b. Bone Matrix

The effects of diphosphonates on matrix synthesis in tissue culture have been discussed in Section VI.A.8. Proline incorporation into collagen is either stimulated[104] or slightly inhibited[85] by Cl_2MDP and HEDP with APD being much more inhibitory[85,104] than Cl_2MDP. Synthesis of the other major constituent of the matrix, proteoglycan, is generally inhibited by diphosphonates, particularly by HEDP.[101,102] Whether these metabolic effects are important in the mineralization process is still unresolved.

c. Inhibition of Resorption Leading to Effects on Mineralization

It is difficult to separate the effects that inhibition of resorption has on the mineralization process from the direct effects diphosphonates may have on mineralization as described above. Tissue culture work with calvaria suggest that concentrations of diphosphonate which inhibit resorption correlate with concentrations which inhibit osteoid formation.[151] Obviously, a decrease in osteoid must ultimately result in decreased mineralization.

When HEDP is used in the treatment of Paget's disease, which is a hyperremodeling process, a reduction in urinary hydroxyproline (a measure of bone turnover) generally preceeds the fall in alkaline phosphatase (a measure of bone formation). However, this may be a result of bone formation continuing at an elevated rate as long as there are excess resorption spaces to fill or that the resorption process is more sensitive to diphosphonate effects. It is likely that the reduction in bone turnover which is seen with low doses of diphosphonates is primarily a result of an inhibition of resorption and that direct effects on mineralization only become significant when excess osteoid begins to form.

2. Hard Tissue Mineralization

Effects on mineralization often attend the use of high doses (>15 mg/kg) of diphosphonates in growing animals, the mechanisms of which are probably a combination of the three mechanisms discussed above.[72,154-159] These doses are generally between 10- and 100-fold higher than are necessary for effects on bone resorption. For HEDP, the higher doses result in a widening of the epiphyseal plate, the appearance of unmineralized cartilage in the growth plate, and a disturbance in the longitudinal growth rate of the bones; all of which return to normal when the drug is discontinued.[154] In contrast to HEDP, only a small widening in the epiphyseal plate is seen with extremely elevated doses of Cl_2MDP. Unmineralized cartilage does not accumulate although longitudinal growth is reduced. Like HEDP, APD causes a widening of the epiphyseal plate, and a retardation of longitudinal growth at the same dose that results in incomplete mineralization.[158] Cl_2MDP is the only well-studied diphosphonate whose usage at high doses is not associated with an accumulation of unmineralized osteoid. It is interesting to note that Cl_2MDP inhibits neither cartilage calcification nor cartilage derived matrix vesicle induced mineralization. Conversely, HEDP inhibits both.

A study of the fluid from the growth plate of rats with an HEDP induced "rickets" revealed a reduction of proteoglycan.[160] The sedimentation coefficient was abnormally high indicating a state of aggregation. Consistent with the hypothesis that proteoglycan aggregates regulate cartilage calcification, the sedimentation coefficient decreased 2.5-fold in rats re-mineralizing after HEDP was withdrawn. During HEDP treatment, the calcium phosphate concentration product was reduced in aspirates from the growth plate but was sufficient to promote mineral growth in vitro.

3. Soft Tissue Calcification

Soft tissue calcification (calciphylaxis) can be experimentally induced by many techniques and the diphosphonates have been found to be effective at preventing calcification in most of these models. The joint ankylosis and heteretopic calcification associated with the Freund's adjuvant model of arthritis are controlled by HEDP and Cl_2MDP.[131,132] These protective effects may be a result of the inhibition of bone resorption so that the local calcium and phosphorous concentrations in the soft tissues of the paw never increase to a concentration that would induce precipitation; conversely, crystal growth inhibition effects cannot be ruled out with current data. A number of diphosphonates, including Cl_2MDP, HEDP, and APD, are effective at inhibiting aortic and kidney calcification in animals receiving high doses of vitamin D.[32,161-163] Cl_2MDP and HEDP also inhibit calciphylaxis of the skin following the administration of lead acetate or polymyxin B.[164] A single oral dose of dihydrotachysterol (DHT) will induce myocardial and vascular degeneration and calcification in rats. Pretreatment or treatment after the administration of DHT with HEDP and to a lesser extent Cl_2MDP inhibits the degeneration and calcification.[165] Similarly, HEDP reduces the number of lesions and amount of calcified arteriosclerotic plaque induced in rabbits by a hypercholesterolemic diet in conjunction with high doses of vitamin D and nicotine.[166] A model of ectopic osteoneogenesis can be developed by the implantation of devitalized bone homografts. HEDP does not affect the initial recellularization of the matrix or the concomitant rise in alkaline phosphatase activity, but does block the remodeling and mineralization process.[167,168]

4. Stones

Bladder stones composed of calcium hydrogen phosphate dihydrate, calcium oxalate, or magnesium ammonium phosphate hexahydrate (struvite) can be induced in rats by dietary manipulations and by implanting zinc pellets in the bladder. The diphosphonate HEDP reduces the size of the calcium oxalate and calcium phosphate stones but the size of the magnesium ammonium phosphate calculi are unaffected.[169] These results are consistent with the finding that HEDP inhibits the precipitation of the first two crystal types from solution but has little effect on magnesium ammonium phosphate precipitation.

5. Osteogenic Sarcoma

The property of diphosphonates to concentrate in bone and at high doses to inhibit hydroxyapatite formation suggests that they might be useful in the treatment of osteogenic sarcoma (an osteoblast-like tumor often characterized by calcification). In one case report, the effects of HEDP on an osteogenic sarcoma in the right distal radius of a Saint Bernard dog were studied.[170] After extended s.c. or oral treatment with HEDP (107 days) there was a marked reduction in active osteoblasts and a complete blockage of the calcific growth of the tumor. In a mixed chondrofibrosarcoma of the rat, HEDP, and to a lesser extent, Cl_2MDP, impaired the tumor associated mineralization but not the cellular growth.[171] Further experimentation is required before any clear efficacy can be established for the use of diphosphonates in this particular type of tumor.

D. Other Effects Related to the Metabolism of Calcium and Phosphate

1. The Relationship Between HEDP, Mineralization, and $1,25(OH)_2D_3$ Synthesis

Dose-related effects are seen on intestinal calcium absorption and $1,25(OH)_2D_3$ production following HEDP administration to rats or chicks. Low doses (1 mg P/kg) stimulate calcium absorption and intestinal $1,25(OH)_2D_3$ content while they inhibit bone resorption and leave cartilage and bone mineralization unimpaired.[172,173] Doses of HEDP on the order of 10 mg P/kg cause hypercalcemia, inhibit mineralization and resorption, and inhibit intestinal calcium transport.[172,174] This reduction in calcium transport is associated with a reduction in 1-hydroxylase[175-177] activity of the kidney but not an inhibition of activity when kidney cells were cultured with HEDP in vitro.[178] Supplementation with $1,25(OH)_2D_3$ but not $25(OH)D_3$ reverses the high dose HEDP effect on intestinal absorption,[175] but does not correct the impaired mineralization.[180] Histologic measurements show a decrease in the broadening of the epiphyseal plate produced by HEDP alone. Vitamin D deficiency and high dose HEDP are similar in that they both produce a widening of the epiphyseal plate but differ in that the former has a greater effect on proliferating cells while the latter affects the hypertrophic cells.[181] Finally, it should be noted that the mineralization defect precedes the inhibition of $1,25(OH)_2D_3$ synthesis. This evidence and the lack of direct effect of HEDP on the kidney cells in vitro[178] suggests that the impairment in mineralization is not due directly to reduced $1,25(OH)_2D_3$ and consequently the reduced calcium absorption may be a homeostatic response to a bone derived "messenger substance" which signals the kidney of a reduced calcium need by this organism. This hypothesis needs to be tempered by the observation that impaired mineralization can be partially overcome with $1,25(OH)_2D_3$ in the chick.[182] The experimental design was not the same in the two studies, however. In addition to a species difference, the dose of HEDP was fourfold lower than that used by Bonjour et al.[180] In another study using rats, $1,25(OH)_2D_3$ prevented or accelerated the reversal of HEDP-induced widening of the epiphyseal plate.[183] Again, it is difficult to compare this experiment to that of Bonjour et al.[180] since calcium balance techniques were not used, the rats were younger, lower doses of HEDP were used, and the source and route of administration of the $1,25(OH)_2D_3$ were different. Taken as a whole, this data suggests that the response to $1,25(OH)_2D_3$ is related to the dose of HEDP; 10 mg P/kg HEDP (s.c.) may be a systemic dose at which the effect on mineralization cannot be overcome, at least with $1,25(OH)_2D_3$.

Relatively less is known about the effects of other diphosphonates on vitamin D metabolism. In one study using rats, Cl_2MDP reduced chromatographically identified $1,25(OH)_2D_3$.[184] In other studies, Cl_2MDP did not reduce the intestinal absorption of calcium as would be expected if $1,25(OH)_2D_3$ levels were decreased. Thus, the original observation has not been corroborated.[172,185]

2. Renal Handling of Phosphate as a Function of HEDP Administration

Doses of HEDP which inhibit bone resorption and mineral deposition and, therefore, the demand for phosphate, markedly reduce the capacity of the tubule of young thyroparathyroidectomized rats to reabsorb phosphate.[186] Inhibition of $1,25(OH)_2D_3$ synthesis by HEDP is not responsible for this phosphaturic effect since HEDP or HEDP and $1,25(OH)_2D_3$ in combination resulted in similar clearance rates for phosphate. Thyroparathyroidectomized rats on a low phosphate diet have a blunted phosphaturic response to PTH which is returned to normal by HEDP. On a high phosphate diet, the phosphaturic response to PTH is enhanced by HEDP. Doses of Cl_2MDP which inhibit bone resorption but not mineralization do not affect the renal handling of phosphate. Neither diphosphonate has an effect on the PTH stimulated rise in renal cAMP. This suggests that the decreased tubular reabsorption of phosphate under chronic HEDP treatment but not Cl_2MDP represents a homeostatic response of the organism to a decreased skeletal demand for phosphate possibly similar to that observed for calcium (see preceding Section VI.D.1). Alternatively, the phosphaturia may be due to a direct effect on the kidney.

Interestingly enough, in man, high doses of HEDP result in a phosphatemic rather than a phosphaturic effect.[187-192] A rise in serum phosphate can be seen within 2 weeks of oral administration of HEDP at 20 mg/kg followed by a return to normal when the drug is discontinued. Hyperphosphatemia is a result of increased tubular reabsorption of phosphorus. Urine cAMP excretion and phosphaturia in response to i.v. bovine PTH (200 units) is unaltered by HEDP administration. Although the direction of phosphate movement at the kidney is opposite in man and rats, in neither case is the effect of PTH opposed or production of cAMP altered. The mechanism of the phosphate effect in man following HEDP treatment remains unknown. Hyperphosphatemia has been observed in pigs with high doses of HEDP[193] and recently has been induced in dogs as a result of high i.v. doses (16 and 32 mg/kg given in a 2 min interval).[42] With the development of an animal model, subsequent studies may shed light on the mechanism of hyperphosphatemia following treatment with HEDP.

VII. MEDICAL APPLICATIONS OF THE DIPHOSPHONATES

A. Osteoporosis

1. Background

Osteoporosis is an abnormal rarefaction of the bone, seen most commonly in the elderly. In advanced cases it may be accompanied by pain, particularly of the lower back, deformities such as loss of stature (curving of the spine and weight-bearing long bones) and pathological fractures. It has been generally accepted that the remaining bone tissue is normal, therefore the fundamental feature of the various osteoporotic syndromes is a loss of bone usually with reduced turnover but not a disordered mineral metabolism.[194] More recent findings suggest that this classical definition of osteoporosis may need to be amended. There is a pronounced difference in trabecular bone volume in osteoporotic females having at least one crushed vertebra as compared to controls. A significant difference in bone density has also been observed.[195] Another deviation from normal bone relates to the fact that within an age matched population of female osteoporotics there is a spectrum of bone loss. "Rapid losers" and "slow losers" are separated by a three- to sixfold difference in the rate of loss depending on whether the measurement is taken at the distal or midshaft of the radius, respectively.[196] Common types of osteoporosis include localized or generalized immobilization, hypogonadism as in estrogen deficient postmenopausal women, bone loss occurring in the presence of excess catabolic agents like corticoids, bone loss in patients with hyperthyroidism, and those receiving cytotoxic chemotherapy[194]

2. Disuse Osteoporosis

Published clinical observations on the use of diphosphonates in osteoporosis are limited to studies involving HEDP.[197-202] In a 20-week study, HEDP reduced the loss of hydroxyproline that attends bedrest osteoporosis.[197] Calcium balance techniques showed that the negative mineral balance had disappeared by the end of the study. Absorptiometry using the ^{125}I photon revealed that there had been no attenuation of mineral loss of the calcaneus. In immobilization osteoporosis, due to spinal cord injury, HEDP inhibited the loss of tibial diaphyseal bone mass as measured by ^{125}I absorptiometry but did not prevent the loss of metaphyseal trabecular bone measured radiographically.[198]

3. Postmenopausal Osteoporosis

Ten patients with postmenopausal osteoporosis responded to a 6 to 12 month treatment of oral disodium HEDP at 20 mg/kg/day with a 50% reduction in bone resorption and a somewhat smaller reduction in mineralization.[200] From this uncoupling of resorption and mineralization, the calcium balance shifted slightly but significantly in the positive direction. At 20 mg/kg, oral HEDP reduces the bone mass of the iliac crest by interfering with mineralization.[202] The production of osteoid may be useful in the treatment of osteoporosis if it occurred from an accelerated *de novo* synthesis which could then mineralize, producing new bone. The accumulation of osteoid attending prolonged administration of HEDP at 20 mg/kg probably stems from an inhibition of mineralization however. In this same study there was no change in the bone density of the distal radius.[202] In one study, 5 mg/kg/day oral HEDP was more effective than higher doses at decreasing urinary hydroxyproline excretion.[202] Doses of HEDP above 5 mg/kg reversibly elevate PTH and serum phosphorus.[197,198,200,202] From these few studies, conflicting results are seen when HEDP is administered at 20 mg/kg. In the study of postmenopausal patients, there was a shift to a positive calcium balance[200] while in a study of immobilization osteoporosis diaphyseal bone mass was spared while metaphyseal was not.[198] These differences might be related to the type of measurements which were taken, the type of osteoporosis which was being treated or the stage at which the particular type of osteoporosis had reached when treatment began. Other dose levels may have generated different results. Whether HEDP can be used to treat any of the forms of osteoporosis has yet to be determined.

B. Hypercalcemia of Malignancy

Active and progressive multiple myeloma is characterized by an increased number of plasma cells in the bone marrow, a homogeneous serum, or urinary myeloma protein, and is often accompanied by extensive skeletal disease. Bone destruction can lead to pain, pathologic fractures, hypercalciuria, and hypercalcemia. The production of OAF by plasma cells from patients with multiple myeloma provides a mechanism for the skeletal destruction associated with this disease.[138] In tissue culture, diphosphonates suppress osteoclastic bone resorption stimulated by OAF suggesting that they might be effective in controlling bone erosion due to multiple myeloma. Phase II clinical trials have supported this belief. In one

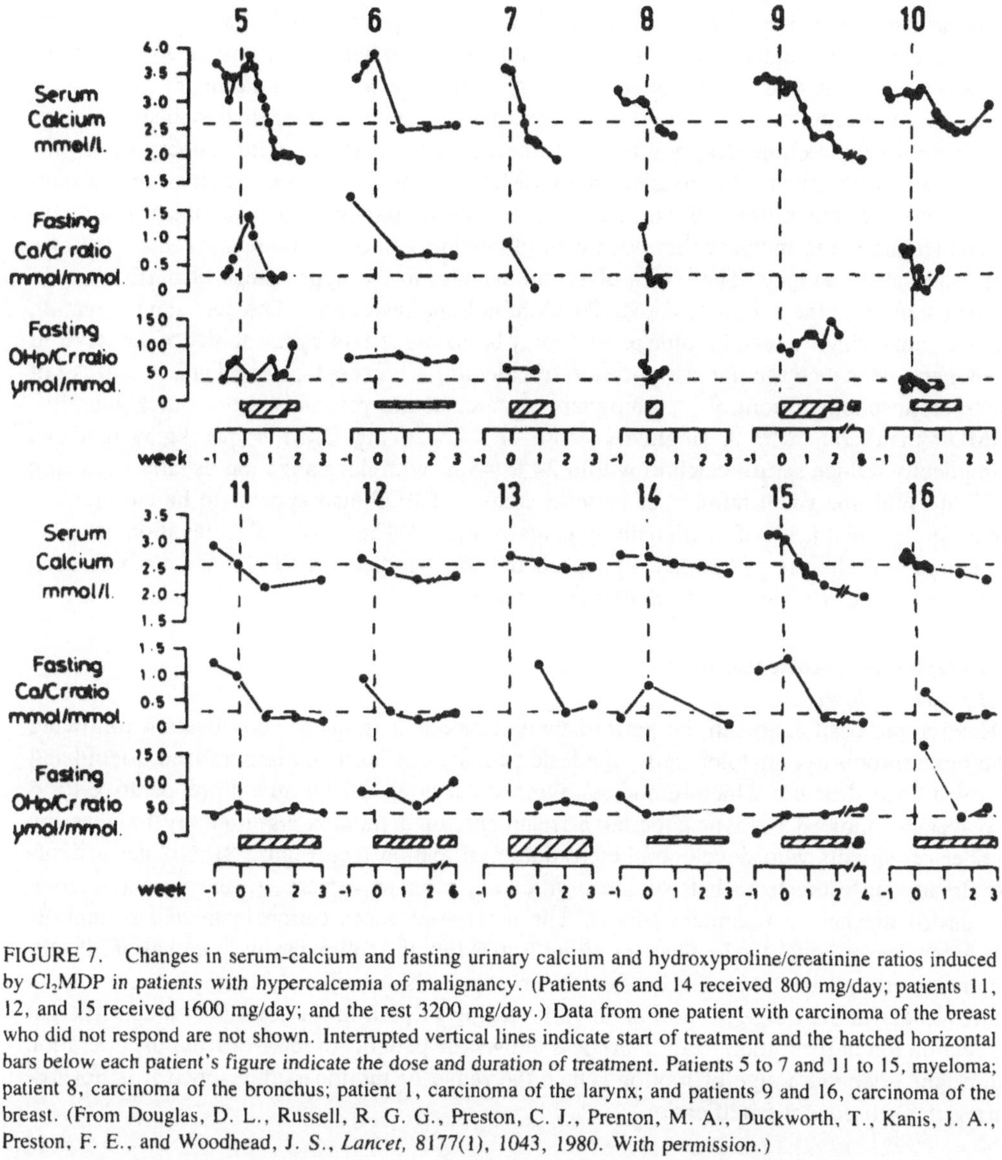

FIGURE 7. Changes in serum-calcium and fasting urinary calcium and hydroxyproline/creatinine ratios induced by Cl_2MDP in patients with hypercalcemia of malignancy. (Patients 6 and 14 received 800 mg/day; patients 11, 12, and 15 received 1600 mg/day; and the rest 3200 mg/day.) Data from one patient with carcinoma of the breast who did not respond are not shown. Interrupted vertical lines indicate start of treatment and the hatched horizontal bars below each patient's figure indicate the dose and duration of treatment. Patients 5 to 7 and 11 to 15, myeloma; patient 8, carcinoma of the bronchus; patient 1, carcinoma of the larynx; and patients 9 and 16, carcinoma of the breast. (From Douglas, D. L., Russell, R. G. G., Preston, C. J., Prenton, M. A., Duckworth, T., Kanis, J. A., Preston, F. E., and Woodhead, J. S., *Lancet*, 8177(1), 1043, 1980. With permission.)

double-blind, placebo-controlled, crossover study, seven out of eight patients with hypercalciuria of myeloma responded during an 8 week oral dose of Cl_2MDP at 3200 mg/day with a significant reduction in urinary calcium.[203] Some patients responded within 7 days. There was a less remarkable drop in urinary hydroxyproline with six out of ten patients showing a decline. Five patients reported a lessening of skeletal pain. Oral Cl_2MDP seems to be effective at controlling the hypercalcemia as well as the hypercalciuria in active multiple myeloma.[204] Furthermore, withdrawal of the drug can be followed by extended periods of normocalcemia. Preliminary trials with APD have shown it to be an effective agent in controlling hypercalcemia, hypercalciuria, and increased urinary hydroxyproline excretion in malignant bone disease due to myeloma or breast cancer.[205] At the present time, Cl_2MDP has been used effectively in a number of clinical trials[203,204,206,207] in controlling malignant bone erosion as a consequence of multiple myeloma, breast cancer, lung cancer, primary hyperparathyroidism, and renal carcinoma (see Figure 7).[204] In general, Cl_2MDP has been more effective at controlling abnormally high calcium values than in producing large effects on the urinary excretion of hydroxyproline. In one five-patient study, while on Cl_2MDP, three patients experienced a rise in serum alkaline phosphatase activity relative to their prestudy values. It was suggested that this rise may be a bone building response as a consequence of reduction in osteoclast activity.[206] In contrast to other malignancies, urinary hydroxyproline excretion rapidly returned to normal when patients with primary hyperparathyroidism were given 1000 to 2000 mg Cl_2MDP per day.[204] Serum calcium values only returned to the upper limit of normal. The actions of parathyroid hormone resulting in hypercalcemia are to increase the movement of calcium out of the bone and reduce the loss of calcium at the kidney. The return of serum calcium to the upper range of normal would be consistent with the action of Cl_2MDP to inhibit bone resorption. The persistent elevation above normal which is seen in some patients may be accounted for by the ability of parathyroid hormones to enhance tubular reabsorption of calcium, a homeostatic mechanism that is not under diphosphonate control.[204] Malignant hypercalcemia responds more quickly to i.v. Cl_2MDP than oral.[208,209] A single i.v. dose of 4 to 20 mg Cl_2MDP per kg weight can dramatically reduce serum calcium within 24 to 48 hr with normalization usually occurring in 72 hr. With the small number of patients studied, HEDP also appears to be an effective agent in the treatment of malignant hypercalcemia.[209] The reversible elevation in PTH associated with the use of Cl_2MDP in Paget's disease (see Section VII.D) has not been seen in the treatment of patients with malignant disease.

C. Heterotopic Ossification

1. Hip Arthroplasty

Heterotopic ossification in the periprosthetic space is a frequent complication following total hip arthroplasty. Histologically, the lesion consists of normally mineralized osteoid and a woven bone structure which remodels to trabecular bone. HEDP given preoperative for 4 to 6 weeks followed by a postoperative treatment for 3 months significantly reduces the number of patients who developed heterotopic ossification (see Figure 8).[38] If the patients are already suffering from heterotopic ossification prior to surgery, further mineralization is retarded during the treatment period. The heterotopic ossification in treated patients is less extensive or equal to the placebo group 6 to 9 months following withdrawal of HEDP. While mineralization takes place after the treatment is terminated, there is a significant improvement in the range of motion and a lessening of pain in the joint.[38,39] It appears that the retardation in the mineralization process allows the patients to recover from the operation and regain adequate range of motion. Once the range of motion is developed it is not lost during the subsequent calcification.

2. Spinal Cord Injury

Heterotopic ossification is also a frequent complication of spinal cord injury. Extensive involvement may cause a severe limitation of motion due to ankylosing of the joints. An inflammatory reaction precedes the deposition of amorphous calcium phosphate which is gradually replaced by enlarged hydroxyapatite crystals. HEDP is effective at reducing the number of patients who get heterotopic ossification and its severity following spinal cord injury (see Figure 9).[37] If HEDP is administered after there is radiographic evidence for heterotopic ossification, further ossification is retarded. One year after termination of treatment, a follow-up study showed that patients who had received HEDP still had significantly less heterotopic ossification than the placebo group.[37] The best results are obtained if the drug is given before there is any evidence of heterotopic ossification.

These data in hip arthroplasty and spinal cord injury suggest that HEDP may be useful in the prevention of heterotopic ossification due to head injury, athletic injuries and severe burns.

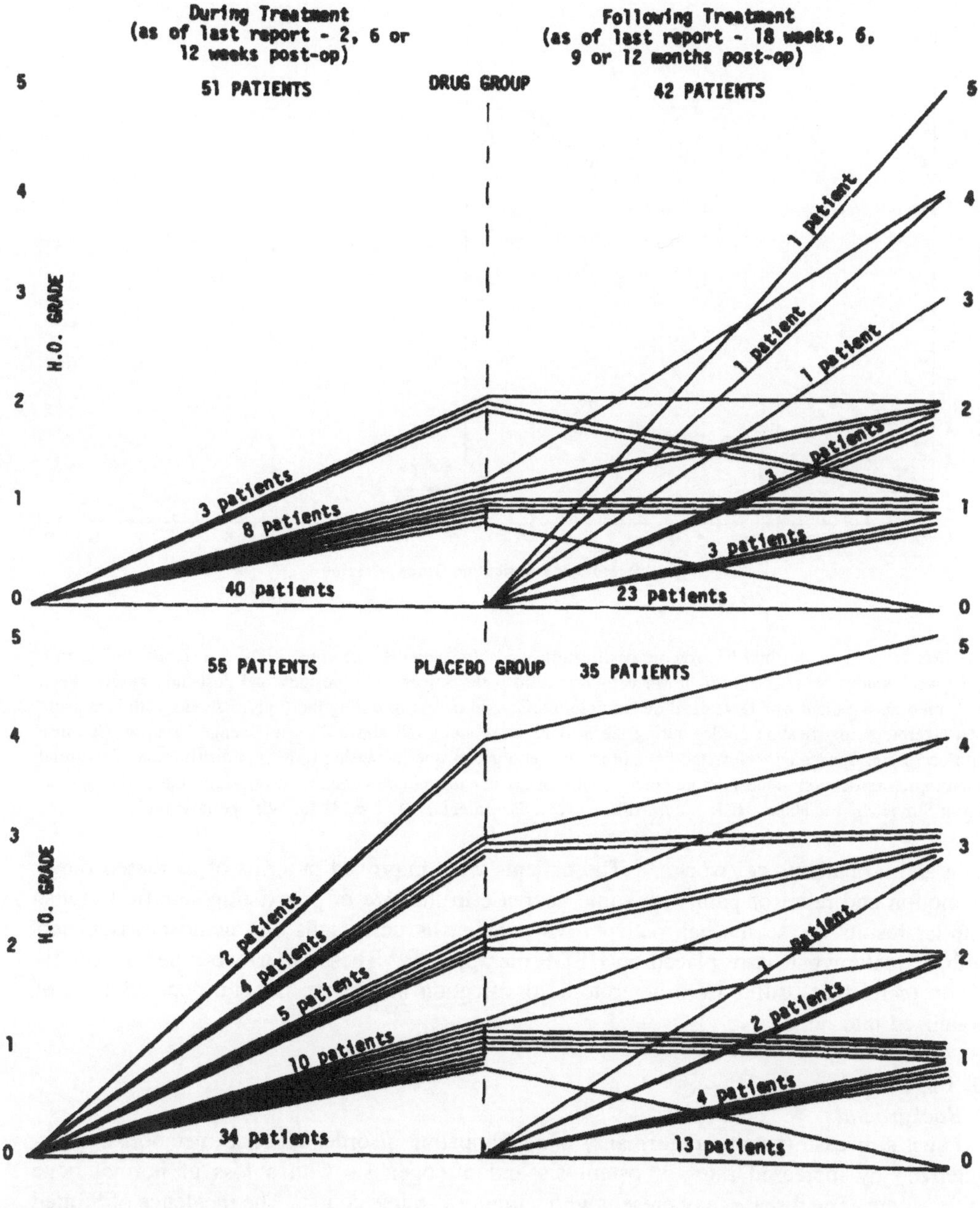

FIGURE 8. The progression of heterotopic ossification (H.O.) during and after treatment with HEDP has been shown for drug-treated and placebo-treated patients. One can see that there has been inhibition of H.O. in the drug group while on medication. Additional H. O. does occur after drug therapy; however, it is not as severe as in the placebo group of patients. (From Finerman, G. A. M., Krenzel, W. F., Jr., Lowell, J. D., Murray, W. R., Volz, R. G., Bowerman, J. W., and Gold, R. H., in *The Hip Proceedings*, Vol. 5, Open Scientific Meeting of the Hip Society, C. V. Mosby, St. Louis, 1977, 222. With permission.)

3. *Myositis Ossificans*

Myositis ossificans progressiva (MOP) is a rare inherited disorder of connective tissue characterized by digital abnormalities and progressive ossification within muscle. In a number of studies, HEDP has been found effective in the prevention of ossification in this disease.[210-212] In one study 44/52 patients were stabilized or improved while 8 were not affected by the

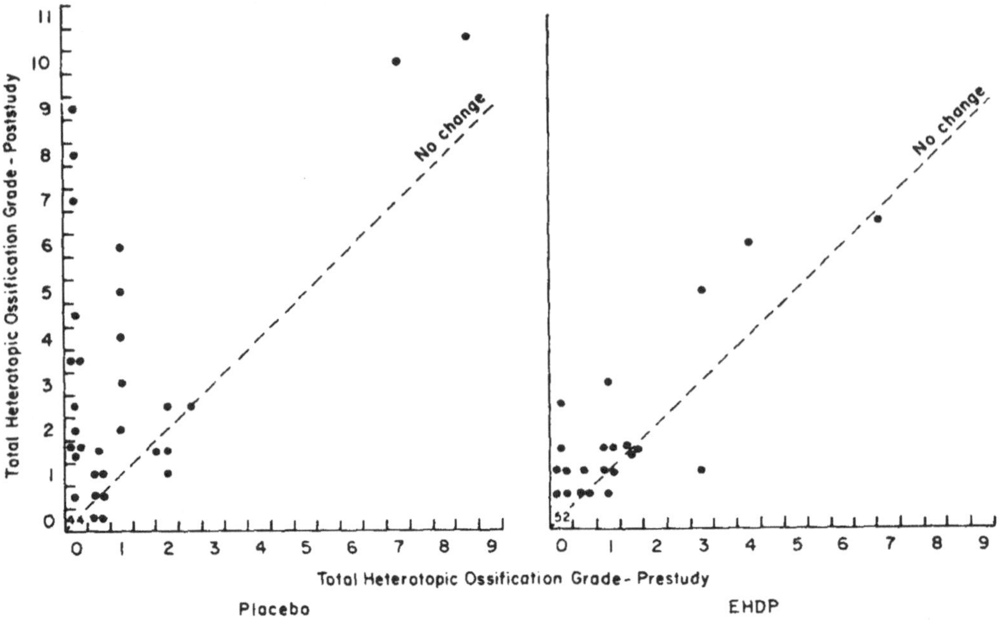

FIGURE 9. Change in total heterotopic ossification grade for patients receiving HEDP or placebo during an 8 to 12 week study. Numbers in left lower cells represent patients with zero prestudy and poststudy grades. Each dot represents a patient who developed heterotopic ossification before or during the study. Patients with increasing total heterotopic ossification grades during the 8 to 12 week study fall above the "no change" line, while those with decreasing grades are below the line. In the placebo group, the increasing (positive) distribution of the total heterotopic ossification grade is more frequent and of greater magnitude compared to those in the EHDP group. (From Stover, S. L., Hahn, H. R., and Miller, J. M., *Paraplegia*, 14, 146, 1976. With permission.)

drug and no patients got worse.[213] The patients were improved in terms of increased range of motion and relief of pain with some decrease in the size of preexisting calcific lesions. Similar results are seen when patients with calcinosis universalis or calcinosis associated with dermatomyositis are placed on HEDP therapy.[213,214] These latter two types of calcification (calcinosis) differ from heterotopic ossification in that the mineral deposits are not organized into bone.

D. Paget's Disease

1. Background

Paget's disease (osteitis deformans) is an idiopathic disorder of bone metabolism characterized by increased rates of osteolysis and osteogenesis with a loss of normal bone architecture. The disease may present with a highly variable course. The incidence of limited and asymptomatic Paget's disease is estimated at 3 to 4% of the population over 40 years of age. Rather than remain quiescent, the disease may progress to involve much of the skeleton. This disorder in bone metabolism can be present with pain, deformity, fractures, increased cardiac output, and nerve compression.[215,216] While serum and urinary calcium and phosphate levels are usually normal in Paget's disease of bone, elevations in blood alkaline phosphatase and excretion of hydroxyproline can be used to monitor the increased bone formation and resorption, respectively.

2. Effects of Diphosphonates

At the present time, HEDP is the only diphosphonate which has been approved for the treatment of Paget's disease.[187,190,217-227] Generally, a dose dependent suppression of the biochemical parameters is seen. The most rapid response is seen with 20 mg/kg/day oral

HEDP disodium salt but significant reductions in hydroxyproline and alkaline phosphatase are obtained with 2.5 to 5 mg/kg/day.[218] Hydroxyproline usually returns to normal or near normal in 1 to 3 months with an additional 1 month delay in the normalization of alkaline phosphatase. A rate difference in the reduction of these two parameters suggests a temporary dissociation of resorption and mineralization. An inhibition of bone resorption (suggested by a normalization of hydroxyproline) probably precedes the suppression in bone formation (reduction in alkaline phosphatase) by the period of time necessary for osteoblasts to mineralize the resorption spaces left by the osteoclasts. Quantitative bone histology has shown that both the number of osteoclasts and osteoblasts are reduced following treatment with HEDP.[225] These changes are more pronounced in Pagetic bone. Symptomatic relief is also good. Most investigators report a reduction in pain in Pagetic lesions and increased mobility following a 3 to 6 month course of HEDP.[190,218-223,225] While the high dose of HEDP, 20 mg/kg/day, brings about the most rapid suppression of biochemical parameters, prolonged use can be associated with a defect in mineralization.[218,223] This mineralization defect, if it occurs, is reversed upon withdrawing the drug. Histological examination of Paget's patients has shown that therapeutic doses of HEDP convert the osteoid from the woven type associated with Paget's disease to a lamellar type associated with normal bone.[190,220] Doses of oral HEDP above 10 mg/kg/day result in a hyperphosphatemia which normalizes when treatment is discontinued. A lower dose of HEDP (5 mg/kg/day) is not associated with hyperphosphatemia or osteomalacia and is now the most common dosage for the treatment of Paget's disease.[226,227]

Curiously enough, bone scans show a normalization of Pagetic lesions following HEDP treatment, however, improvement in X-rays are not always seen.[218,222,223] This may be due to the bone scan's ability to pick up less involved areas which are more responsive to treatment. In areas where the Pagetic lesion is described by an "osteolytic cone" of resorption and can be seen on X-ray, HEDP may slow the rate of bone turnover but the lesion may not heal within the time frame of the study. In this case the scan would appear more normal but the X-ray would not improve. Recently, lytic Paget bone lesions have shown radiographic improvement when patients were treated with HEDP at 5 mg/kg.[228]

Some investigators have combined HEDP and calcitonin in the treatment of Paget's disease.[221,222] In one study, the rate at which hydroxyproline and alkaline phosphatase returned to normal was essentially identical suggesting an even more rapid return to normal biochemical parameters than seen with HEDP alone.[221]

One of the most gratifying aspects of the use of HEDP in Paget's disease of bone has been the long-term relief even after the drug is discontinued.[190,218-221,223,225-227] In one study, 21 of 49 patients had a sustained suppression of their serum alkaline phosphatase and urinary hydroxyproline for 12 months following cessation of HEDP.[218] Symptomatic relief can last as long as two years in 50% of the patients (see Figure 10).[226]

Two additional diphosphonates, Cl_2MDP and APD, are not commercially available but have been tested for clinical efficacy in Paget's disease.[229-231] Both drugs are potent inhibitors of the hyperremodeling state of Paget's disease. Elevated hydroxyproline and alkaline phosphatase return to normal within 1 to 6 months in a dose-related fashion after treatment with Cl_2MDP (Figure 11).[204] Published data is somewhat limited, but the available histology indicates that highly effective doses of Cl_2MDP and APD do not result in a suppression of mineralization. Excess woven bone is replaced by normal amounts of well-mineralized lamellar bone following 6 months of treatment with Cl_2MDP at 1600 mg/day orally.[230] Good improvement in the clinical symptoms of pain and immobility have been reported following the oral administration of Cl_2MDP and APD. These potent inhibitors of resorption lower urinary calcium and reduce the TmP/GFR ratio [transport maximum for phosphate (mg/min) $\div$ glomerular filtration rate (mℓ/min)]. Serum calcium is lowered but generally stays within the normal range. Reversible elevations in iPTH, ostensibly as a result of the depression in

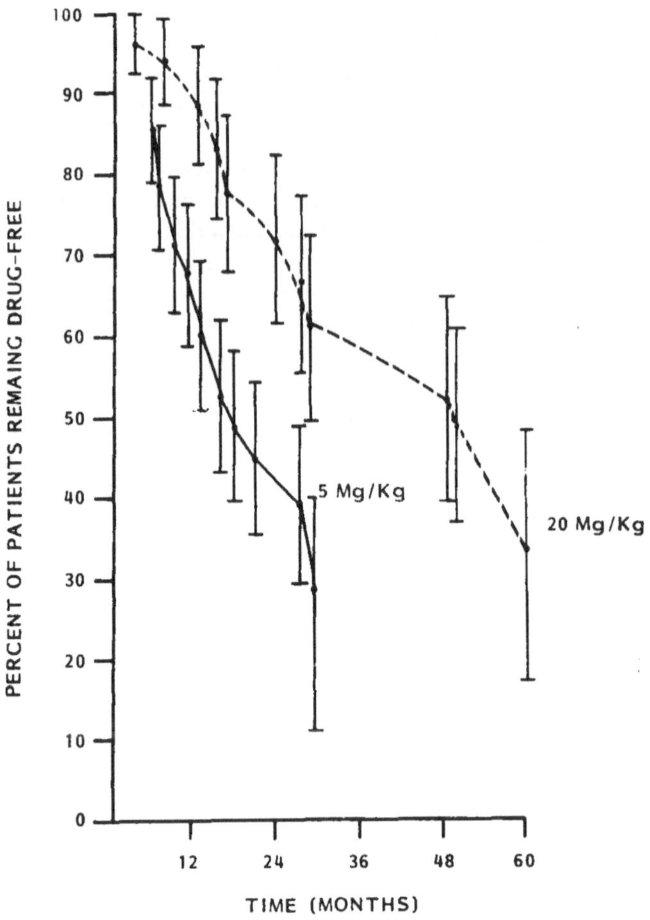

FIGURE 10. A life table analysis of the drug-free interval on two doses of etidronate. Of Paget's patients, 50% have symptomatic relief for 18 or 48 months after a 6 months course of treatment with etidronate at 5 or 20 mg/kg/day, respectively. (From Johnston, C. C., Khairi, M. R. A., and Meunier, P. J., *Arthritis Rheum.*, 23, 1172, 1980. With permission.)

serum calcium, has not been associated with any untoward effects. As with HEDP, a posttreatment remission in biochemical indices (up to 9 months at present) is a positive aspect of the use of Cl_2MDP for Paget's disease.

Currently, the therapeutic index of Cl_2MDP is being reassessed by the manufacturer prior to further human clinical trials.

3. Nuclear Inclusions

Nuclear and cytoplasmic filamentous inclusions described as the nucleocapsid of a paramyxovirus have been seen with the electron microscope in the osteoclasts of Pagetic bone.[231-233] It is thought that this viral infection may play a role in the etiology of the disease. Treatment for 6 months with a dose of Cl_2MDP that is 800 mg/day or higher is associated with a complete disappearance of the osteoclastic inclusions.[231] Osteoclasts are absent in bone dosed with 1600 mg/day Cl_2MDP as the disodium salt and inactive and degenerated osteoclasts characterize the lower doses. It has been postulated that the osteoclastic suppression induced by Cl_2MDP treatment may result in the mounting of an immunological attack upon the released nucleocapsid or delay the infection of new osteoclasts.[231] This could

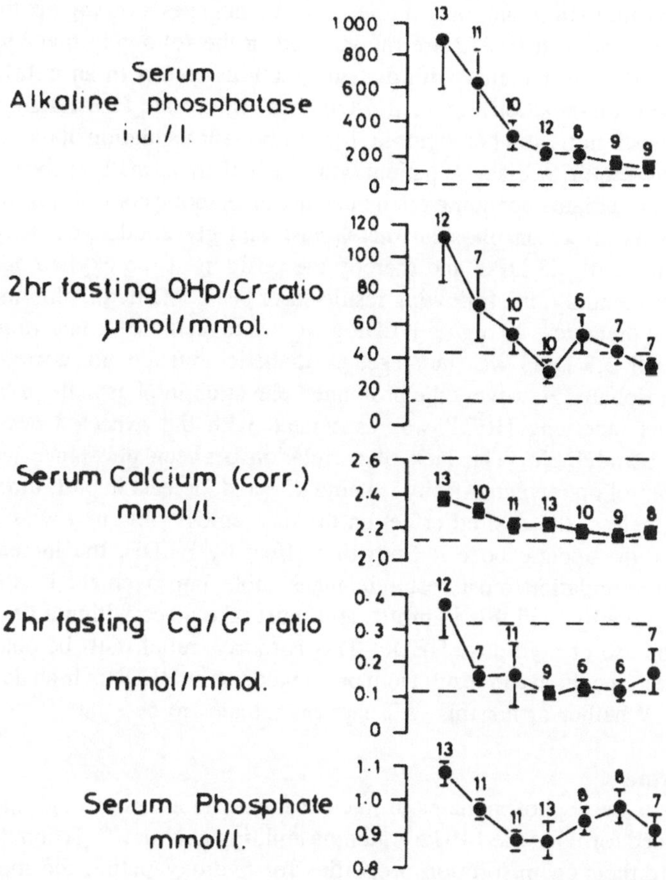

FIGURE 11. Biochemical changes in serum and urine (mean and SEM) in 13 patients with Paget's disease treated with 1600 mg/day of Cl_2MDP. (OHp/Cr = urinary hydroxyproline/creatinine; Ca/Cr = urinary calcium/creatinine; TmP/GFR = maximum tubular resorption of phosphate/glomerular filtration rate.) (From Douglas, D. L., Russell, R. G. G., Preston, C. J., Prenton, M. A., Duckworth, T., Kanis, J. A., Preston, F. E., and Woodhead, J. S., *Lancet*, 8177(1), 1043, 1980. With permission.)

explain the long-lasting, posttreatment clinical effects of diphosphonates in Paget's disease of bone.

E. Diabetes

An imaginative use of the hyperphosphatemic effect of HEDP is being pursued for the treatment of diabetic angiopathy.[234-236] In diabetics, insufficient oxygen release leading to hypoxia in the microcirculation may play a role in the pathogenesis of angiopathy. Diabetic hypoxia and its treatment with HEDP are rationalized in the following manner. Increased glycohemoglobin (HbA_{1c}) in patients with diabetes mellitus results in an unfavorable shift in the oxyhemoglobin dissociation curve. 2,3-Diphosphoglycerate (2,3-DPG), through its effect on oxygen dissociation from hemoglobin, is an important regulation of oxygen transport in the red cell. Insulin-requiring diabetic patients vary more than normally in the concentration of this glycolytic intermediate depending on their state of metabolic control. Plasma inorganic phosphate, through its effect on phosphofructokinase and glyceraldehyde dehydrogenase, may influence the red cell 2,3-DPG and thereby the position of the oxyhemoglobin dissociation curve. Experimentally, the following results have been obtained. Administered orally to normal controls or diabetics, 20 mg/kg HEDP gave rise to an increase in serum phosphate. Interestingly, red cell 2,3-DPG was increased in diabetics but did not correlate with the increased serum phosphate. However, the prolonged elevation in phosphate in both diabetic and healthy subjects receiving HEDP was associated with the expected decrease in the oxygen affinity of hemoglobin. The lack of correlation between phosphate and 2,3-DPG suggests that the control on oxygen affinity is being affected, at least in part, through another mechanism. Nevertheless, the desired effect on the oxyhemoglobin curve was obtained. In order to derive any therapeutic benefit from this effect by HEDP, the increased oxygen release in the microcirculation must result in measurable improvement in some diabetic malady such as angiopathy, and this improvement must take place within a time frame that is acceptable for the use of high dose HEDP. This is to say, relief must be obtained before defects in mineralization which can result from prolonged use of HEDP at high doses becomes a contraindication. Whether or not this will happen, remains to be seen.

F. Nuclear Medicine

The application of the diphosphonates to nuclear medicine arises through their ability to complex with reduced forms of the 140 keV gamma emitting nuclide, ^{99m}Technetium (^{99m}Tc), while still retaining their chemisorption properties for hydroxyapatite, the main inorganic constituent of bone. The most common form of ^{99m}Tc is the commercially available $^{99m}Tc(VII)O_4^-$ obtained either from elution of an alumina column of ^{99}molybdate or as the free extract $^{99m}TcO_4^-$. In this form when given intravenously to animals or man it concentrates heavily in the stomach and thyroid glands[237] and distributes more or less uniformly in other soft tissues. However, when the $^{99m}TcO_4^-$ is reduced electrolytically[238] or chemically by such means as $SnCl_2$,[239] HBr,[240] or $NaBH_4$[240] the reduced forms of ^{99m}Tc can be complexed with diphosphonates such as HEDP,[59,60,237,239,240] MDP,[241] HMDP,[28,242] and others.[243] In complexed form, these agents adsorb to bone[27,237,244] providing excellent imaging agents for the skeleton and particularly for sites of abnormal bone metabolism such as primary and secondary bone tumors[59,60,170,242,245] sites of abnormal calcification,[28,242] and inflammatory osseous disease.[246] The structure of the diphosphonate seems to play an important role in the amount of uptake on normal bone and the discriminatory power of the scanning agent to differentiate between normal and abnormal bone. Significant differences in whole body retention of ^{99m}Tc among HEDP, MDP, and HMDP (skeletal imaging agents) were shown to be 18.4, 30.3, and 36.6% of dose, respectively. Differences in adsorption of ^{99m}Tc between bone lesion (target) and normal bone (nontarget), however, suggest that both HMDP and HEDP have the greatest discrimination.[242] Clinical data on HEDP[248] and HMDP[249] support this difference. This discrimination is thought to be related to the highly specific structural coordination of the 1-hydroxy diphosphonates for the trigonal face of a calcium center in the 001 face (Figure 4) of the hydroxyapatite growing crystal. The research on different diphosphonates in this area of medical application is currently in a dynamic state.

VIII. CONCLUSIONS

The primary mechanism for the localization of the diphosphonates in the mammalian skeleton is chemisorption to deposits of solid calcium phosphate. It can be argued that the extracellular deposition of diphosphonates can account for all of the biological effects on resorption and mineralization. The simplest way to explain the mechanism of action of diphosphonates involving only extracellular adsorption would be an interference with osteoclastic resorption leading to a homeostatic reduction in osteoblastic activity and, consequently, a reduction in bone turnover. This inhibition in osteoclastic bone resorption may be the result of an interference by the diphosphonate with calcium and phosphate fluxes in the vicinity of the cell through moderation of the solubility rate of adjacent solid calcium phosphate. Another possibility for altered osteoclastic function may be the change in the surface of the hydroxyapatite due to the binding of the diphosphonate and hence attachment of the osteoclast to the surface of bone. These possible changes in the milieu around the osteolcasts could then account for its inability to recognize or act on the new surface or function in its normal capacity.

Equally important to the known extracellular adsorption of diphosphonates on bone tissue may be their intracellular actions on osteoclasts and osteoblasts. These actions could involve the direct inhibitions of enzymes or control mechanisms, or more likely the chemisorption to organelles containing calcium phosphate solids (such as mitochondria) causing alterations in intracellular calcium and phosphate fluxes. The potential of diphosphonates to form soluble chelates with intracellular calcium, magnesium, or other metal ions could also bring about alterations in cell activity. The inability to distinguish between the relative importance of the intra- and extracellular effects of the diphosphonates as they relate to bone turnover has been one of the biggest stumbling blocks in the understanding of the mechanism of action of these agents. A solution to this problem would do much toward providing a rational basis for the development of therapeutically more effective diphosphonates.

Through its action on bone remodeling, the diphosphonate HEDP has found effective therapeutic application in the treatment of Paget's disease and prophylactic uses in heterotopic ossification resulting from hip arthroplasty or spinal cord injury and in dental calculus prevention. Laboratory and clinical studies are currently underway to evaluate the diphosphonates for their medical use in osteoporosis, arthritis, and primary and secondary tumors of the bone.

REFERENCES

1. **Harvey, R. G. and DeSombre, E. R.,** The Michaelis-Arbuzov and related reactions, *Top. Phosphorus Chem.*, 1, 57, 1964.
2. **Canavan, A. E., Dowden, B. F., and Eaborn, C.,** The thermal decomposition of alkylphosphonates: a new route to alkylphosphonic acids, *J. Chem. Soc.*, 331, 1962.
3. **Curry, J. D. and Nicholson, D. A.,** Oligophosphonates, *Top. Phosphorus Chem.*, 7, 37, 1972.
4. **Von Blaser, B., Worms, K.-H., Germscheid, H.-G., and Wollmann, K.,** Uber 1-hydroxyalkan-1,1-diphosphonsaüren, *Z. Anorg. Allg. Chem.*, 381, 247, 1971.
5. **Sekine, M. and Hata, T.,** Convenient synthesis of unesterified acylphosphonic acids, *J. Chem. Soc., Chem. Commun.*, 285, 1978.
6. **Worms, K.-H. and Blum, H.,** Unsetzungen von 1-aminoalkan-1,1-diphosphonsaüren mit saltpetriger saüre, *Z. Anorg. Allg. Chem.*, 457, 209, 1979.
7. **Kosolapoff, G. M. and Maier, L.,** *Organic Phosphorus Compounds*, Vol. 7, John Wiley & Sons, New York, 1976, chap. 18.

8. **Schroeder, J. P. and Sopchak, W. P.,** The reaction of phosphorus trichloride and oxygen with polymers, *J. Polym. Sci.*, AI 47, 417, 1960.
9. **Rochlitz, F. and Vilcsek, H.,** β-Chloroethylphosphonic dichloride: its synthesis and use, *Angew. Chem., Int. Ed.*, 74, 940, 1962.
10. **Anbar, M., St. John, G. A., and Scott, A. C.,** Organic polymeric polyphosphonates as potential preventive agents of dental caries: in vitro experiments, *J. Dent. Res.*, 53, 867, 1974.
11. **Cabasso, I., Jagur-Grodzinski, J., and Vofsi, D.,** Synthesis and characterization of polymers with pendant phosphonate groups, *J. Appl. Polym. Sci.*, 18, 1969, 1974.
12. **Bartels, T., Schuthof, J., and Ahrends, J.,** The adsorption of two polyphosphonates on hydroxyapatite and their influence on the acid solubility of whole bovine enamel, *J. Dent.*, 7, 221, 1979.
13. **Anbar, M., St. John, G. A., and Edward, T. E.,** Organic polymeric polyphosphonates as potential preventive agents of dental caries. In vivo experiments, *J. Dent. Res.*, 53, 1240, 1974.
14. **Francis, M. D. and Briner, W. W.,** The effect of phosphonates on dental enamel in vitro and calculus formation in vivo, *Calcif. Tissue Res.*, 11, 1, 1973.
15. **Briner, W. W. and Francis, M. D.,** In vitro and in vivo evaluation of anti-calculus agents, *Calcif. Tissue Res.*, 11, 10, 1973.
16. **Meyer, J. L. and Nancollas, G. H.,** The influence of multidentate organic phosphonates on the crystal growth of hydroxyapatite, *Calcif. Tissue Res.*, 13, 295, 1973.
17. **Kabachnik, M. I., Lastovskii, R. P., Medvad, T. Y., Medyntser, V. V., Kolpakova, I. D., and Dyatlova, N. M.,** The complexing properties of hydroxyethylidenediphosphonic acid in aqueous solutions, *Dokl. Akad. Nauk., SSSR*, 177, 582, 1967.
18. **Grabenstetter, R. J., Quimby, O. T., and Flaut, T. J.,** The acid dissociation constants of substituted methanediphosphonic acids: a correlation with P^{31} magnetic resonance chemical shift and with Taft σ^*, *J. Phys. Chem.*, 71, 4194, 1967.
19. **Carroll, R. L. and Irani, R. R.,** On the acidity of substituted methylenediphosphonates and their interaction with alkali metal ions, *Inorg. Chem.*, 6, 1994, 1967.
20. **Francis, M. D. and Centner, R. L.,** The development of diphosphonates as significant health care products, *J. Chem. Ed.*, 55, 760, 1978.
21. **Tobias, R. S.,** Studies on the hydrolysis of metal ions. 21. The hydrolysis of tin (II) ion, Sn^{2+}, *Acta Chem. Scand.*, 12, 198, 1958.
22. **Grabenstetter, R. J. and Cilley, W. A.,** Polynuclear complex formation in solutions of calcium ion and ethane-1-hydroxy-1,1-diphosphonic acid. 1. Complexometric and pH titrations, *J. Phys. Chem.*, 75, 676, 1971.
23. **Wiers, B. H.,** Polynuclear complex formation in solution of calcium ion and ethane-1-hydroxy-1,1-diphosphonic acid. II. Light scattering, sedimentation, mobility and dialysis measurements, *J. Phys. Chem.*, 75, 682, 1971.
24. **Yesinowski, J. P. and Benedict, J. J.,** Unpublished work, 1978.
25. **Deutsch, E. and Barnett, B. L.,** Synthetic and structural aspects of technetium chemistry as related to nuclear medicine, in *ACS Symposium Series, No. 140, Inorganic Chemistry in Biology and Medicine*, Martell, A. E., Ed., American Chemical Society, Washington, D.C., 1980, 103.
26. **Barnett, B. L. and Strickland, L. C.,** Structure of disodium dihydrogen 1-hydroxyethylidenediphosphonate tetrahydrate: a bone growth regulator, *Acta Crystallogr.*, B35, 1212, 1979.
27. **Francis, M. D., Tofe, A. J., Benedict, J. J., and Bevan, J. A.,** Imaging the skeletal system, in *Radiopharmaceuticals II*, Proc. 2nd Int. Symp. on Radiopharm., Society of Nuclear Medicine, New York, 1979, 603.
28. **Bevan, J. A., Tofe, A. J., Francis, M. D., Barnett, B. L., and Benedict, J. J.,** Tc-99m HMDP (hydroxymethylenediphosphonate): a new radiopharmaceutical for skeletal and acute myocardial infarct imaging. Part I. Synthesis and biodistribution in animals. Part II. Comparison of Tc-99m HMDP with other technetium labelled bone imaging agents in a canine model, *J. Nucl. Med.*, 21, 961, 1980.
29. **Francis, M. D.,** The inhibition of calcium hydroxyapatite crystal growth by polyphosphonates and polyphosphates, *Calcif. Tissue Res.*, 3, 151, 1969.
30. **Francis, M. D. and Webb, N. C.,** Hydroxyapatite formation from a hydrated calcium monohydrogen phosphate precursor, *Calcif. Tissue Res.*, 6, 335, 1971.
31. **Francis, M. D.,** Composition for inhibiting anomalous deposition and mobilization of calcium phosphate in animal tissue, U.S. Patent 3,683,080, 1972.
32. **Francis, M. D., Russell, R. G. G., and Fleisch, H.,** Diphosphonates inhibit formation of calcium phosphate crystals in vitro and pathological calcification in vivo, *Science*, 165, 1264, 1969.
33. **Francis, M. D., Slough, C. L., Briner, W. W., and Oertel, R. P.,** An investigation of melitate and ethane-1-hydroxy-1,1-diphosphonate in calcium phosphate systems, *Calcif. Tissue Res.*, 23, 53, 1977.
34. **Sturzenburger, O. P., Swancar, J. R., and Reiter, G.,** Reduction of dental calculus in humans through the use of a dentrifice containing a crystal-growth inhibitor, *J. Periodont.*, 42, 416, 1971.

35. **Conroy, C. W., Sturzenburger, O. P., Bollmer, B. W., Swancar, J. R., and Zimmerman, E. R.,** The effect of a sodium etidronate dentifrice on calculus and gingivitis in adults, *J. Dent. Res.*, (Suppl. 208), 100, 1972.
36. **Herforth, V. A., Fligge, U., and Strossburg, M.,** Clinical investigation on the reduction of calculus formation by use of a toothpaste containing 1% HEDP, *Dtsch. Zahnaertztl. Z.*, 32, 757, 1977.
37. **Stover, S. L., Hahn, H. R., and Miller, J. M.,** Disodium etidronate in the prevention of heterotopic ossification following spinal cord injury, *Paraplegia*, 14, 146, 1976.
38. **Finerman, G. A. M., Krenzel, W. F., Jr., Lowell, J. D., Murray, W. R., Volz, R. G., Bowerman, J. W., and Gold, R. H.,** Role of diphosphonate (EHDP) in the prevention of heterotopic ossification after total hip arthroplasty, in *The Hip Proceedings*, Vol. 5, Open Scientific Meeting of the Hip Society, C. V. Mosby, St. Louis, 1977, 222.
39. **Bijvoet, O. L. M., Nollen, A. J. G., Slooff, T. J. J. H., and Feith, R.,** Effect of a diphosphonate on para-articular ossification after total hip replacement, *Acta Orthop. Scand.*, 45, 926, 1974.
40. **Nixon, G. A., Buehler, E. V., and Newman, E. A.,** Preliminary safety assessment of disodium etidronate as an additive to experimental oral hygiene products, *Toxicol. Appl. Pharmacol.*, 22, 661, 1972.
41. **Tofe, A. J. and Francis, M. D.,** The author's reply, *J. Nucl. Med.*, 16, 444, 1975.
42. **Francis, M. D. and Slough, C. L.,** Acute IV infusion of etidronate disodium(EHDP): mechanism of toxicity, *Acad. Pharm. Sci., Abstr.* 5, 1981.
43. **Francis, M. D. and Slough, C. L.,** Unpublished work, 1975.
44. **King, W. R., Francis, M. D., and Michael, W. R.,** Effect of disodium ethane-1-hydroxy-1,1-diphosphonate on bone formation, *Clin. Orthop.*, 78, 251, 1971.
45. **Nolen, G. A. and Buehler, E. V.,** The effects of disodium etidronate on the reproductive functions and embryogeny of albino rats and New Zealand rabbits, *Toxicol. Appl. Pharmacol.*, 18, 548, 1971.
46. **Nixon, G. A. and Buehler, E. V.,** Unpublished work, 1972.
47. **Flora, L., Hassing, G. S., Parfitt, A. M., and Villanueva, A. R.,** Comparative skeletal effects of two diphosphonates in dogs, in *Bone Histomorphometry*, Jee, W. S. S. and Parfitt, A. M., Eds., Société de Publications Médicales et Dentaires, Paris, 1981, 389.
48. **Flora, L.,** Unpublished work, 1980.
49. **Vignery, A. and Baron, R.,** Comparative effects of APD and Cl_2MDP on bone in the rat: in vivo and in vitro studies, in *Bone Histomorphology*, Jee, W. S. S. and Parfitt, A. M., Eds., Société de Publications Médicales at Dentaires, Paris, 1981, 381.
50. **Michael, W. R., King, W. R., and Wakim, J. M.,** Metabolism of disodium ethane-1-hydroxy-1,1-diphosphonate (disodium etidronate) in the rat, rabbit, dog, and monkey, *Toxicol. Appl. Pharmacol.*, 21, 503, 1972.
51. **Saville, P. D. and Recker, R. R.,** Unpublished work, 1973.
52. **Michael, W. R.,** Unpublished work, 1970.
53. **Varns, C. L., Harvey, W. J., and King, W. R.,** Unpublished work, 1974.
54. **Ginn, D. L. and Hedglin, W. L.,** Unpublished work, 1974.
55. **Hedglin, W. L. and Ginn, D. L.,** Unpublished work, 1974.
56. **King, W. R.,** Unpublished work, 1974.
57. **Wasserman, R. H., Bonjour, J. P., and Fleisch, H.,** Ileal absorption of disodium ethane-1-hydroxy-1,1-diphosphonate (EHDP) and disodium dichloromethylene diphosphonate (Cl_2MDP) in the chick, *Experientia*, 29, 1110, 1973.
58. **Liggett, J. S.,** Unpublished work, 1974.
59. **Wellman, H. N., Browne, A., Kavula, M., Khairi, R., Anger, R., Tofe, A. J., and Francis, M. D.,** Optimization of a new kit — prepared skeletal imaging agent, ^{99m}Tc-Sn-EHDP, compared with ^{18}F, in *Radiopharmaceuticals and Labeled Compounds*, International Atomic Energy Agency, Vienna, 1973, 1.
60. **Silberstein, E., Saenger, E., Tofe, A., Alexander, G., and Park, H.,** Imaging of bone metastases with ^{99m}Tc-Sn-EHDP (diphosphonate), ^{18}F, and skeletal radiography — a comparison of sensitivity, *Radiology*, 197, 551, 1973.
61. **Troehler, U., Bonjour, J. P., and Fleisch, H.,** Renal secretion of diphosphonates in rats, *Kidney Int.*, 8, 6, 1975.
62. **La Nauze, J. M., Rosenberg, H., and Shaw, D. C.,** The enzymatic cleavage of the carbon-phosphorus bond: purification and properties of the phosphonatase, *Biochim. Biophys. Acta*, 212, 332, 1970.
63. **Floyd, B., Francis, S., and Benedict, J. J.,** Unpublished work, 1978.
64. **Preston, D., Dittert, L. W., and Foster, T.,** Unpublished work, 1973.
65. **Reynolds, J. J., Minkin, C., Morgan, D. B., Spychen, D., and Fleisch, H.,** The effects of two diphosphonates on the resorption of mouse calvaria in vitro, *Calcif. Tissue Res.*, 10, 302, 1972.
66. **Russell, R. G. G., Muhlbauer, R. C., Bisaz, S., Williams, D. A., and Fleisch, H.,** The influence of pyrophosphate, condensed phosphates, phosphonates and other phosphate compounds on the dissolution of hydroxyapatite in vitro and on bone resorption induced by parathyroid hormone in tissue culture and in thyroparathyroidectomized rats, *Calcif. Tissue Res.*, 6, 183, 1970.

67. **Fleisch, H., Russell, R. G. G., and Francis, M. D.,** Diphosphonates inhibit hydroxyapatite dissolution in vitro and bone resorption in tissue culture and in vivo, *Science*, 165, 1262, 1969.
68. **Hausman, E., Weinfeld, N., and Michaels, S.,** Lipopolysaccharide-diphosphonate interaction on bone in tissue culture, *Arch. Oral Biol.*, 17, 1381, 1972.
69. **Goldberg, L. S.,** Unpublished work, 1979.
70. **Gebauer, U. and Fleisch, H.,** The effects of two diphosphonates on the resorption of mouse calvaria induced by 1,25-dihydroxy-cholecalciferol and prostaglandin E_1, in culture, *Miner. Electrolyte Metab.*, 4, 172, 1980.
71. **Rowe, D. J. and Hausman, E.,** The alteration of osteoclast morphology by diphosphonates in bone organ culture, *Calcif. Tissue Res.*, 20, 53, 1976.
72. **Schenk, R., Merz, W. A., Mühlbauer, R., Russell, R. G. G., and Fleisch, H.,** Effect of ethane-1-hydroxy-1,1-diphosphonate (EHDP) and dichloromethylene diphosphonate (Cl_2MDP) on the calcification and resorption of cartilage and bone in the tibial epiphysis and metaphysis of rats, *Calcif. Tissue Res.*, 11, 196, 1973.
73. **Miller, S. C. and Jee, W. S. S.,** Ethane-1-hydroxy-1,1-diphosphonate (EHDP): Effects on growth and modeling of the rat tibia *Calcif. Tissue Res.*, 18, 215, 1975.
74. **Miller, S. C., Jee, W. S. S., Kimmel, D. B., and Woodbury, L.,** Ethane-1-hydroxy-1,1-diphosphonate (EHDP) effects on incorporation and accumulation of osteoclast nuclei, *Calcif. Tissue Res.*, 22, 243, 1977.
75. **Plasmans, C. M. T., Jap, P. H. K., Kuijpers, W., and Slooff, T. J. J. H.,** Influence of a diphosphonate on the cellular aspect of young bone tissue, *Calcif. Tissue Int.*, 32, 247, 1980.
76. **Rosenblum, I. Y., McCuskey, R. S., McNeal, N. C., Kirckaert, G. A., Flora, L., and Metzger, C. A.,** The effects of EHDP on regenerating trabecular bone using in vivo microscopic, light and electron microscopic and electron microprobe techniques, *Calcif. Tissue Res.*, 20, 91, 1976.
77. **Plasmans, C. M. T., Kuijpers, W., and Slooff, T. J. J. H.,** The effect of ethane-1-hydroxy-1,1-diphosphonic acid (EHDP) on matrix induced ectopic bone formation, *Clin. Orthop.*, 132, 233, 1978.
78. **Weisbrode, S. E., Capen, C. C., and Pendley, C. B.,** Effect of dichloromethylene diphosphonate on morphology, enzyme activity, and ash content of bones of thyroparathyroidectomized rats, *Calcif. Tissue Res.*, 25, 119, 1978.
79. **Ende, J. J.,** Some effects of EHDP and Cl_2MDP on the metabolism of isolated bone cells, *Proc. K. Ned. Akad. Wet. Ser. C. Biol. Med. Sci.*, 81, 252, 1978.
80. **Fast, D. K., Felix, R., Dowse, C., Neuman, W. F., and Fleisch, H.,** The effects of diphosphonates on the growth and glycolysis of connective-tissue cells in culture, *Biochem. J.*, 172, 97, 1978.
81. **Felix, R., Fast, D. K., Sallis, J. D., and Fleisch, H.,** Effects of diphosphonates on glycogen content of rabbit ear cartilage cells in culture, *Calcif. Tissue Int.*, 30, 163, 1980.
82. **Morgan, D. B., Monod, A., Russell, R. G. G., and Fleisch, H.,** Influence of dichloromethylene diphosphonate (Cl_2MDP) and calcitonin on bone resorption, lactate production and phosphatase and pyrophosphatase content of mouse calvaria treated with parathyroid hormone in vitro, *Calcif. Tissue Res.*, 13, 287, 1973.
83. **Ende, J. J. and van Rooijen, H. J. M.,** Some effects of EHDP and Cl_2MDP on the metabolism of mouse calvaria in tissue culture, *Proc. K. Ned. Akad. Wet. Ser. C. Biol. Med. Sci.*, 82, 43, 1979.
84. **Shinoda, H., Felix, R., and Fleisch, H.,** Structure-activity relationship of diphosphonates with alkyl groups of various length, *Miner. Electrolyte Metab.*, 2, 268, 1979.
85. **Lemkes, H. H. P. J., Reitsma, P. H., Frijlink, W., Verlinden-Ooms, H., and Bijvoet, O. L. M.,** A new diphosphonate: dissociation between effects on cells and mineral in rats and a preliminary trial in Paget's disease, *Adv. Exp. Med. Biol.*, 103, 459, 1978.
86. **Pilczyk, R., Sutcliffe, H., and Martin, T. L.,** Effects of pyrophosphate and diphosphonates on parathyroid hormone- and fluoride-stimulated adenylate cyclase activity, *FEBS Lett.*, 24, 225, 1972.
87. **Eisman, J. A., Martin, T. J., Pilczyk, R., Legge, D. G., and Sutcliffe, H. S.,** Influence of pyrophosphate and diphosphonates on rat liver adenylate cyclase, *Clin. Exp. Pharmacol. Physiol.*, 1, 13, 1974.
88. **Goulding, A., McChesney, R., and Malthus, R. S.,** Diminished urinary excretion of cyclic AMP by rats given disodium ethane-1-hydroxy-1,1-diphosphonate, *J. Endocrinol.*, 65, 145, 1975.
89. **Gebauer, U., Russell, R. G. G., Touabi, M., and Fleisch, H.,** Effect of diphosphonates on adenosine 3′:5′cyclic monophosphate in mouse calvaria after stimulation by parathyroid hormone in vitro, *Clin. Sci. Mol. Med.*, 50, 473, 1976.
90. **Felix, R., Russell, R. G. G., and Fleisch, H.,** The effect of several diphosphonates on acid phosphohydrolases and other lysosomal enzymes, *Biochem. Biophys. Acta*, 429, 429, 1976.
91. **Guilland, D. E., Sallis, J. D., and Fleisch, H.,** The effect of two diphosphonates on the handling of calcium by rat kidney mitochondria in vitro, *Calcif. Tissue Res.*, 15, 303, 1974.
92. **Guilland, D. F. and Fleisch, H.,** The effect of in vivo treatment with EHDP and/or 1,25-DHCC on calcium uptake and release in isolated kidney mitochondria, *Biochem. Biophys. Res. Commun.*, 61, 906, 1974.

93. **Felix, R. and Fleisch, H.,** The effects of pyrophosphate and diphosphonates on calcium transport in red cells, *Separatum Experientia*, 33, 1003, 1977.
94. **Felix, R. and Fleisch, H.,** Increase in alkaline phosphatase activity in calvaria cells cultured with diphosphonates, *Biochem. J.*, 183, 73, 1979.
95. **Weisbrode, S. E., Capen, C. C., and Pendley, C. B. II,** Interaction of dichloromethylene diphosphonate and vitamin D on bone of thyroparathyroidectomized rats, *Am. J. Pathol.*, 89, 137, 1977.
96. **Felix, R., Herrmann, W., and Fleisch, H.,** Stimulation of precipitation of calcium phosphate by matrix vesicles, *Biochem. J.*, 170, 681, 1978.
97. **Woltgens, J. H. M., Bervoets, T. H. J. M., and DeVries, W.,** Influence of diphosphonates on the pyrophosphatase activity in *Streptococcus mutans*, *J. Periodontal Res.*, 12, 462, 1977.
98. **Herczog, J. I. B., DeVries, W., and Woltgens, J. H. M.,** Influence of ethane-1-hydroxy-1,1-diphosphonate on anaerobic growth and acid formation by *S. mutans*, *Caries Res.*, 11, 90, 1977.
99. **Vreven, J. and Frank, R. M.,** Activities pyrophosphatiques acide et alcaline de la plaque dentaire humaine, *Arch. Oral. Biol.*, 19, 203, 1974.
100. **Ritz, H. L.,** Microbial populationship in developing human dental plaque, *Arch. Oral. Biol.*, 12, 1561, 1967.
101. **Guenther, H. L., Guenther, H. E., and Fleisch, H.,** Effects of 1-hydroxyethane-1,1-diphosphonate and dichloromethanediphosphonate on rabbit articular chondrocytes in culture, *Biochem. J.*, 184, 203, 1979.
102. **Palmoski, M. and Brandt, K.,** Effects of diphosphonates on glycosaminoglycan synthesis and proteoglycan aggreagation in normal adult articular cartilage, *Arthritis Rheum.*, 21, 942, 1978.
103. **Spengler, D. M., Baylink, D. J., Rosenquist, J. B., and Evans, R.,** Mechanical properties of connective tissue: effect of diphosphonate, *Clin. Res.*, 25, 368A, 1977.
104. **Guenther, H. L., Guenther, H. E., Fast, D. K., and Fleisch, H.,** The effect of two diphosphonates on the collagen metabolism of mouse and rat bone and cartilage, *Experientia*, 3, 792, 1977.
105. **Miller, S. C. and Jee, W. S. S.,** The effect of dichloromethylene diphosphonate, a pyrophosphate analog, on bone and bone cell structure in the growing rat, *Anat. Rec.*, 193, 439, 1979.
106. **Miller, S. C. and Jee, W. S. S.,** The comparative effects of dichloromethylene diphosphonate (Cl_2MDP) and ethane-1-hydroxy-1,1-diphosphonate (EHDP) on growth and modeling of the rat tibia, *Calcif. Tissue Res.*, 23, 207, 1977.
107. **Larsson, A. and Larsson, S. E.,** The effects of ethane-1-hydroxy-1,1-diphosphonate on cellular transformation and organic matrix of the epiphyseal growth plate of the rat: a light microscopic and ultrastructural study, *Acta Pathol. Microbiol. Scand.*, 86, 211, 1978.
108. **Russell, J. E., Termine, J. D., and Avioli, L. V.,** Experimental renal osteodystrophy. The response to 25-hydroxycholecalciferol and dichloromethylene diphosphate therapy, *J. Clin. Invest.*, 56, 548, 1975.
109. **Reynolds, J. J., Murphy, H., Mühlbauer, R. C., Morgan, D. B., and Fleisch, H.,** Inhibition by diphosphonates of bone resorption in mice and comparison with grey lethal osteopetrosis, *Calcif. Tissue Res.*, 12, 59, 1973.
110. **Yarrington, J. T., Capen, C. C., Black, H. E., Potts, J. T., Jr., and Geho, W. B.,** Experimental parturient hypocalcemia in cows following prepartal chemical inhibition of bone resorption, *Am. J. Pathol.*, 83, 569, 1976.
111. **Yarrington, J. T., Capen, C. C., Black, H. E., Nagode, L. A., and Geho, W. B.,** Effect of dichloromethane diphosphonate on calcium homeostasis mechanisms in pregnant cows, *Am. J. Pathol.*, 87, 615, 1977.
112. **Bonjour, J. P., Fleisch, H., and Copp, D. H.,** Influence of a diphosphonate on serum calcium homeostasis, *Proc. Exp. Biol. Med.*, 143, 404, 1973.
113. **Yarrington, J. T., Capen, C. C., Black, H. E., Potts, J. T., Jr., and Geho, W. B.,** Effect of ethane-1-hydroxy-1,1-diphosphonate (EHDP) on the ultrastructure of parathyroid glands and plasma immunoreactive parathyroid hormone in pregnant cows fed a low calcium diet, *Lab. Invest.*, 36, 402, 1977.
114. **Michael, W. R., King, W. R., and Francis, M. D.,** Effectiveness of diphosphonates in preventing "osteoporosis" of disuse in the rat, *Clin. Orthop. Relat. Res.*, 78, 271, 1971.
115. **Mühlbauer, R. C., Russell, R. G. G., Williams, D. A., and Fleisch, H.,** The effects of diphosphonates, polyphosphates, and calcitonin on "immobilisation osteoporosis" in rats, *Eur. J. Clin. Invest.*, 1, 336, 1971.
116. **Goulding, A. and McChesney, R.,** Comparison of effects of parathyroidectomy and ethane-1-hydroxy-1,1-diphosphonate (EHDP) administration upon bone synthesis of hydroxyproline and urinary excretion of hydroxyproline in rats, *Calcif. Tissue Res.*, 23, 115, 1977.
117. **Fleisch, H., Russell, R. G. G., Simpson, B., and Mühlbauer, R. C.,** Prevention by a diphosphonate of immobilization "osteoporosis" in rats, *Nature (London)*, 223, 211, 1969.
118. **Lane, J. M. and Steinberg, M. E.,** The role of diphosphonates in osteoporosis of disuse, *J. Trauma*, 13, 863, 1973.

119. **Cabanela, M. E. and Jowsey, J.**, The effects of phosphonates on experimental osteoporosis, *Calcif. Tissue Res.*, 8, 114, 1971.
120. **Cates, J., Sheets, D. D., and Johnston, C. C., Jr.**, Prevention of disuse osteoporosis in the rat, *Clin. Res.*, 19, 473, 1971.
121. **Ellsasser, J. C., Moyer, C. F., Lesker, P. A., and Simmons, D. J.**, Effect of low doses of disodium ethane-1-hydroxy-1,1-diphosphonate on disuse osteoporosis in the denervated cat tail, *Clin. Orthop. Relat. Res.*, 91, 235, 1973.
122. **Black, H. E. and Jee, W. S. S.**, A histomorphometric and biochemical evaluation of the effects of a diphosphonate in corticosteroid-treated rabbits, in *Bone Histomorphometry*, P. J. Meunier, Ed., Armour Montagu, Paris, 1977, 157.
123. **Hahnel, H., Muhlbach, R., Lindenhayn, K., Schaetz, P., and Schmidt, U. J.**, Zum Einfluss von Diphosphonat auf die experimentelle Heparinosteopathie, *Z. Alternsforsch.*, 27, 289, 1973.
124. **Hahnel, H., Lindenhayn, K., Muhlbach, R., Schmidt, U. J., and Kalbe, I.**, Antikoagulantienosteopathie durch Kumarin und deren Beeinfluseng mit Diphosphonat. Eine tierexperimentelle Studie, *Dtsch. Gesundheitswes.*, 31, 856, 1976.
125. **Kaye, M.**, The effect of a diphosphonate on uremic hyperparathyroidism, *Nephron*, 10, 188, 1973.
126. **Jowsey, J. and Holley, K. E.**, Influence of diphosphonates on progress of experimentally induced osteoporosis, *J. Lab. Clin. Med.*, 82, 567, 1973.
127. **Morgan, D. B., Gasser, A., Largiader, U., Jung, A., and Fleisch, H.**, Effects of a diphosphonate on calcium metabolism in calcium-deprived rats, *Am. J. Physiol.*, 228, 1750, 1975.
128. **Lindenhayn, K., Schmidt, U. J., Hirthe, D., Wegner, G., and Kalbe, I.**, Zur Wirkung von Dexamethason und Diphosphonat auf den Knochen, *Dtsch. Gesundheitswes.*, 28, 202, 1973.
129. **Gasser, A. B., Morgan, D. B., Fleisch, H. A., and Richelle, L. J.**, The influence of two diphosphonates on calcium metabolism in rat, *Clin. Sci.*, 43, 31, 1972.
130. **Weisbrode, S. E., Capen, C. C., and Pendley, C. B.**, Effect of dichloromethylene diphosphonate on morphology, enzyme activity, and ash content of bones of thyroparathyroidectomized rats, *Calcif. Tissue Res.*, 25, 119, 1978.
131. **Flora, L.**, Comparative antiinflammatory and bone protective effects of two diphosphonates in adjuvant arthritis, *Arthritis Rheum.*, 22, 340, 1979.
132. **Francis, M. D., Flora, L., and King, W. R.**, The effects of disodium ethane-1-hydroxy-1,1-diphosphonate on adjuvant induced arthritis in rats, *Calcif. Tissue Res.*, 9, 109, 1972.
133. **Baker, B. L. and Ebert, M.**, Unpublished work, 1981.
134. **Komissarenko, S. V., Zhuravskii, N. I., Karlova, N. P., and Gulyi, M. F.**, Inhibition of hypersensitivity of delayed and immediate types in guinea pigs by methylenediphosphonic acid, *Bull. Exp. Biol. Med.*, 84, 1322, 1977.
135. **Galasko, C. S. B.**, Mechanism of bone destruction in the development of skeletal matastases, *Nature (London)*, 263, 507, 1976.
136. **Chopra, D. and Clerkin, E. P.**, Hypercalcemia and malignant disease, *Med. Clin. N. Am.*, 59, 441, 1975.
137. **Besarab, A. and Caro, J. F.**, Mechanisms of hypercalcemia in malignancy, *Cancer*, 41, 2276, 1978.
138. **Gordan, G. S.**, Osteolytic sterol in human breast cancer, *Science*, 151, 1226, 1966.
139. **Mundy, G. R.**, Evidence for the secretion of an osteoclast stimulating factor in myeloma, *N. Engl. J. Med.*, 291, 1041, 1974.
140. **Voelkel, E. F., Tashjian, A. H., Jr., Franklin, R., Wasserman, E., and Levine, L.**, Hypercalcemia and tumor prostaglandins: the VX_2, carcinoma in the rabbit, *Metab. Clin. Exp.*, 24, 973, 1975.
141. **Wolfe, H. J., Bitman, W. R., Voelkel, E. F., Griffiths, H. J., and Tashjian, A. H.**, Systemic effects of the VX_2, carcinoma on the osseous skeleton: a quantitative study of trabecular bone, *Lab. Invest.*, 38, 208, 1978.
142. **Doppelt, S., Slovik, D., Neer, R., and Potts, J.**, Dichloromethane diphosphonate treatment of hypercalcemia in the VX_2 carcinoma bearing rabbit, in *VII Int. Conf. on Calcium Regulating Hormones*, Talmalge, R. V., Matthews, J. L., and Cohn, D., Eds., 1980, 165.
143. **Minne, H., Streibl, W., Staneslaus, S., and Arnaud, C. D.**, Paraneoplastic parathyroid hormone secretion by the transplantable Walker carcinosarcoma 256 of the rat, *Acta Endocrinol. Suppl.*, 215, 129, 1978.
144. **Minne, H., Raue, F., Klar, H., Ziegler, R., and Bijvoet, O. L. M.**, Treatment of paraneoplastic hypercalcemia with a new diphosphonate (APD) in the animal experiment, in *Symposium CEMO III, Bones and Tumors*, Donath, A. and Courvoisier, B., Eds., Editions Medecine et Hygiene, Geneva, 1980, 166.
145. **Jung, A., Bornand, J., van Ouwenaller, C., Depierr, D., Chantraine, A., Donath, A., Seematte, R., and Courvoisier, B.**, Effect of diphosphonate on tumour-induced osteolysis in animals and in humans, in *Symposium CEMO III, Bone and Tumors*, Donath, A. and Courvoisier, B., Eds., Editions Medecine et Hygiene, Geneva, 1980, 170.
146. **Jung, A., Barras, C., Mermillod, B., and Courvoisier, B.**, The effect of two diphosphonates on osteolysis induced by tumor extracts, *Schweiz. Med. Wochenschr.*, 109, 1888, 1979.

147. **Galasko, C. S. B., Samuel, A. W., Rushton, S., and Lacey, E.,** The effects of prostaglandin synthesis inhibitors and diphosphonates on tumor mediated osteolysis, *Br. J. Surg.*, 67, 493, 1980.
148. **Eilon, G. and Mundy, G. R.,** Direct resorption of bone by human breast cancer cells in vitro, *Nature (London)*, 276, 726, 1978.
149. **Martodam, R. R.,** Unpublished work, 1981.
150. **Anderson, H. C.,** Matrix vesicles of cartilage and bone, in *The Biochemistry and Physiology of Bone*, Vol. IV, Bourne, G. H., Ed., Academic Press, New York, 1976, 135.
151. **Felix, R., Herrmann, W., and Fleisch, H.,** Stimulation of precipitation of calcium phosphate by matrix vesicles, *Biochem. J.*, 170, 681, 1978.
152. **Boskey, A. L., Goldberg, M. R., and Posner, A. S.,** Effect of diphosphonates on hydroxyapatite formation induced by calcium-phospholipid-phosphate complexes, *Calcif. Tissue Int.*, 27, 83, 1979.
153. **Minkin, C., Rabadjija, L., and Goldhaber, P.,** Bone remodeling in vitro: the effects of two diphosphonates on osteoid synthesis and bone resorption in mouse calvaria, *Calcif. Tissue Res.*, 14, 161, 1978.
154. **King, W. R., Francis, M. D., and Michael, W. R.,** Effect of disodium ethane-1-hydroxy-1,1-diphosphonate on bone formation, *Clin. Orthop. Relat. Res.*, 78, 251, 1971.
155. **Russell, R. G. G., Kislig, A. M., Casey, P. A., Fleisch, H., Thornton, J., Schenk, R., and Williams, D. A.,** Effect of diphosphonates on calcitonin on the chemistry and quantitative histology of rat bone, *Calcif. Tissue Res.*, 11, 179, 1973.
156. **Rosenblum, I. Y.,** The effects of disodium ethane-1-hydroxy-1,1-diphosphonate (EHDP) on bone and serum chemistry in rabbits, *Calcif. Tissue Res.*, 16, 145, 1974.
157. **Larsson, A. and Larsson, S. E.,** Light microscopic and ultrastructural observations on the short term effects of ethylene-1-hydroxy-1,1-diphosphonate (EHDP) on rat tibia epiphysis, *Acta Pathol. Microbiol. Scand.*, 84, 17, 1976.
158. **Reitsma, P. H., Bijvoet, O. L. M., Verlinden-Ooms, H., and van der Wee-Pals, L. J. A.,** Kinetic studies of bone and mineral metabolism during treatment with (3-amino-1-hydroxy-propylidene)-1,1-bisphosphonate (APD) in rats, *Calcif. Tissue Int.*, 32, 145, 1980.
159. **Evans, R. A., Baylink, D. J., and Wergedal, J.,** The effects of two diphosphonates on bone metabolism in the rat, *Metab. Bone Dis. Relat. Res.*, 2, 39, 1979.
160. **Howell, D. S., Muniz, O. E., Blanco, L. N., and Pita, J. C.,** A micropuncture study of growth cartilage in phosphonate (EHDP) induced rickets, *Calcif. Tissue Int.*, 30, 35, 1980.
161. **Potokar, M. and Schmidt-Dunker, M.,** The inhibitory effect of new diphosphonic acids on aortic and kidney calcification in vivo, *Atherosclerosis*, 30, 313, 1978.
162. **Fleisch, H., Russell, R. G. G., Bisaz, S., Casey, P. A., and Mühlbauer, R. C.,** The influence of pyrophosphate analogues (diphosphonates) on the precipitation and dissolution of calcium phosphate in vitro and in vivo, *Calcif. Tissue Res.*, 2(Suppl.), 10, 1968.
163. **Fleisch, H. A., Russell, R. G. G., Bisaz, S., Mühlbauer, R. C., and Williams, D. A.,** The inhibitory effect of phosphonates on the formation of calcium phosphate crystals in vitro and on aortic and kidney calcification in vivo, *Eur. J. Clin. Invest.*, 1, 12, 1970.
164. **O'hara, J., Jowsey, J., and Williams, G. A.,** Effects of dichloromethylene diphosphonate, ethane-1-hydroxy-1,1-diphosphonate, and rickets in rats, *J. Lab. Clin. Med.*, 86, 793, 1975.
165. **Rosenblum, I. Y., Black, H. E., and Ferrell, J. F.,** The effects of various diphosphonates on a rat model of cardiac calciphylaxis, *Calcif. Tissue Res.*, 23, 151, 1977.
166. **Rosenblum, I. Y., Flora, L., and Eisenstein, R.,** The effects of disodium ethane-1-hydroxy-1,1-diphosphonate (EHDP) on a rabbit model of atherosclerosis, *Atherosclerosis*, 22, 411, 1975.
167. **Plasmans, C. M. T., Kuypers, W., and Slooff, T. J. J. H.,** The effect of ethane-1-hydroxy-1,1-diphosphonic acid (EHDP) on matrix induced ectopic bone formation, *Clin. Orthop. Relat. Res.*, 132, 233, 1978.
168. **Strates, B. S., Firschein, H. E., and Urist, M. R.,** Alkaline phosphatase and failure of calcification under the influence of a diphosphonate, *Biochim. Biophys. Acta*, 224, 121, 1971.
169. **Fraser, D., Russell, R. G. G., Pohler, O., Robertson, W. G., and Fleisch, H.,** The influence of disodium ethane-1-hydroxy-1,1-diphosphonate (EHDP) on the development of experimentally induced urinary stones in rats, *Clin. Sci.*, 42, 197, 1972.
170. **Francis, M. D., Slough, C. L., Black, H. E., Tofe, A. J. and Cloyd, G. G.,** Diphosphonate treatment of a primary osteogenic sarcoma in a dog; a case report, *Vet. Radiol.*, 21, 168, 1980.
171. **Francis, M. D. and Slough, C. L.,** Unpublished work, 1975.
172. **Gasser, A. B., Morgan, D. B., Fleisch, H. A., and Richelle, L. J.,** The influence of two diphosphonates on calcium metabolism in the rat, *Clin. Sci.*, 43, 31, 1972.
173. **Guilland, D., Trechsel, U., Bonjour, J. P., and Fleisch, H.,** Stimulation of calcium absorption and apparent increased intestinal 1,25 dihydroxycholecalciferol in rats treated with low doses of ethane-1-hydroxy-1,1-diphosphonate, *Clin. Sci. Mol. Med.*, 48, 157, 1975.

174. **Bonjour, J. P., Russell, R. G. G., Morgan, D. B., and Fleisch, H. A.,** Intestinal calcium absorption, Ca-binding protein, and Ca-ATPase in diphosphonate-treated rats, *Am. J. Physiol.*, 224, 1011, 1973.
175. **Bonjour, J. P., DeLuca, H. F., Fleisch, H., Trechsel, U., Matejowec, L. A., and Omdahl, J. L.,** Reversal of the EHDP inhibition of calcium absorption by 1,25-dihyroxycholecalciferol., *Eur. J. Clin. Invest.*, 3, 44, 1973.
176. **Baxter, L. A., DeLuca, H. F., Bonjour, J. P., and Fleisch, H.,** Inhibition of vitamin D metabolism by ethane-1-hydroxy-1,1-diphosphonate, *Arch. Biochem. Biophys.*, 164, 655, 1974.
177. **Hill, L. F., Lumb, G. A., Mawer, E. B., and Stanbury, S. W.,** Indirect inhibition of the biosynthesis of 1,25-dihydroxycholecalciferol in rats treated with a diphosphonate, *Clin. Sci.*, 44, 335, 1973.
178. **Trechsel, U., Bonjour, J. P., and Fleisch, H.,** Regulation of the metabolism of 25-hydroxyvitamin D_3 in primary cultures of chick kidney cells, *J. Clin. Invest.*, 64, 206, 1979.
179. **Bonjour, J. P., DeLuca, H. F., Fleisch, H., Trechsel, U., Matejowec, L. A., and Omdahl, J. L.,** Reversal of the EHDP inhibition of calcium absorption by 1,25-dihydroxycholecalciferol,, *Eur. J. Clin. Invest.*, 3, 44, 1973.
180. **Bonjour, J. P., Trechsel, U., Fleisch, H., Schenk, R., DeLuca, H. F., and Baxter, L. A.,** Action of 1,25-dihydroxyvitamin D_3 and diphosphonate on calcium metabolism in rats, *Am. J. Physiol.*, 229, 402, 1975.
181. **Bisaz, S., Schenk, R., Kunin, A. S., Russell, R. G. G., Mühlbauer, R., and Fleisch, H.,** The comparative effects of vitamin D deficiency and ethane- 1-hydroxy-1,1-diphosphonate administration on the histology and glycolysis of chick epiphyseal and articular cartilage, *Calcif. Tissue Res.*, 19, 139, 1975.
182. **Baxter, L. A., Canty, D. J., Bedner, G. J., Stern, L., DeLuca, H. F., Ginn, D. L., Flora, L., and Hassing, G. S.,** Effect of ethane-1-hydroxy-1,1-diphosphonate and vitamin D on bone mineralization, *Calcif. Tissue Int.*, 28, 73, 1979.
183. **Boris, A., Hurley, J. F., Tomal, T., Mallon, J. P., and Matuszewski, D. S.,** Evidence for the promotion of bone mineralization by 1,25-dihydroxy-cholecalciferol in the rat unrelated to the correction of deficiencies in serum calcium and phosphorus, *J. Nutr.*, 108, 1899, 1978.
184. **von Herrath, D., Kraft, D., and Schaefer, K.,** Influence of disodium dichloromethylene diphosphonate on 25-hydroxy-cholecalciferol metabolism in rats, *Naunyn-Schmiedeberg's Arch. Pharmacol.*, 278, 435, 1973.
185. **Trechsel, U., Schenk, R., Bonjour, J. R., Russell, R. G. G., and Fleisch, H.,** Relation between bone mineralization, Ca absorption, and plasma Ca in phosphonate-treated rats, *Am. J. Physiol.*, 232, E298, 1977.
186. **Bonjour, J. P., Troehler, U., Preston, C., and Fleisch, H.,** Parathyroid hormone and renal handling of Pi: effect of dietary Pi and diphosphonates, *Am. J. Physiol.*, 234, F487, 1978.
187. **Guncaga, J., Lauffenburger, T., Lentner, C., Dambacher, M. A., Haas, H. G., Fleisch, H., and Olah, A. J.,** Diphosphonate treatment of Paget's disease of bone. A correlated metabolic, calcium kinetic and morphometric study, *Horm. Metab. Res.*, 6, 62, 1974.
188. **Jowsey, J., Riggs, B., Kelly, P. J., Hossman, D. L., and Bordier, P.,** The treatment of osteoporosis with disodium ethane-1-hydroxy-1,1-diphosphonate, *J. Lab. Clin. Med.*, 78, 574, 1971.
189. **Recker, R. R., Hassing, G. S., Lau, J. R., and Saville, P. D.,** The hyperphosphatemic effect of disodium ethane-1-hydroxy-1,1-diphosphonate (EHDP𝒯ℳ): renal handling of phosphorus and the renal response to parathyroid hormone, *J. Lab. Clin. Med.*, 81, 258, 1973.
190. **Russell, R. G. G., Smith, R., Preston, C., Walton, R. J., and Woods, C. G.,** Diphosphonates in Paget's disease, *Lancet*, 7863(1), 894, 1974.
191. **Smith, R., Russell, R. G. G., and Bishop, M.,** Diphosphonates and Paget's disease of bone, *Lancet*, 7706(1), 945, 1971.
192. **Citrin, D. L.,** Toxicity of EHDP in humans, *J. Nucl. Med.*, 17, 417, 1976.
193. **Fleisch, H., Bisaz, S., Care, A. D., Mühlbauer, R. G., and Russell, R. G. G.,** A review of the effects of phosphonates on calcium metabolism, *Calcitonin*, Proc. 2nd Int. Symp., Taylor, S., Ed., Wm. Heinemann Medical Books, London, 1970, 409.
194. **Gordan, G. S. and Vaughan, C.,** *Clinical Management of the Osteoporoses*, Publishing Sciences Group, Acton, Mass., 1976.
195. **Meunier, P. J., Courpron, P., Edouard, C., Alexandre, C., Bressot, C., Lips, P., and Boyce, B. F.,** Bone histomorphometry in osteoporotic states, *Osteoporosis*, Vol. 11, Barzel, U.S., Ed., Grune & Stratton, New York, 1979, 27.
196. **Khairi, M. R. A. and Johnston, C. C.,** Exponential bone loss in postmenopausal women, *Clin. Res.*, 26(3), 107, 1978.
197. **Lockwood, D. R., Vogel, J. M., Schneider, V. S., and Hulley, S. B.,** Effect of the diphosphonate EHDP on bone mineral metabolism during prolonged bed rest, *J. Clin. Endocrinol. Metab.*, 41, 533, 1975.
198. **Arnstein, A. R., Blumenthal, F. S., and McCann, D. S.,** The effect of diphosphonate (EHDP) therapy on immobilization osteoporosis, in *Proc. 55th Ann. Meet. Endocr. Soc.*, Chicago, 1973, A183.

199. **Lauffenburger, T., Olah, A. J., Dambacher, M. A., Guncaga, J., Lentner, C., and Hass, H. G.,** Bone remodeling and calcium metabolism: a correlated histomorphometric, calcium kinetic, and biochemical study in patients with osteoporosis and Paget's disease, *Metab. Clin. Exp.*, 26, 589, 1977.
200. **Heaney, R. P. and Saville, P. D.,** Etidronate disodium in postmenopausal osteoporosis, *Clin. Pharmacol. Ther.*, 20, 593, 1976.
201. **Gennari, C.,** Disodium etidronate treatment of osteoporosis, in *Etidronate,* Proc. 1st Int. Symp. Diphosphonate Ther. Caniggia, A., Ed., Instituto Gentili, Pisa, Italy, 1979, 133.
202. **Jowsey, J., Riggs, B. L., Kelly, P. J., Hoffman, D. L., and Bordier, P.,** The treatment of osteoporosis with disodium ethane-1-hydroxy-1,1-diphosphonate, *J. Lab. Clin. Med.*, 78, 574, 1971.
203. **Siris, E. S., Sherman, W. H., Baquiran, D. C., Schlatterer, J. P., Osserman, E. F., and Canfeld, R. E.,** Effects of dichloromethylene diphosphonate on skeletal mobilization of calcium in multiple myeloma, *N. Engl. J. Med.*, 302, 310, 1980.
204. **Douglas, D. L., Russell, R. G. G., Preston, C. J., Prenton, M. A., Duckworth, T., Kanis, J. A., Preston, F. E., and Woodhead, J. S.,** Effect of dichloromethylene diphosphonate in Paget's disease of bone and in hypercalcemia due to primary hyperparathyroidism or malignant disease, *Lancet*, 8177(1), 1043, 1980.
205. **van Breukelen, F. J. M., Bijvoet, O. L. M., and van Oosterom, A. T.,** Inhibition of osteolytic bone lesions by (3-amino-1-hydroxypropylidene)-1,1-biphosphonate (A.P.D.), *Lancet*, 8120(1), 803, 1979.
206. **Chapuy, M. C., Meunier, P. J., Alexander, C. M., and Vignon, E. P.,** Effects of disodium dichloromethylene diphosphonate on hypercalcemia produced by bone metastases, *J. Clin. Invest.*, 65, 1243, 1980.
207. **Jacobs, T. P., Siris, E. S., Bilezikian, J. P., Baquiran, D. C., Shane, E., and Canfield, R. E.,** Hypercalcemia of malignancy: treatment with intravenous dichloromethylene diphosphonate, *Ann. Intern. Med.*, 94, 312, 1981.
208. **Cohen, A. I., Koeller, J., Davis, T. E., and Citrin, D. L.,** Intravenous dichloromethylene diphosphonate in cancer associated hypercalcemia-A phase I/II Evaluation. *Cancer Treat. Rep.*, 65, 651, 1981.
209. **Jung, A., Bornand, J., van Ouwenaller, C., Chantraine, A., and Donath, A.,** Effects of diphosphonates on tumor-induced osteolysis in animals and in humans, in *VII Int. Conf. on Calcium Regulating Hormones*, Talmage, R. V., Matthews, J. L., and Cohn, D., Eds., Excerpta Medica, Amsterdam, 1980, 1103.
210. **Bassett, C. A. L., Donath, A., Magagno, F., Preisig, R., Fleisch, H., and Francis, M. D.,** Diphosphonates in the treatment of Myositis Ossificans, *Lancet*, 7625(2), 845, 1969.
211. **Russell, R. G. G., Smith, R., Bishop, M. C., Price, D. A., and Squire, C. M.,** Treatment of Myositis Ossificans Progressiva with a diphosphonate, *Lancet*, 7740(1), 10, 1972.
212. **Weiss, I. W., Fisher, L., and Phang, J. M.,** Diphosphonate therapy in a patient with Myositia Ossificans Progressiva, *Ann. Intern. Med.*, 74, 933, 1971.
213. **Geho, W. B. and Whiteside, J. A.,** Experience with disodium etidronate in diseases of ectopic calcification, in *Clinical Aspects of Metabolic Bone Disease*, Frame, B., Parfitt, A., and Duncan, H., Eds., International Congress Series, Amsterdam, 1973, 506.
214. **Cram, R. L., Barrmada, R., Geho, W. B., and Ray, R. D.,** Diphosphonate treatment of calcinosis universalis, *N. Engl. J. Med.*, 285, 1012, 1971.
215. **Barry, H. C.,** Paget's Disease of Bone, *Practitioner*, 210, 340, 1973.
216. **Nagant de Deuxchaisnes, C. and Krane, S. M.,** Paget's disease of bone. Clinical and metabolic observations, *Medicine*, 43, 233, 1964.
217. **Altman, R. D., Johnston, C. C., Khari, M. R. A., Wellman, H., Serafini, A. N., and Sankey, R. R.,** Influence of disodium etidronate on clinical and laboratory manifestations of Paget's disease of bone (Osteitis deformans), *N. Engl. J. Med.*, 289, 1379, 1973.
218. **Canfield, R., Rosner, W., and Skinner, J., McWhorter, J., Resnick, L., Feldman, F., Kammerman, S., Ryan, K., Kunegonis, M., and Bohne, W.,** Diphosphonate therapy of Paget's disease of bone, *J. Clin. Endocrinol. Metab.*, 44, 96, 1977.
219. **Khairi, M. R. A., Altman, R. D., DeRosa, G. P., Zimmerman, J., Schenk, R. K., and Johnston, C. C.,** Sodium etidronate in the treatment of Paget's disease of bone, *Ann. Intern. Med.*, 87, 656, 1977.
220. **Meunier, P., Chapuy, M. C., Courpron, P., Vignon, E., Edouard, C., and Bernard, J.,** Effets cliniques, biologiques et histologiques de l'ethane-1-hydroxy-1,1-diphosphonate (EHDP) dans la maladie de Paget, *Rev. Rhum.*, 42, 699, 1975.
221. **Hosking, D. J., Bijvoet, O. L. M., van Aken, J., and Will, E. J.,** Paget's bone disease treated with diphosphonate and calcitonin, *Lancet*, 7960(1), 615, 1976.
222. **Nagant de Deuxchaisnes, C., Rombouts-Lindemans, C., Huaux, J. P., Devogelaer, J. P., Maglhem, J., and Maldague, B.,** Roentgenologic evaluation of the action of the diphosphonate EHDP and of combined therapy (EHDP and Calcitonin) in Paget's disease of bone, in *Molecular Endocrinology*, Macintyre, I. and Szelke, M., Eds., Elsevier/North Holland and Biomedical Press, Amsterdam, 1979, 405.
223. **Finerman, G. A. M., Gonick, H. C., Smith, R. K., and Mayfield, J. M.,** Diphosphonate treatment of Paget's disease, *Clin. Orthop. Relat., Res.*, 120, 115, 1976.

224. **Funcaga, J., Lauffenburger, T., Lentner, C., Darnbacher, M. A., Haas, H. G., Fleisch, H., and Oleh, A. J.**, Diphosphonate treatment of Paget's disease of bone, *Horm. Metab. Res.*, 6, 62, 1974.
225. **Khairi, M. R. A., Meunier, P., Edouard, C., Courpron, P., Bernard, J., Derosa, G. P., and Johnston, C. C., Jr.**, Quantitative bone histology in Paget's disease of bone: influence of sodium etidronate (EHDP) therapy, *Calcif. Tissue Res.*, 22, 355, 1977.
226. **Johnston, C. C., Khairi, M. R. A., and Meunier, P. J.**, Use of Etidronate (EHDP) in Paget's disease of bone, *Arthritis Rheum.*, 23, 1172, 1980.
227. **Siris, E. S., Thomas, P. J., and Canfield, R. E.**, Paget's disease of bone, *Bull. N. Y. Acad. Med.*, 56(3), 285, 1980.
228. **Murphy, W. A., Whyte, M. P., and Haddad, J. G.**, Healing of lytic Paget bone disease with diphosphonate therapy, *Diag. Radiol.*, 134, 635, 1980.
229. **Frijlink, W. B., Bijvoet, O. L. M., te Velde, J., and Heynen, G.**, Treatment of Paget's disease with (3-amino-1-hydroxypropylidene)-1,1-bisphosphate (APD), *Lancet*, 1(8120), 799, 1979.
230. **Meunier, P. J., Chapuy, M. C., Alexandre, C., Bressot, C., Edouard, C., Vignon, E., Mathieu, L., and Trechsel, U.**, Effects of disodium dichloromethylene diphosphonate on Paget's disease of bone, *Lancet*, 8141(2), 489, 1979.
231. **Alexandre, C. M., Matthews, J. L., Meunier, P. J., Martin, J. H., and Petit, H.**, Ultrastructural aspects of osteoclasts and osteoblasts in Paget's disease of bone treated with dichloromethylene disphosphonate (Cl_2MDP), in *VII Int. Conf. on Calcium Regulating Hormones*, Talmage, R. V., Matthews, J. L., and Cohn, D., Eds., Excerpta Medica, Amsterdam, 1980, 120.
232. **Mills, B. G. and Singer, F. R.**, Nuclear inclusions in Paget's disease of bone, *Science*, 194, 201, 1976.
233. **Rebel, A., Malkani, K., Basle, M., and Bregeon, C.**, Nuclear inclusions in osteoclasts in Paget's bone disease, *Calcif. Tissue Res.*, 21(Suppl.), 113, 1976.
234. **Ditzel, J.**, Oxygen transport impairment in diabetics, *Diabetes*, 25, (Suppl. 2) 832, 1976.
235. **Ditzel, J., Hau, C., and Daugaard, N.**, Effect of the diphosphonate ethane-1-hydroxy-1,1-diphosphonate (EHDP) on hemoglobin oxygen affinity of diabetic and healthy subjects, *Microvasc. Res.*, 13, 355, 1977.
236. **Ditzel, J., Hau, C., and Daugaard, N.**, Effects of the diphosphonate EHDP on plasma inorganic phosphate and hemoglobin oxygen affinity of diabetic and healthy subjects, in *Advances in Experimental Medicine and Biology*, Massry, S. G. and Ritz, E., Eds., Plenum Press, New York, 1977, 423.
237. **Francis, M. D., Slough, C. L., Tofe, A. J., and Silberstein, E. B.**, Factors affecting uptake and retention of technetium-99m-diphosphonate and 99m-pertechnetate in osseous, connective, and soft tissues, *Calcif. Tissue Res.*, 20, 303, 1976.
238. **Steigman, J., Chin, E. V., and Solomon, N. A.**, Scintiphotos in rabbits made with Tc-99m preparations reduced by electrolysis and by $SnCl_2$, *J. Nucl. Med.*, 20, 766, 1979.
239. **Tofe, A. J. and Francis, M. D.**, Optimization of the ratio of stannous tin: ethane-1-hydroxy-1,1-diphosphonate for bone scanning with ^{99m}Tc-pertechnetate, *J. Nucl. Med.*, 15, 69, 1974.
240. **Deutsch, E., Libson, K., Becker, C. B., Francis, M. D., Tofe, A. J., Ferguson, D. L., and McCreary, L. D.**, Preparation and biological distribution of technetium diphosphonate radiotracers synthesized without stannous ion, *J. Nucl. Med.*, 21, 859, 1980.
241. **Subramanian, G., McAfee, J. G., Blair, R. J., Kallfelz, F. A., and Thomas, F. D.**, Technetium-99m-methylene diphosphonate — a superior agent for skeletal imaging. Comparison with other technetium complexes, *J. Nucl. Med.*, 16, 744, 1975.
242. **Francis, M. D., Ferguson, D. L., Tofe, A. J., Bevan, J. A., and Michaels, S. E.**, Comparative evaluation of three diphosphonates: in vitro adsorption (C-14 labelled) and in vivo osteogenic uptake (Tc-99m complexed), *J. Nucl. Med.*, 21, 1185, 1980.
243. **Davis, M. A. and Jones, A. G.**, Comparison of ^{99m}Tc-labeled phosphate and phosphonate agents for skeletal imaging, *Semin. Nucl. Med.*, 6, 19, 1976.
244. **Jones, A. G., Francis, M. D., and Davis, M. A.**, Bone scanning: radionuclidic reaction mechanisms, *Semin. Nucl. Med.*, 6, 3, 1976.
245. **Tofe, A. J., Francis, M. D., and Harvey, W. J.**, Correlation of neoplasms with incidence and localization of skeletal metastases: an analysis of 1,355 diphosphonate bone scans, *J. Nucl. Med.*, 16, 986, 1975.
246. **Handmaker, H. and Leonards, R.**, The bone scan in inflammatory osseous disease, *Semin. Nucl. Med.*, 6, 95, 1976.
247. **Fogelman, I., Tofe, A. J., and Francis, M. D.**, Whole body retention: comparison of three diphosphonates, *J. Nucl. Med.*, 22, P78, 1981.
248. **Fogelman, I., Citrin, D. L., McKillop, J. H., Turner, J. G., Bissent, R. G., and Greig, W. R.**, A clinical comparison of Tc-99m HEDP and Tc-99m MDP in the detection of bone metastases: concise communication, *J. Nucl. Med.*, 20, 98, 1979.
249. **Arnold, J. S., Barnes, W. E., and Khedkar, N.**, Kinetic studies of a new and superior Tc-99m diphosphonate bone imaging agent, *J. Nucl. Med.*, 20, 653, 1979.

Chapter 5

PHOSPHONIC ACIDS AND PHOSPHONATES AS ANTIMETABOLITES

Robert Engel

TABLE OF CONTENTS

I. INTRODUCTION

The identification of an antimetabolite, some agent which inhibits or perturbs a given metabolic reaction, is a relatively simple, virtually trivial matter. The profound difficulty arises in the development of antimetabolites which will perform in a *specific* manner and be capable of being delivered intact to a specific reaction domain. That is, the difficulty is in finding an antimetabolite which will perturb only one reaction of many occurring in the same reaction domain, or will affect only one of numerous organisms which may be co-existing. In this chapter, consideration will be given to the use and potential use of phosphonic acids as specific antimetabolites.

It has been increasingly recognized that phosphonic acids structurally related to natural phosphate metabolites possess particularly intriguing potential as antimetabolites. Since the author's review[1] of several years ago, numerous new synthetic procedures for phosphonic acids have been developed as have new approaches to metabolic regulation by use of C–P bonded molecules. Fundamentally, multifunctional molecules are involved wherein, ideally, one of the functionalities is modified in some manner, preferably subtle, while the others remain the same as in the natural metabolite. Several modes by which phosphonic acids and phosphonates may act as antimetabolites may be envisioned. That which is most commonly found involves the introduction of a phosphonic acid linkage in place of the normal phosphate *ester* portion of a metabolite while the remaining functional groups are allowed to remain the same. Another involves the replacement of an acidic hydroxyl function on phosphorus by a methyl or other alkyl group. A third approach involves the substitution of a phosphonic acid function for some other acidic linkage, such as a carboxylate. For optimization or variability of the effect of the antimetabolite it may be necessary that other or additional structural variations be made.

For each of the above mentioned structural variations several means may be envisioned by which the phosphonic acid, substituting for a natural metabolite, may act as an antimetabolite. The simplest of these would be by the generation of an "unnatural product" which perturbs, by product or feedback inhibition, the biosynthesis of precursors to the "natural" product. Another possibility is that the phosphonic acid or phosphonate ester might be capable of direct involvement in the specific or nonspecific inhibition of one or more enzymatic processes.

For the *design* of specific antimetabolites based on the "phosphonic-acid-for-phosphate" concept, several "architectural" aspects must be given detailed consideration. The first of these is correspondence of steric factors between the natural phosphate and the postulated phosphonic acid antimetabolite.

It might appear that a steric factor of prime importance would be the gross correspondence of lineal relationships between the natural phosphate and the potential antimetabolite. In particular, it might be presumed that a good antimetabolite or "metabolic mimic" would possess similar distances between its phosphorus acid site and other functionalities when compared to those in the natural phosphate. For certain systems this would appear to be the situation. For example, 2,3-dihydroxypropyl-1-phosphonic acid, generated in both racemic[2] and chiral[3] forms as a potential antimetabolite for processes normally utilizing *sn*-glycerol-3-phosphate, was found[4] to be totally without inhibitory activity toward L-glycerol-3-phosphate:NAD oxidoreductase. For comparison one may consider the phosphonic acid which is "isosteric" *(vide infra)* with the natural phosphate, i.e., 3,4-dihydroxybutyl-1-phosphonic acid;[5-9] this material has been found[5-10] to be oxidized by the L-glycerol-3-phosphate:NAD oxidoreductase at approximately the same rate as the natural substrate and to have nearly the same value of K_m. Obviously the original inference[4] of the necessity of the phosphate ester oxygen for activity is invalid here.

However, for some other systems the lineal relationship does not appear to be a critical factor. For example, the "shortened" nonisosteric phosphonic acid analogue of UDP-glucose[11] (phosphorus bound directly to the 5′-carbon of the nucleoside sugar) has been found to substitute for the natural material as substrate for UDP-glucose dehydrogenase. Moreover, while both isosteric and nonisosteric phosphonic acid analogues of 3-deoxy-D-arabinoheptulosonate-7-phosphate and D-gluco-heptulosonate-7-phosphate are competitive inhibitors of the 3-dehydroquinate synthetase of *Escherichia coli*, the nonisosteric systems are far better inhibitors than the isosteric systems.[12] Thus it may be seen that the issue of steric correspondence is not equivalent for all systems of interest.

The implication of an isosteric relationship between a natural phosphate and a phosphonic acid wherein a methylene group is present in place of the normal esteratic oxygen requires close consideration. In the strictest sense, the term "isosteric" refers to compounds of *identical* size and shape. However, crystallographic data[13-17] for simple related phosphate esters and phosphonic acids indicate there to be nontrivial differences in the following comparisons: C–O–P and C–C–P bond angles; O–P–O and C–P–O bond angles; C–O and C–C bond distances; and O–P and C–P bond distances. In spite of these nontrivial differences it is found that there is only very minor variation in distances between phosphoryl or phosphonyl oxygen and distant functional groups for corresponding conformations. This is simply a fortuitous consequence of effects being in opposite directions.

With methylenediphosphonic acids, which are structurally related to pyrophosphate, the correspondence is not so favorable. As a result of the P–C–P bond angle being significantly greater than that for P–O–P,[18] and with the absence of any compensating factor of direction or distance, there are sizable deviations in interatomic nonbonded distances. One might expect resultant notable differences in chelating ability for these systems.

A second factor which is expected to be significant in the design of phosphonic acids as antimetabolites of natural phosphate processes is the difference in the acidities of phosphonic acids and phosphate monoesters. Both phosphonic acids and phosphate monoesters are dibasic acids, the first pK_a of each representing a strong acid. Variations are found, however, in the values for the second pK_a, that for the phosphonic acids generally being in the range 7.0 to 8.2[19,20] and approximately 0.5 to 1.0 pK_a units higher than that for the structurally related monophosphate esters.[21]

There have also been efforts to substitute phosphonic acid linkages for other nonphosphorus acid functions in molecules of biological interest. Here deviations in fundamental character of the antimetabolite from the natural material are anticipated to be quite significant, since there will be differences of a sizable nature in pK_a and (depending on the natural material) in the charged state of the species at physiological pH.

A further factor of "architectural" difference between natural phosphate metabolites and related phosphonic acid antimetabolites is that of binding capability which is lost by the structural elimination of the esteratic oxygen of the natural phosphate linkage. The significance of this binding capability will of course vary among diverse metabolic processes; at this point it may be evaluated only empirically. For a given phosphonic acid antimetabolite, this factor may be seen to be of no consequence with a particular metabolic process whereas, with another process, it may exclude the postulated agent from activity. For several systems where this appears to be a significant factor *(vide infra)* introduction of a new substituent bearing binding capability results in antimetabolic activity.

In the design of antimetabolites for in vitro enzymatic investigations, consideration of the above mentioned factors is usually sufficient. However, for in vivo antimetabolic activity a further consideration need be made, that of transport of the potential agent into the organism or to a specific site of the organism. While it would seem unreasonable, on intuitive grounds, to anticipate transport of the analogue when transport of the natural material is absent, it would be a reasonable venture to attempt transport of the analogue when transport of the natural material is present. Of course, there is no guarantee for such transport although to date the isosteric systems investigated have exhibited transport characteristics corresponding, in at least a general way, to those of the natural materials. In the absence of a transport system for natural materials or of site selectivity for a species in a complex organism, coupling of the potential antimetabolite to a "chemical vector" might be of value. This technique has been used successfully in several systems discussed below.

In the following sections the preparation and use of phosphonic acids as antimetabolites will be considered on a functional basis. That is, potential antimetabolites will be reviewed under the heading of the metabolic process involved. We will begin with a very brief consideration of pyrophosphatases (the phosphonic acids involved are considered in other regards in greater detail in another section) followed by a much more detailed consideration of the glycolytic and other carbohydrate pathways, and glycerol related processes. This leads quite reasonably to a consideration of antimetabolites of phospholipid metabolism. Finally, consideration will be given to antimetabolites related to nucleotides, those phosphonic acids with demonstrated antiviral characteristics, and phosphonic acids generated as potential antimetabolites of carboxylate and other miscellaneous systems.

II. ANTIMETABOLITES OF PYROPHOSPHATASES

The fundamental structural class of compounds that would be anticipated to be of utility as antimetabolites of pyrophosphatases are those related to methylenediphosphonic acid (Structure 1). A number of syntheses of this material have been reported with widely varying yield.[22-30]

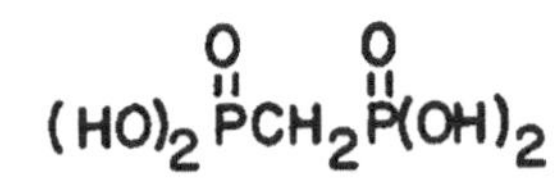

1

In the laboratory, the most reliable method has been found to be a modification of that of Kosolapoff[26] using triisopropyl phosphite with methylene bromide refluxed under a 2-m-long packed column followed by pyrolysis of the isopropyl ester functions from the resultant tetraisopropyl methylenediphosphonic acid.[31]

In spite of the geometric variations of this material from inorganic pyrophosphate,[18] this material and related structures bind to divalent cations in much the same manner as the natural material, but without the possibility of hydrolysis.[32-53] As a result, these materials are useful in the treatment of metabolic disorders where simple chelation with divalent metal cations is involved. This topic is treated in detail in Chapter 4 and other reviews.[53,54]

As regards antimetabolic activity where further involvement is concerned, the results are variable. Several systems might be noted wherein (1) serves reasonably as a "substitute" for inorganic pyrophosphate. For example, (1) replaces inorganic pyrophosphate in ATP upon interaction with *E. coli* lysyl t-RNA synthetase[55,56] and is an inhibitor of both the formation and phosphorolysis of polyadenylic acid by bacterial polynucleotide phosphorylase.[57] Also, (1) has been found to be a competitive inhibitor of nonspecific alkaline phosphatases from *E. coli* and bovine intestine.[58,59] This latter result is of particular interest in light of the total lack of activity of (1) toward a variety of other pyrophosphatases, including those from *Aspergillus* sp.,[60] yeast,[49,58,60-62] and rat liver microsomes.[60]

For several of these systems other analogues related to (1) and inorganic pyrophosphate *do* exhibit activity. For example, the compound (Structure 2),[63] a bisphosphoramide structure, inhibits yeast pyrophosphatase where (1) is without effect.[58] Also, the hydroxymethylene compound (Structure 3)[64] has been bound to be an effective inhibitor of yeast pyrophosphatase where (1) is without effect.[49,62]

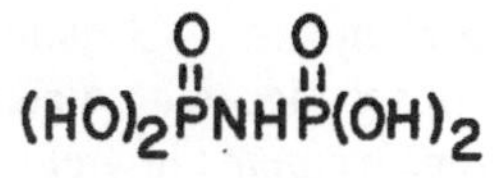

2

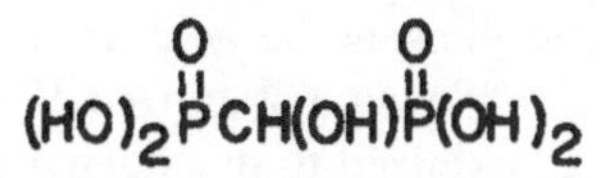

3

These results have several complications, the most obvious being a fundamental difference in the nature of pyrophosphatases from various sources. As such, the species (1) to (3) have the potential of serving as useful probes for details of enzyme function. Also, one may see here the potential importance of a binding capability associated with the atom, oxygen in the natural system, directly bound to the phosphorus. It is of great significance that this binding capability may be accommodated, at least in some systems, by a functionality *attached* to that atom rather than by an electron pair directly on the atom. Utilization of this situation is noted later for other systems (Section IV).

III. ANTIMETABOLITES OF THE GLYCOLYTIC PATHWAY

The glycolytic pathway has been the object of intensive synthetic efforts for phosphonic acid preparation. Indeed, the glycolytic pathway provides a host of targets for which phosphonic acids might be utilized as substitutes for normal phosphate esters. For organizational purposes we will consider antimetabolites in the approximate order that their natural counterparts occur in the pathway. Several approaches to the synthesis of phosphonic acid analogues of glucose-6-phosphate have been reported. For the nonisosteric species (Structure 4) wherein the esteratic oxygen is simply "absent" and the 6-carbon site is directly bound to the phosphorus, Griffin and Burger[65] attempted a synthesis proceeding through the diethyl ester (Structure 5) although all attempts to hydrolyze these ester linkages resulted in destruction of the molecule. Synthesis of (4) was then accomplished[65] with the generation of the diphenyl ester, which could be converted to the free acid via hydrogenolysis.[66] Although (4) may be generated by this latter route, it is rather inconvenient in that phosphonylation is accomplished via the Arbuzov reaction[67] using diphenyl ethyl phosphite, which must be prepared for the purpose rather than by using commercially available trialkyl phosphites. With currently available techniques, dialkyl esters of (4) may be converted readily to the free acid, obviating the difficulties previously encountered.[65] Such esters may be cleaved cleanly without damage to other functionalities by using trimethylsilyl halides followed by treatment with water,[68-71] or by heating with sodium iodide to 150°C in dimethylformamide followed by treatment with aqueous acetic acid.[72] A further alternative would be to use tris(trimethylsilyl) phosphite in the Arbuzov reaction, generating the free acid upon addition of water.[73-76] Another synthesis of (4) has been claimed[77] although details have not as yet been published; it was noted that (4) was used in the synthesis of other phosphonic acids.

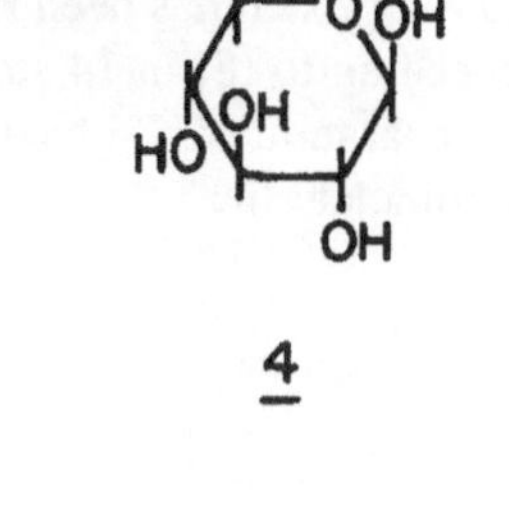

4

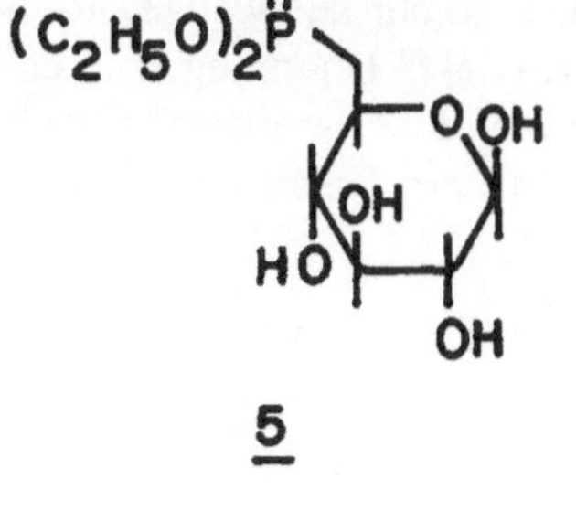

5

The isosteric phosphonic acid analogue of glucose-6-phosphate, i.e., (Structure 6), has been reported[78,79] using a stabilized Wittig-type reagent[80,81] in reaction with a suitably protected glucose, oxidized to the aldehyde at the 6-position. Improvements in yield have been obtained[82] for this general route by modification of the hydroxyl protecting steps and phosphonylation using a phosphonate-anion type Wittig reagent.[31]

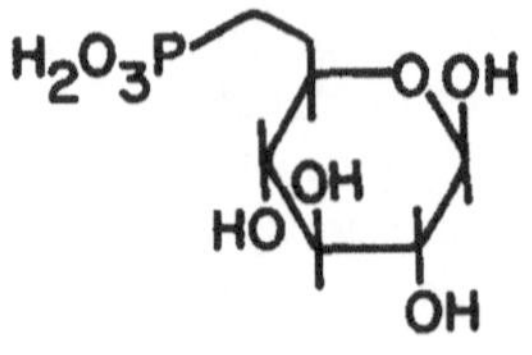

6

Again, another synthesis of (6) has been claimed[77] without details except that it was used in the synthesis of other phosphonic acids.

To date few attempts have been made to study the activity of (6) and (4). Griffin and Burger[65] found (4) to inhibit acetylcholine formation by choline acetylase in the absence of adenosine triphosphate (ATP), although no activity could be found with xanthine oxidase, hyaluronidase, and histidine decarboxylase. Moreover, it was found[77] not to interact with glucose-6-phosphate dehydrogenase. For comparison with this latter result, the isosteric system (6) was found[77] to be a substrate for glucose-6-phosphate dehydrogenase (from yeast), the product acid (Structure 7) also being a substrate for 6-phosphogluconate dehydrogenase. As yet no nonenzymatic synthesis for (7) has been reported.

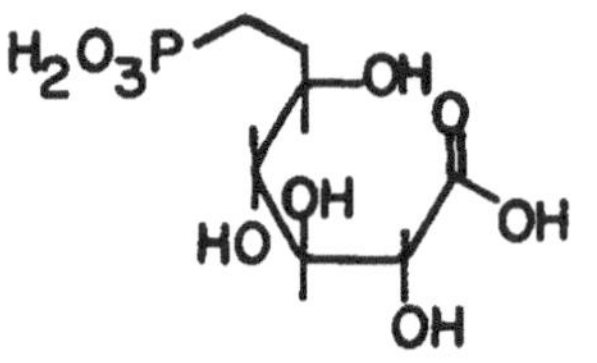

7

Attempts to use (6) as a growth regulator with strains of *E. coli* constitutive for the hexose phosphate transport system have given inconclusive results.[82]

Two approaches to the synthesis of thesosteric phosphonic acid analogue (Structure 8) of glucose-1-phosphate have been reported. An attempt by Paulsen, et al.[83] involved a Horner type[84] reaction for phosphonylation of glucose pentaacetate. The resultant species (Structure 9) could indeed serve as a precursor to (8) by reduction and deacetylation, although the reactions have, to our knowledge, not been performed. Another approach has been reported by BeMiller, et al.[85] beginning with calcium D-gluconate proceeding to (8) in 14 steps.[86] It is their intention that the material (8) be used further in the preparation of C-glycosides,[85] although the material itself bears significant potential as an antimetabolite.

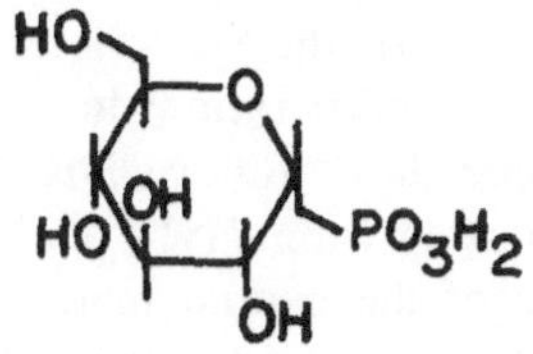

8

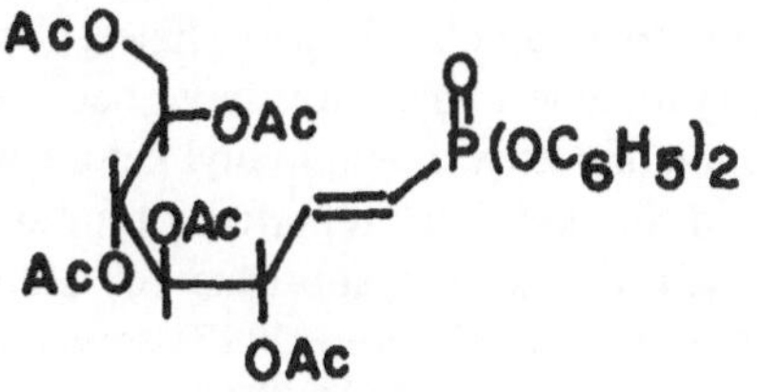

9

Analogues of fructose phosphates have also been the target of synthetic efforts. Paulsen and Bartsch[87] reported an approach to the synthesis of the nonisosteric phosphonic acid analogue of fructose-1-phosphate although final ester cleavage and deprotection was not performed. The synthesis of the isosteric phosphonic acid analogue (Structure 10) of fructose-1-phosphate has been reported[88] using a Horner reaction[84] for phosphonylation of the aldehyde (Structure 11) followed by reduction-hydrogenolysis and deprotection. For metabolic studies a radioactive label has been introduced via reduction of the vinylic phosphonate using tritium.

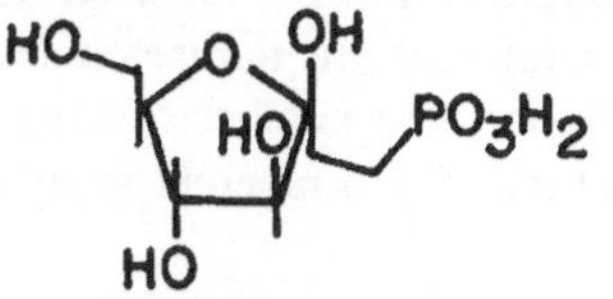

10

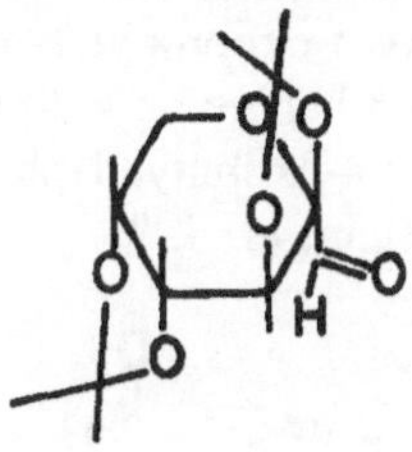

11

As an analogue of fructose-1-phosphate, (10) might be able to serve as an antimetabolite and growth regulator for mutant strains of *E. coli* which are constitutive for the hexose phosphate transport system. This in fact has been found;[88] growth of strains RK 1042 and RK 1435 was completely inhibited at a 2 m*M* external concentration of (10).

One further phosphonic acid related to the fructose phosphate system has been reported[89] as being generated by an enzymatic procedure. The phosphonic acid (Structure 12), related to fructose-1,6-diphosphate, has been generated in the aldolase mediated reaction of D-glyceraldehyde-3-phosphate and 4-hydroxy-3-oxobutyl-1-phosphonic acid (Structure 13), the isosteric phosphonic acid analogue of dihydroxyacetone phosphate.

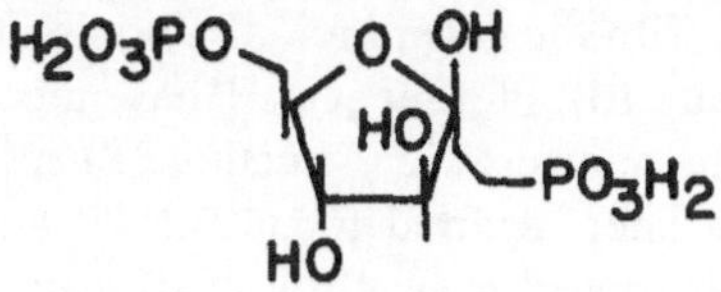

12

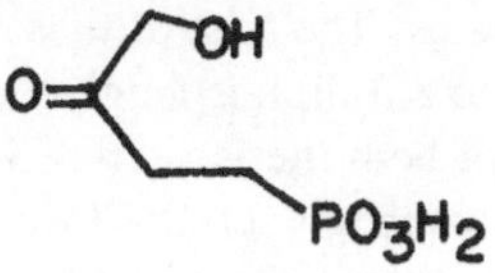

13

This material bears antimetabolic activity in vitro, binds to both aldolase and fructose diphosphatase much as does the natural material, and serves as an inhibitor of fructose diphosphatase.[89] To date there has been no report of an in vivo investigation with (12) or of a nonenzymatic synthesis.

The species (12) also leads naturally to a consideration of phosphonic acid antimetabolites related to "post aldolase" phosphates. Two syntheses of (13), the isosteric analogue of dihydroxyacetone phosphate, have been reported. The first preparation[8] involved phosphonylation of acetoxymethyl vinyl ketone by a standard procedure[90] followed by hydrolysis. Dixon and Sparks[91] later reported a four-step synthesis starting with acrylic acid. As noted above, (13) serves as a substrate for aldolase, in place of the natural material, for the generation of (12). Moreover, (13) serves as a substrate for *sn*-glycerol-3-phosphate dehydrogenase from *E. coli.*[8,10,89,91] However, (13) does not interact with triose phosphate isomerase from several sources.[91]

Initial efforts were not promising for in vivo antimetabolic activity for (13),[8] presumably due to a lack of transport mechanism for dihydroxyacetone phosphate, although it would be anticipated that the species would demonstrate activity if a cell could be induced to accept it. This has recently been accomplished[92] by attachment of (13) to the beta-carboxyl group of the tripeptide alanylaspartylalanine; the growth of a strain of *E. coli* impervious to (13) alone has been thus inhibited. The tripeptide function serves as an "illicit transport" agent for (13) utilizing the tripeptide transport system present in strains of *E. coli* and other organisms.[93-102]

Synthetic routes for both isosteric and nonisosteric phosphonic acid analogues related to glyceraldehyde-3-phosphate have been reported. Baer and Robinson[103] reported the preparation of 2-hydroxy-3-oxopropyl-1-phosphonic acid (Structure 14), the nonisosteric analogue by two routes, each beginning with DL-glyceraldehyde dimer. The synthesis of the isosteric system, 3-hydroxy-4-oxobutyl-1-phosphonic acid (Structure 15), in racemic form was later reported by Goldstein, et al.[104]

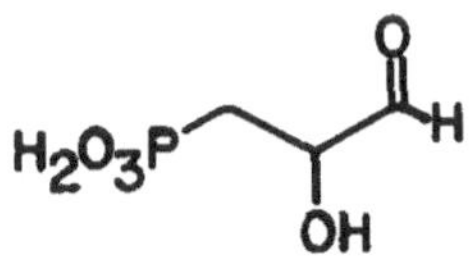

14

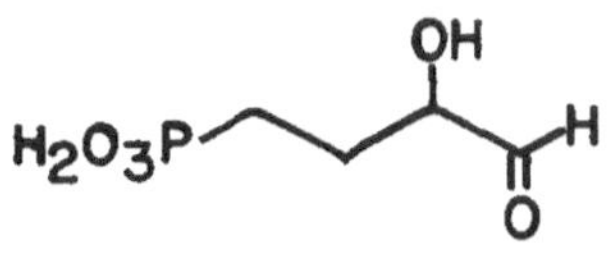

15

The former of these systems has not been tested for antimetabolic activity. The isosteric system (15), however, has been investigated in both enzymatic and intact bacterial systems. Phosphonic acid (15) has been noted[105] to be a competitive inhibitor of glyceraldehyde-3-phosphate dehydrogenase from rabbit muscle and is a competitive inhibitor of *sn*-glycerol-3-phosphate in reactions catalyzed by acyl coenzyme A:*sn*-glycerol-3-phosphate phospha tidyltransferase. With *E. coli,* (15) has been found to enter intact cells via the glycerol-3-phosphate transport system and inhibit growth, having a profound effect on lipid synthesis.[106]

A number of synthetic efforts have been directed at finding phosphonate analogues of the phosphoglycerates. The first of these[107] was concerned with the generation of a nonisosteric system related to 2,3-diphosphoglycerate, that is (Structure 16). Pfeiffer et al.[108] have reported the syntheses of both the isosteric (Structure 17) and the nonisosteric (Structure 18) species related to 3-phosphoglycerate. Two other routes were later reported for (17),[81,104] one of these proceeding through oxidation of (15).[104] Another route,[7] basically a modification of that first used,[108] allows incorporation of a radioactive label via intermediate cyanohydrin formation using 14-C cyanide.

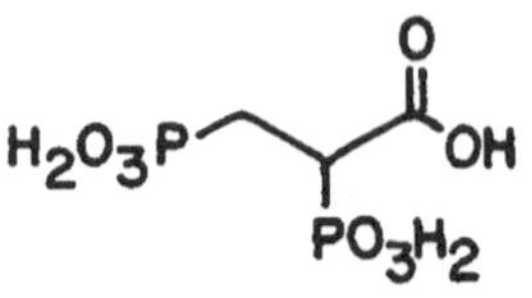

16

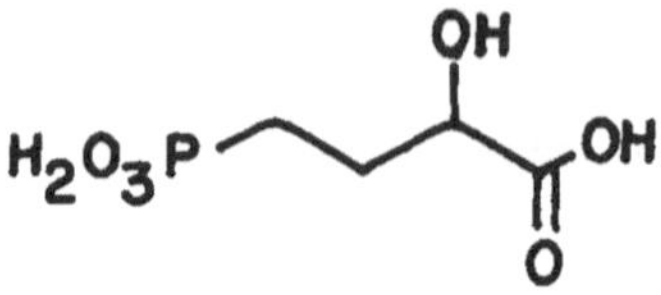

17

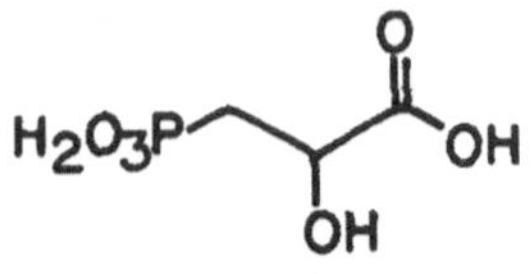

18

Although (16) and (18) were found to be totally without activity in the systems studied,[108,109] the isosteric species (17) was found to be a suitable replacement for the natural material in NADH oxidation[91] and to be a substrate for phosphoglycerate kinase.[110]

Finally, for this section, one considers species which are potential antimetabolites structurally related to phosphoenolpyruvate. Stubbe and Kenyon[111] prepared, among a variety of other structurally related materials, the isosteric phosphonic acid analogue (Structure 19) of phosphoenolypyruvate.

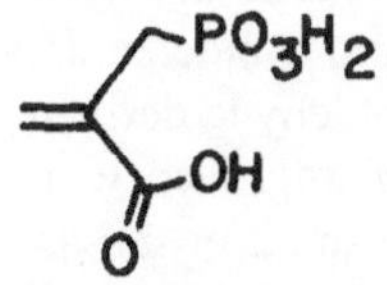

19

The material (19) was found[111-113] to be a very poor inhibitor of pyruvate kinase and a very slow reacting substrate for enolase from rabbit muscle and yeast,[111,114,115] presumably generating the isosteric phosphonic acid analogue of 2-phosphoglycerate, a species which has not, as yet, been reported by independent synthesis.

IV. ANTIMETABOLITES OF GLYCEROL RELATED PROCESSES

Glycerol-3-phosphate provides one of the most interesting species for investigation of intermediary metabolism. It has been separated in the present consideration from both glycolysis (of which it is a branch-product) and lipid biosynthesis (of which it is a precursor) due to its variety of routes of generation and reactivity. For example, the intracellular presence of *sn*-glycerol-3-phosphate may be the result of active transport via one of several transport systems, enzymatic phosphorylation of glycerol, or the enzymatic reduction of dihydroxyacetone phosphate, and may be involved in phospholipid generation via a variety of reactions. Existing as it does at a branch point of intermediary metabolism, it is quite reasonable that it should be a target for antimetabolic activity by structurally related phosphonic acids.

As noted above[2,3] the nonisosteric phosphonic acid analogue (Structure 20) was early prepared in both racemic form (via Arbuzov reaction on allyl bromide followed by hydroxylation and deesterification)[2] and chiral form corresponding to the absolute configuration of *sn*-glycerol-3-phosphate (from D-mannitol diacetonide).[3] The material prepared by either route was found[4] to be inactive as a substitute for *sn*-glycerol-3-phosphate in reaction involving L-glycerol-3-phosphate:NAD oxidoreductase.

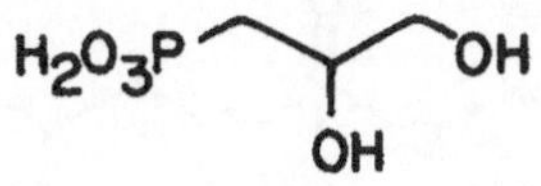

20

Further efforts with this point as a target have involved isosteric species. The primary species in this category is 3,4-dihydroxybutyl-1-phosphonic acid (Structure 21). A variety of syntheses of this material have been developed, generating it in racemic, chiral, and specifically labeled forms.

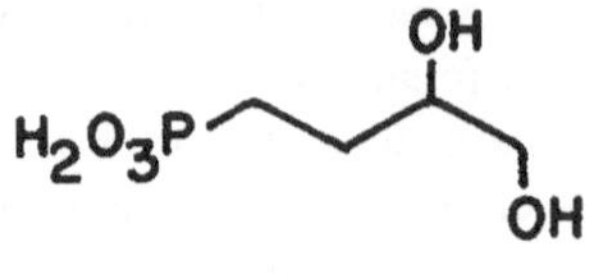

21

The initial synthesis[5] generated the material in racemic form via a route analogous to that used[2] for (20) but starting with 4-bromo-1-butene. It was later[6] generated in chiral form using a Wittig phosphonylation of the aldehyde derived from D-mannitol diacetonide cleavage. This latter process has now been improved to provide higher yields[31] and to allow specific introduction of tritium at the 1- and 2-positions via catalytic reduction of the vinylic phosphonate intermediate.[9] A further chiral synthesis has been reported[7] beginning with optically active malic acid; this route readily provides both enantiomers and a variety of useful intermediates. A further synthesis of racemic material involves sodium borohydride reduction of the immediate precursor to (13), the isosteric phosphonic acid analogue of dihydroxyacetone phosphate.[8] This route is of particular interest as it allows specific introduction of a tritium label at the 3-position. A route vaguely related to those involving D-mannitol diacetonide[6,7,31] has also been reported by Paulsen and Bartsch.[116,117]

The compound (21) was of particular interest as bacterial species were readily available which actively transported and used the natural phosphate. Thus, there was a significant possibility that (21) would be of use in vivo. In fact, the enantiomer of (21) corresponding in absolute configuration to *sn*-glycerol-3-phosphate is transported into a variety of bacterial species and its transport properties have now been thoroughly studied in *E. coli.*[118,119] It is also noteworthy that its antipode does not appear to be transported.[105] Thus it would appear that transport of (21) correlates well with that of the natural material.

For those species which transport (21) growth inhibition is observed, a characteristic not shown by the nonisosteric species (20); presumably (20) is not transported. At low external concentrations of (21) suitable strains of *E. coli* undergo significant perturbation of phospholipid synthesis and growth stasis.[5,120-122] The mode of antimetabolic activity of (21) in *E. coli* has been noted by both in vivo and in vitro efforts to involve perturbation of phosphatidylglycerol synthesis;[123,124] compound (21) replaces *sn*-glycerol-3-phosphate in the reaction catalyzed by CDP-diglyceride:*sn*-glycerol-3-phosphate phosphatidyltransferase and produces a polar lipid material (Structure 22) which prevents phosphatidylglycerol synthesis and thus seriously perturbs the normal lipid composition. Moreover, at low concentration of (21) the synthesis of lipoprotein is strongly inhibited.[125]

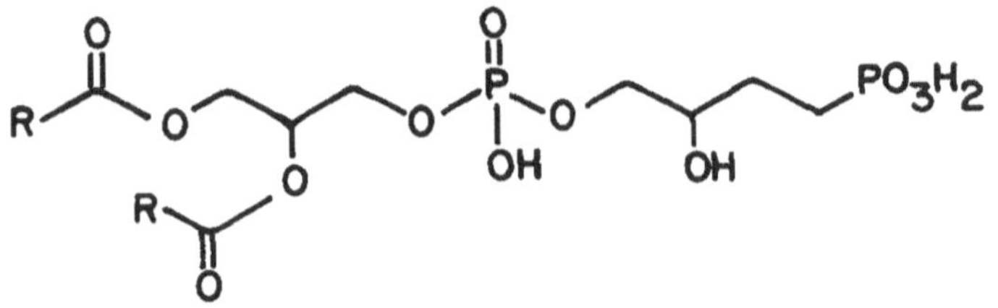

22

The material (21) has also been noted to be active in the inhibition of growth of strains of *Bacillus subtilis.*[126,127] Here it is noted, in similarity to *E. coli,* that phosphatidylglycerol synthesis is inhibited, but also that (21) is incorporated into the cell wall. With strain 168, containing polyglycerolphosphate cell wall teichoic acid, (21) is bacteriostatic; with strain W23, containing polyribitol cell wall teichoic acid, (21) is bactericidal.

In vitro enzymatic studies with (21) indicate a number of interesting points. While (21) serves as a substrate for CDP-diglyceride:*sn*-glycerol-3-phosphate phosphatidyltransferase and is an inhibitor of the anaerobic *sn*-glycerol-3-phosphate:NAD(P) oxidoreductase of *E. coli*,[6,123] it does not appear to interact with the catabolic membrane-bound *sn*-glycerol-3-phosphate dehydrogenase, CDP-diglyceride:L-serine phosphatidyltransferase, or acyl coenzyme A:*sn*-glycerol-3-phosphate acyltransferase.[123] The lack of interaction in these latter systems is attributed to the loss of binding capability resulting from the substitution of a methylene group for the esteratic oxygen of the natural substrate *(vide supra)*.

As an attempt to test this hypothesis and overcome the apparent lack of interaction, further species were synthesized without the esteratic oxygen but still bearing a potential binding functionality. These include the 1,3,4-trihydroxybutyl-1-phosphonic acid (Structure 23) generated from *l*-malic acid with chirality at the 3-position corresponding to that of *sn*-glycerol-3-phosphate;[7] the pair of diastereoisomers that resulted (due to the hydroxyl at the 1-position) in the Abramov reaction have been partially separated by column chromatography.[128] In addition there was synthesized the *N*-(2,3-dihydroxypropyl) phosphoramide (Structure 24)[9] wherein an electron pair on nitrogen could serve in place of those on the esteratic oxygen of the natural substrate.

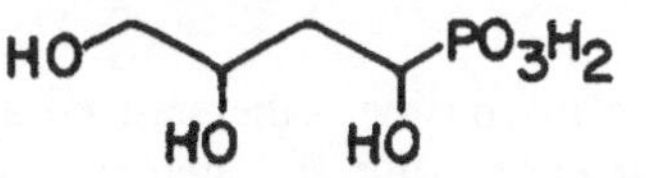

23

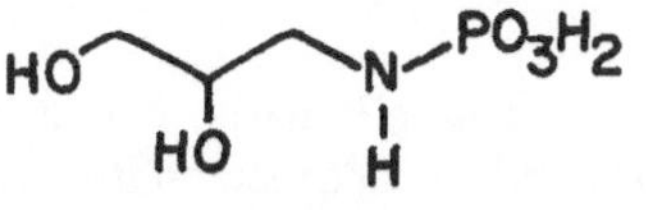

24

When investigated with actyl coenzyme A:*sn*-glycerol-3-phosphate acyltransferase both (23) and (24) were found[128,129] to serve as substrates, supporting the postulate that inactivity of (21) resulted from loss of a binding function. Moreover, (23) serves as a growth inhibitor of strains of *E. coli*.[128,129]

Several other species bearing carbon-phosphorus bonds and structurally related to *sn*-glycerol-3-phosphate were synthesized and investigated.[6] These are the 1,2,3-trihydroxypropyl-1-phosphonic acid (Structure 25) and the phosphinic acid (Structure 26) wherein not only is the normal esteratic oxygen missing but also a methyl group has replaced a normal acidic hydroxyl on phosphorus.

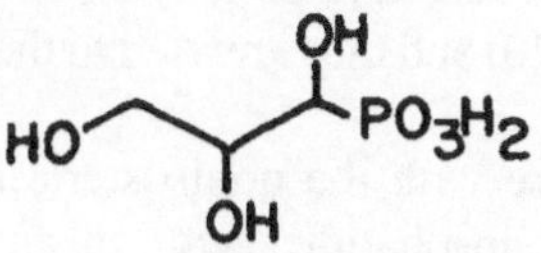

25

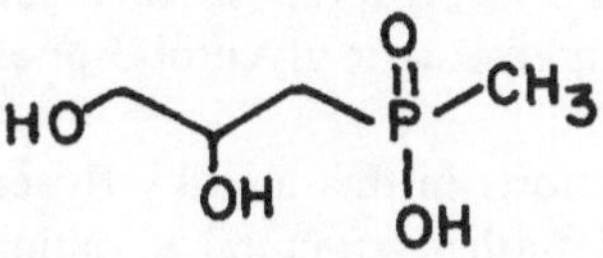

26

Neither of these compounds inhibited the *sn*-glycerol-3-phosphate:NAD(P) oxidoreductase as did (21). A further compound synthesized[7] as a structural analogue of glycerol-3-phosphate is the 1-methyl-3,4-dihydroxybutyl-l-phosphonic acid (Structure 27). This material is inactive as a growth inhibitor of *E. coli*.

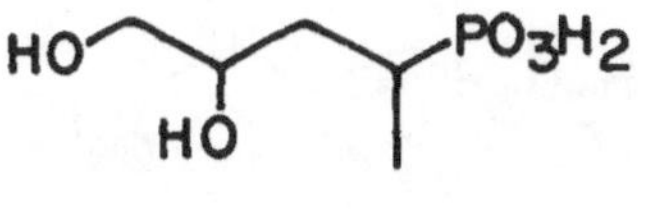

27

A major consideration in the choice of analogues of *sn*-glycerol-3-phosphate for investigation was the availability of species for which an active transport system was known. In the absence of such transport systems, as in mutant strains of *E. coli,* (21) appears to be inactive in inhibiting growth.[105,119] Presumably this problem could be overcome should a system for "illicit transport" be available. When (21) was bound via the 4-position hydroxyl to the tripeptide system noted above for use with (13),[92] inhibition of growth of these *E. coli* strains was noted.[105] Presumably, the antimetabolite-tripeptide conjugate is transported to allow the inhibition to occur.

V. ANTIMETABOLITES OF LIPID PROCESSES

Interest in phosphonic acids as antimetabolites of lipid processes has been partially spurred by the observation[130-137] of a series of natural "phosphonolipids" containing 2-aminoethylphosphonic acid (Structure 28) (see Chapter 3).

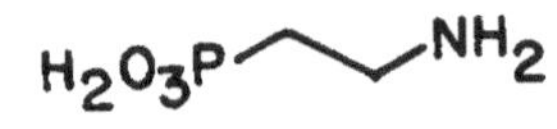

28

With this as a "natural" basis, it is a logical extension to consider other related systems bearing C–P bonds as antimetabolites. Several possibilities for utility of such systems may be considered. Such phosphonolipids structurally related to natural phospholipids may be capable of serving as inhibitors of natural processes, such as phosphate or carboxylate ester cleavage. Also, simple phosphonolipids might be capable of serving as replacements for the normal lipids in enzymatic coupling processes generating more complex systems; or, the fundamental "inert" character of the C–P bond(s) of phosphonolipids might be used for the generation of inert membranes of membrane components.

An extensive synthetic effort has been made in the area of phosphonolipids. As a result of the variety of possible structural modifications, it is best here to proceed in a rather formal fashion, organizing the discussion on the basis of individual structural relationships.

A. Compounds Related to Phosphatidic Acid in which the Normal Phosphate Ester Linkage is Replaced by a C–P Linkage

Within this class, two fundamental subcategories have received attention, those which are related to the nonisosteric glycerol-3-phosphate analogue (20) and those related to the isosteric system (21).

The first efforts in this area by Rosenthal et al.[138] were with the nonisosteric backbone and included further structural variations from normal phosphatidic acid. In an effort to generate truly inert species, compounds of the type (Structure 29) were generated wherein the normal carboxylate ester linkages were replaced by ether linkages in addition to the presence of the C–P bond. Soon after this report, the isosteric systems (Structure 30) were prepared, again with ether linkages in place of the normal carboxylate esters, and a novel lipid analogue (Structure 31) bearing only a hydrocarbon chain attached to the phosphonic acid function.[139,140]

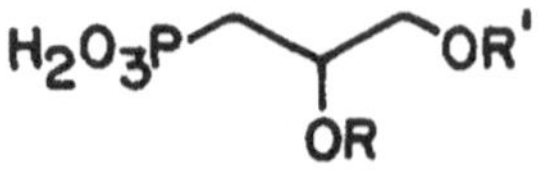

29

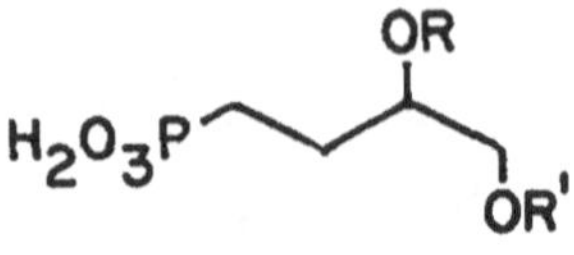

30

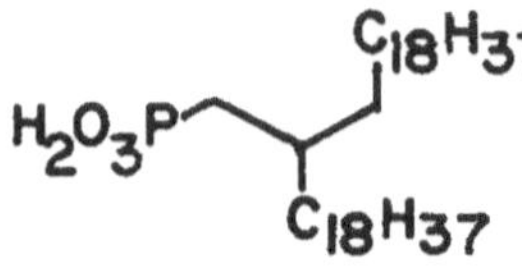

31

These species serve as antimetabolites with several systems.[141] Species (29) serves as a reasonably potent inhibitor of pig kidney phosphatide phosphohydrolase when measured with phosphatide derived from egg lecithin. The species (30) and (31) were also active in this system, but to a lesser extent. These species are specific rather than general enzyme inhibitors and require sonication for dispersal before activitiy may be observed. This is in correspondence with numerous other observations of decreased solubility with the phosphonic acids as compared to the natural phosphate esters, presumably due to a loss of hydrophilic binding characteristic upon removal of the esteratic oxygen. These species were also reported to cause complete growth inhibition of *Staphylococcus epidermidis* and *Streptococcus salivarius*.[141]

The phosphonic acids related to phosphatidic acid bearing normal carboxylate ester linkages have also been reported. The first preparations involved the nonisosteric backbone (Structure 32) and were accomplished by acylation of (20).[142,143] An improvement in the synthesis of (32) was later reported by Rosenthal et al.[75,76] using a reaction of tris(trimethylsilyl) phosphite[73,74] with the appropriate 2,3-diacyloxy-1-iodopropanes. Both routes were applicable for chiral (32), although only the first preparations were actually of chiral species of (32). The phosphatidic acids incorporating an isosteric backbone (Structure 33) were later reported[144] utilizing the Rosenthal approach with the 3,4-diacyloxy-1-bromobutanes.

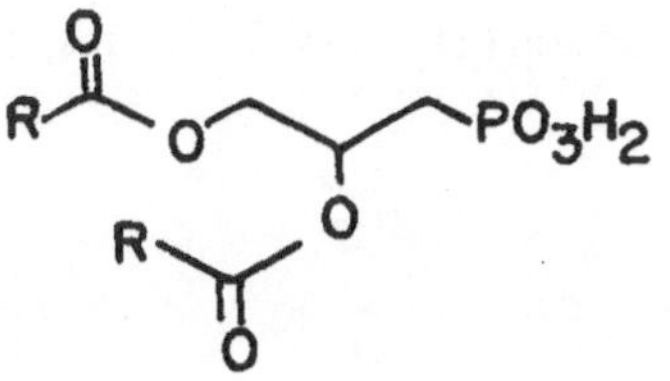

32

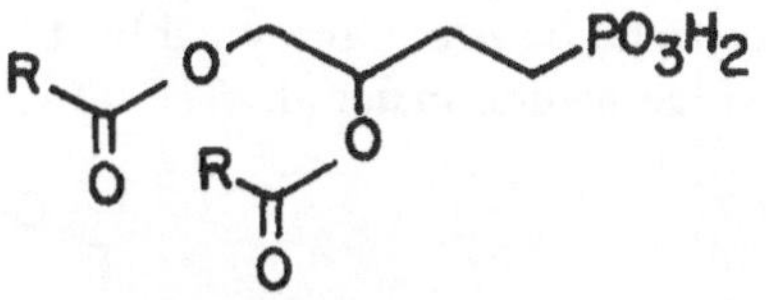

33

Again, a significant difficulty is observed with (33) bearing C-16 and C-18 chains in that they exhibit low solubility compared to the natural systems.[144,145] Increased solubility is found with shorter (C-10) acyl chains.[146] To date, investigations of antimetabolic activity of the type found for (29) to (31) have not been performed for (32) and (33).

The synthetic approaches noted above generated phosphatidic acids wherein the two alkoxy or acyloxy functions attached to the backbone were identical. For differentially substituted compounds of the type (33) another approach is used. The isosteric analogue (13) of dihydroxyacetone phosphate was acylated[144] to generate (Structure 34), the isosteric phosphonic acid analogue of acyldihydroxyacetone phosphate. This material may be reduced either catalytically or with sodium borohydride[147,148] to generate (Structure 35), an isosteric analogue of *lyso*-phosphatidic acid. This material was then acylated differentially to yield the phosphatidic acid. The species (35) serves as a substrate for *lyso*-phosphatidic acid acyl transferase.[148]

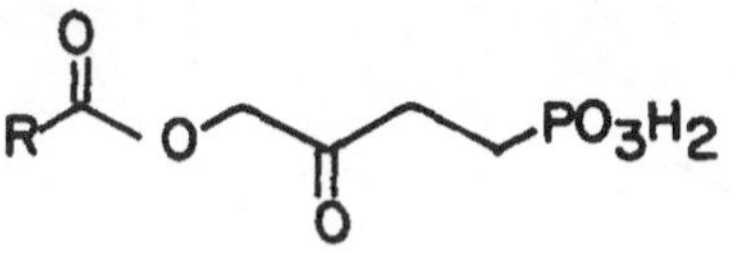

34

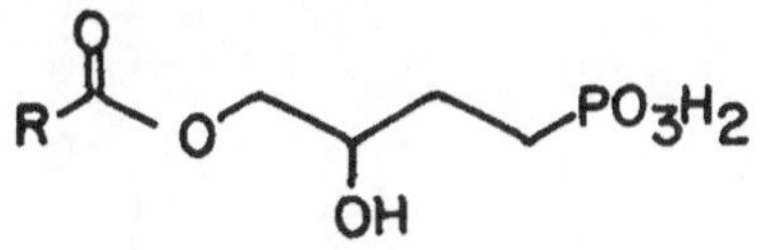

35

B. Compounds Related to Phosphatidylethanolamine

With this general category one may consider several aspects of structural variation from the "natural" system. Of particular interest for synthetic purposes are compounds bearing the C–P bond in the polar head group portion as these materials are, or are related to, natural phosphonolipids. These materials (Structure 36) bearing acyloxy functions in the glycerol backbone have been reported by reaction of the alpha, beta-diglycerides (chiral and racemic) with a variety of N-protected P-activated derivatives of 2-aminoethylphosphonic acid.[149-155] All of these systems involved alpha-beta-diglycerides bearing similar acyloxy functions.

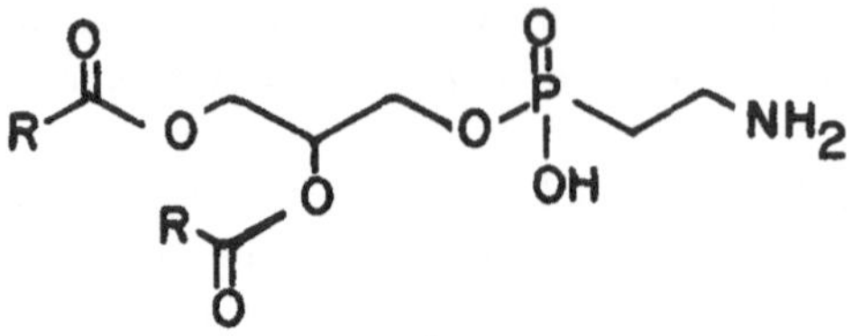

36

Baer later reported[156] the synthesis of a compound of the class (36) bearing a saturated acyloxy function at the alpha-position and an unsaturated function at the beta-position. Turner et al.[155] have reported these compounds to serve as substrates for phospholipase A from snake venom. Using racemic phosphonolipid, one half of the material was degraded to the *lyso*-phosphatidyl system, presumably of absolute configuration (Structure 37) corresponding to that of the natural material.

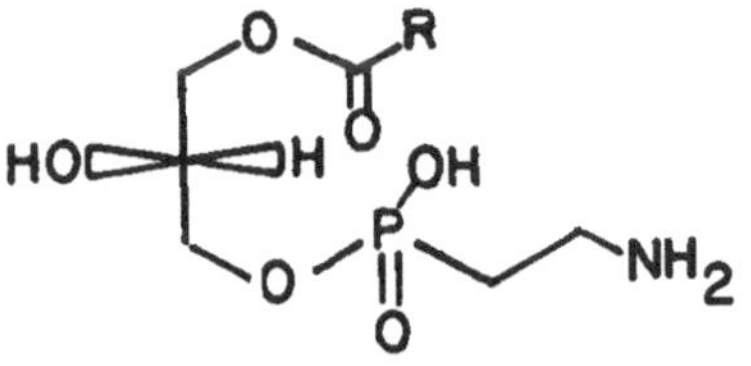

37

An unusual form of (36) was synthesized by Pfeiffer et al.[157] wherein the alkyl functions were adamantyl; this material was considered for its ability to inhibit the renin-catalyzed conversion of renin substrate to angiotensin I and found to be quite active. The diether systems (Structure 38) have also been prepared,[158] as has the ether-ester system (Structure 39).[159]

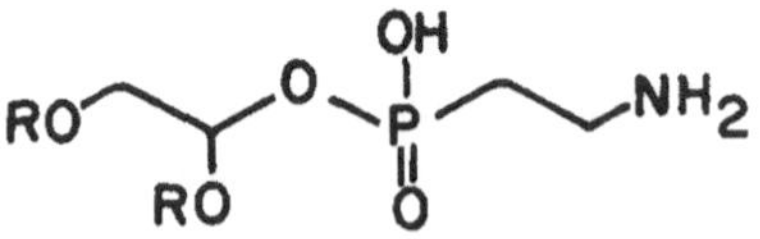

38

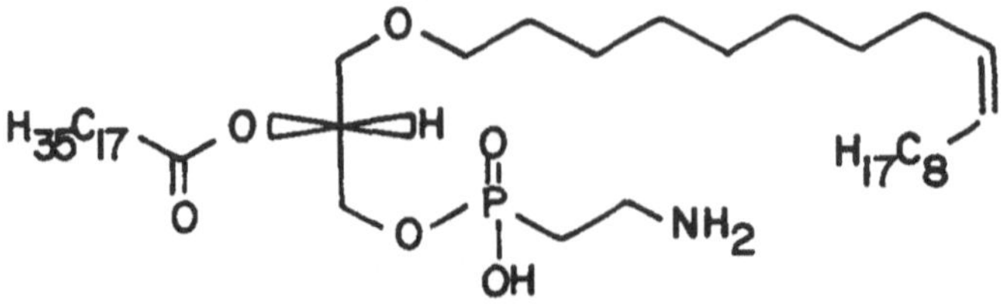

39

The synthesis of sphingolipids related to natural species isolated from sea anemone[163] has also been reported.[160-162] The dihydroceramide aminoethylphosphonate (Structure 40) was prepared by the same technique as mentioned above. Another unusual "shortened" system (Structure 41) has been reported.[164] These materials have not been investigated for anti-metabolic activity.

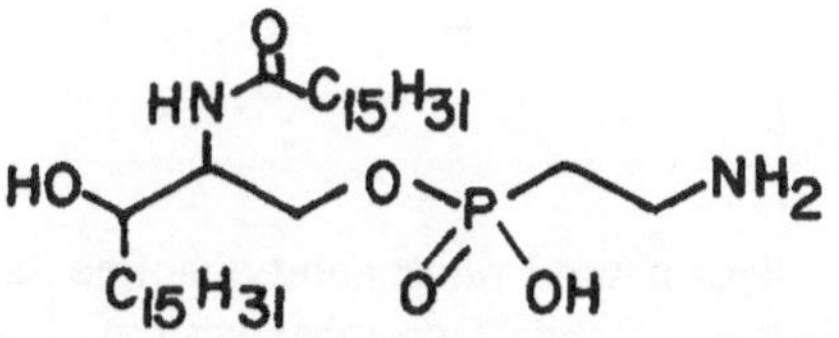

40

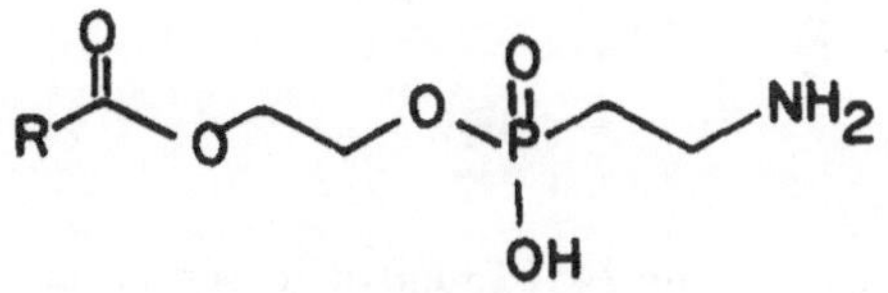

41

Several systems have been reported bearing the C–P linkage in the backbone of the lipid. Rosenthal et al.[140,151] have generated the diether lipids (Structure 42) related to the nonisosteric analogue (20) of glycerol-3-phosphate by coupling of protected 2-aminoethanol with activated forms of the phosphatidic acids (29). The isosteric diester system (Structure 43) related to (21) was later reported[165] via similar coupling processes involving the phosphatidic acids (33). Pfeiffer et al.[157] have also reported the "shortened" system (Structure 44).

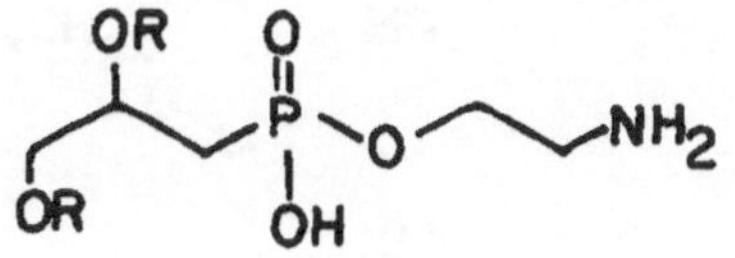

42

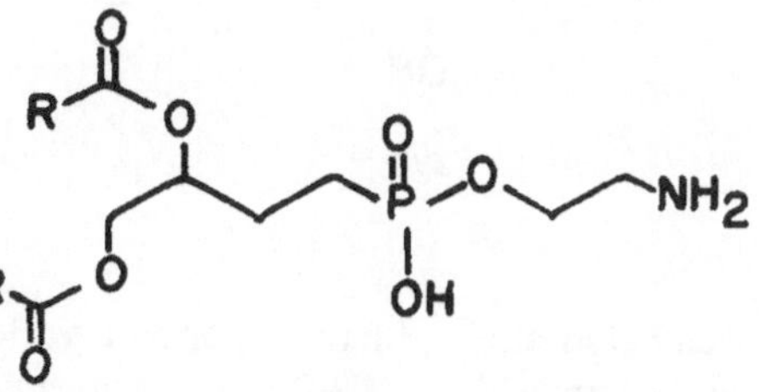

43

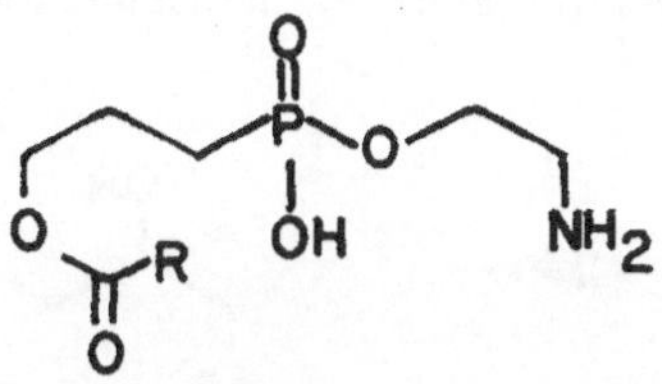

44

C. Compounds Related to Phosphatidylcholine

Again several aspects of structural variation from the natural systems may be considered.

Baer et al. have reported[166-170] the syntheses of a number of chiral species of structures (Structure 45) and (Structure 46) related to naturally occurring phosphonolipids. Although several approaches were used for linking the several parts of these molecules, all proceeded through a 2-bromoethylphosphonate species with introduction of the ammonium function by reaction with trimethylamine.

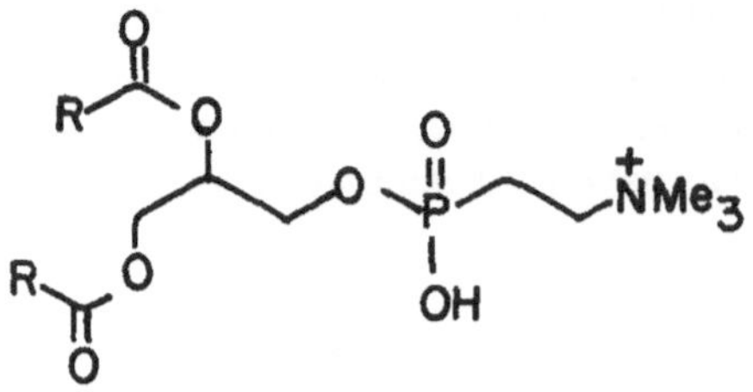

45

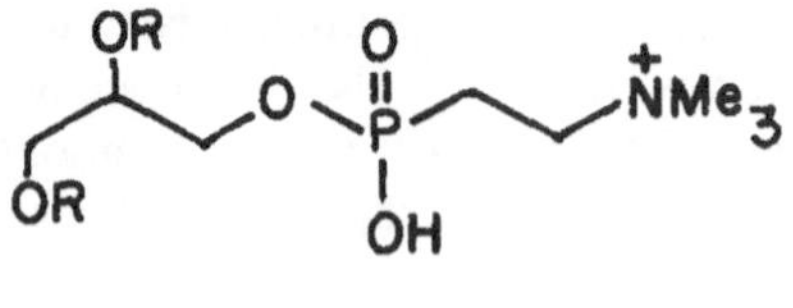

46

Studies[171] using (45) found it to serve nearly as well as natural phosphatidylcholine as a substrate for phospholipase C isolated from *Clostridium welchii*. Two other systems have been synthesized bearing the C–P bond in the polar head group. Baer and Robinson[172] reported the "shortened" species (Structure 47), again proceeding through a 2-bromoethylphosphonate species, as did Chacko and Hanahan[159] for the species (Structure 48). To date no biological data regarding these materials have been reported.

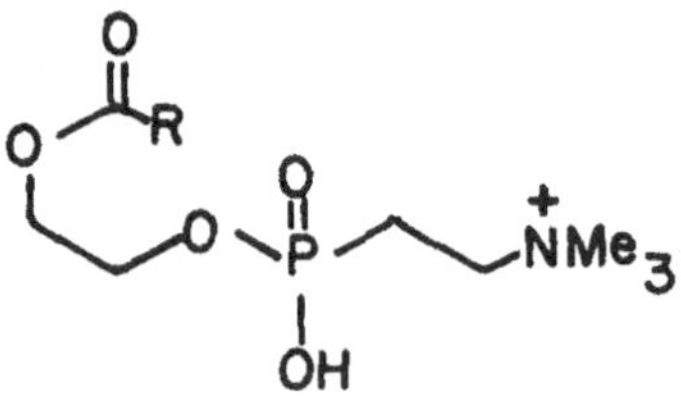

47

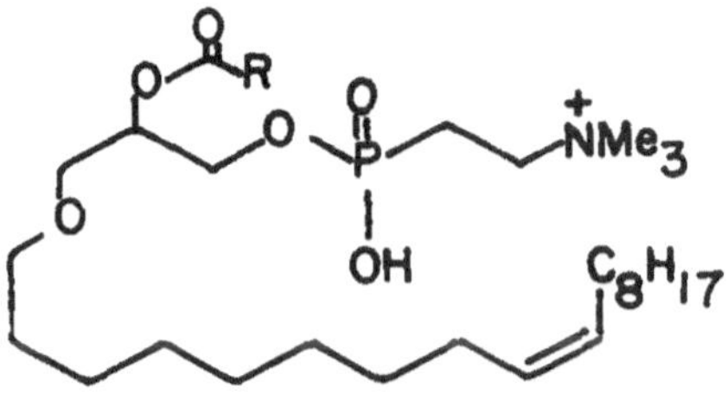

48

Rosenthal et al.[140,172] have reported a variety of phosphonate species related to phosphatidylcholine in which a C–P bond is located in the backbone portion. These include the diether species (Structure 49) and (Structure 50) in racemic form, and later[75,76] the diacyloxy species (Structure 51).

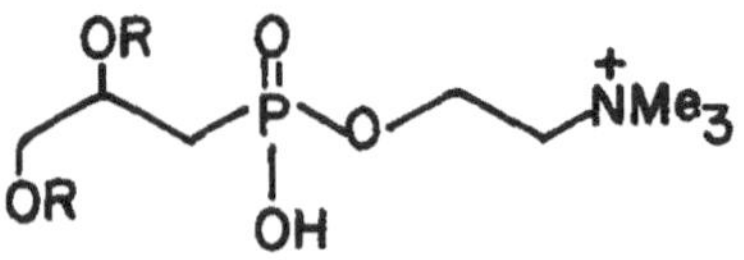

49

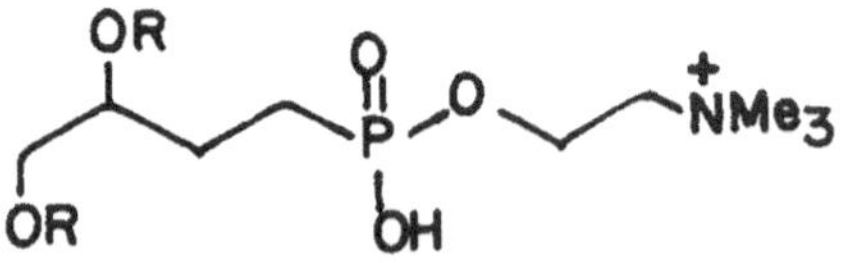

50

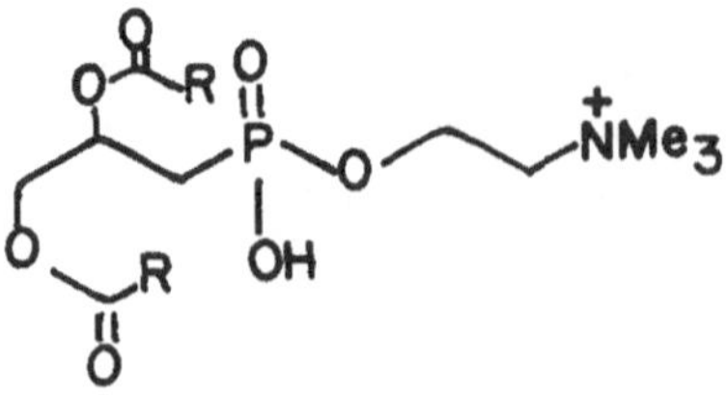

51

The species (49) and (50) have been found[173,174] to be significant inhibitors of phospholipase C from *C. perfringens*. Of these, (50), with the "isosteric" backbone is the best inhibitor, and both (49) and (50) are significantly more inhibitory than is (45). The species (49) also inhibits phospholipase A.[175] The synthesis of the totally isosteric diacyloxy system (Structure 52) has also been reported.[165]

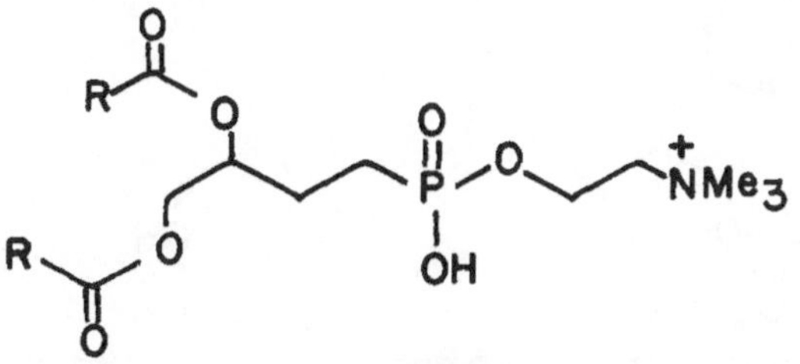

52

In addition to these systems, Rosenthal et al.[176-178] have studied a series of phosphinates. These species, (Structures 53, 54, 55, and 56) bearing ether linkages, as anticipated were found to be completely resistant to hydrolysis catalyzed by any normal phospholipase system.

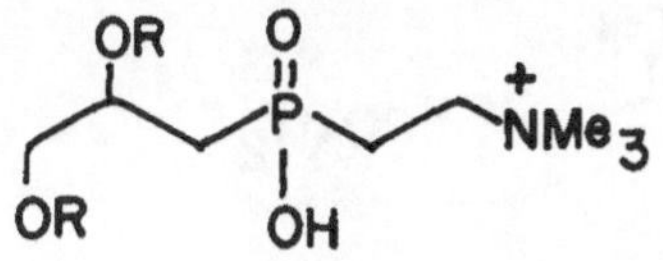

53

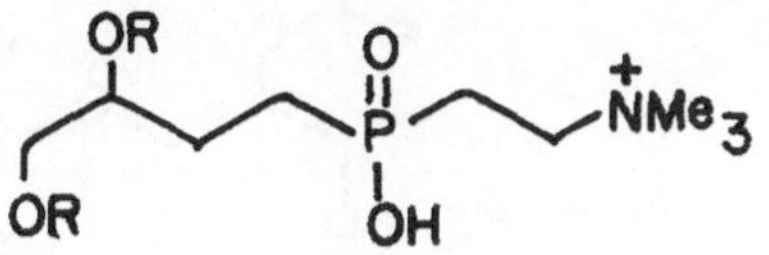

54

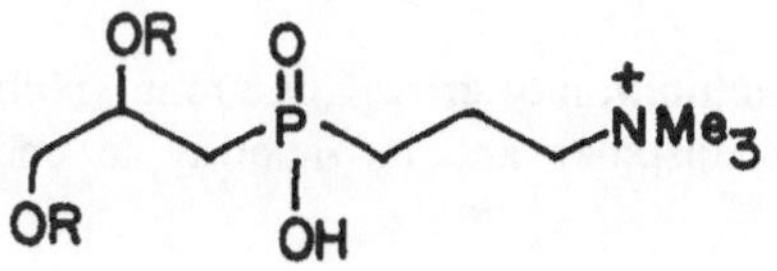

55

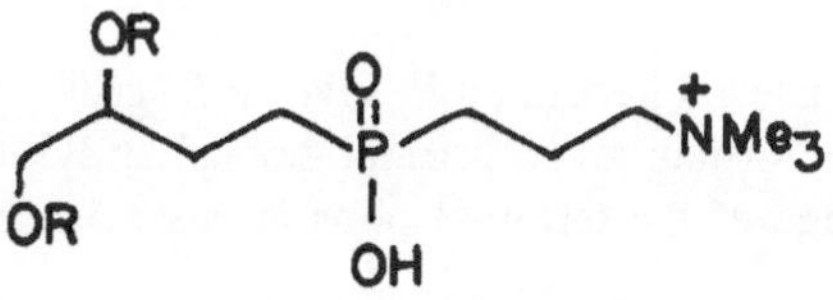

56

These materials have been found[179] to be competitive inhibitors of phospholipase A from *Agkistrodon piscivorus* venom. They have also been found to be inhibitors of phospholipase C,[174] B, and D.[180]

The use of ether linkages in place of normal acyloxy linkages of phospholipids has proven to be a significant substitution. Ether linkages lend an inert character to the molecules without disturbing properties required in normal lipid interactions. Several studies have found a negligible effect of this substitution on the interaction of lipids with sterols;[181-183] the length of the backbone is more critical; the ''isosteric'' systems pack better in liposomes and exhibit sterol interaction.[181] There have also been two studies of the specificity of these analogues prepared by Rosenthal et al. for enzyme activation.[184,185]

D. Compounds Related to Phosphatidylglycerol

To date only the isosteric analogue (Structure 57) has been reported with the C–P bond in the backbone portion.[165] As with many of the other phosphonolipids bearing long alkyl chains, decreased solubility compared to the natural system has made difficult the evaluation of antimetabolic activity.

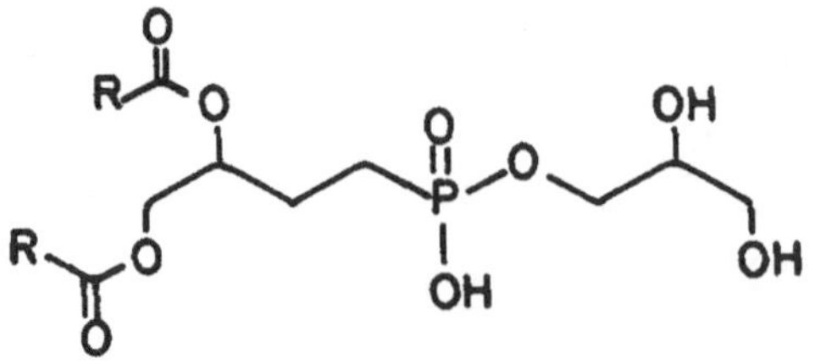

57

The isosteric analogue (Structure 58) of phosphatidylglycerol phosphate has been generated[124] enzymatically via reaction of (21) with CDP-diglyceride:*sn*-glycerol-3-phosphate phosphatidyl transferase.

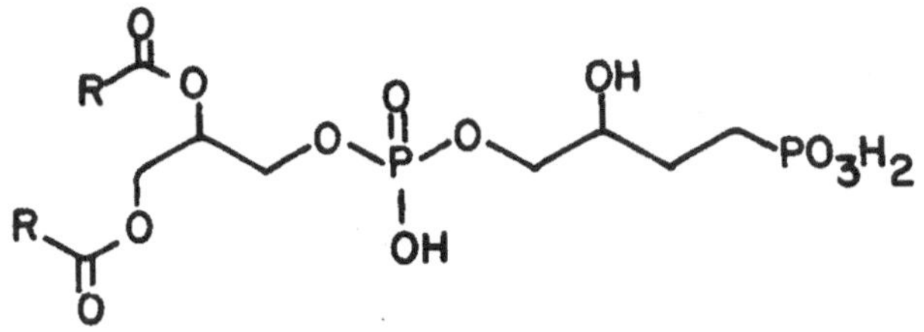

58

This material generated in vivo in *E. coli* results in inhibition of growth due to its inhibition of the formation of normal phosphatidylglycerol phosphate and its inability to undergo cleavage of the terminal phosphorus acid function.

E. Compounds Related to Nucleoside and Serine Phosphatides

Diether (Structure 59) and diacyloxy (Structure 60) phosphonate species, isosteric and nonisosteric in the backbone as compared to the natural material, phosphatidylserine, have been prepared.[186] These materials, which are racemic in the "diglyceride" portion, but chiral in the serine portion, were prepared to be used in studies of lipid metabolism.

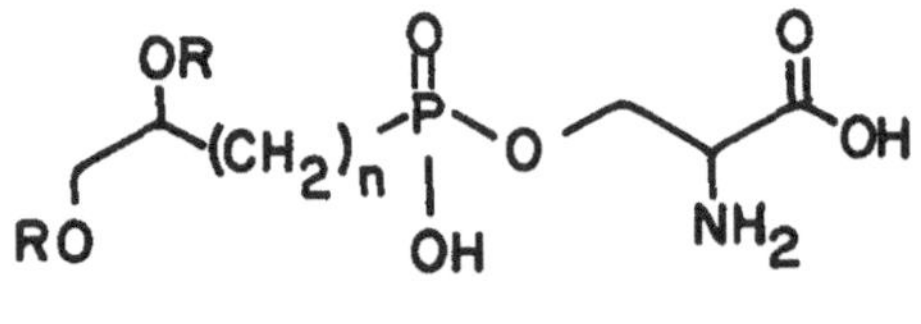

59

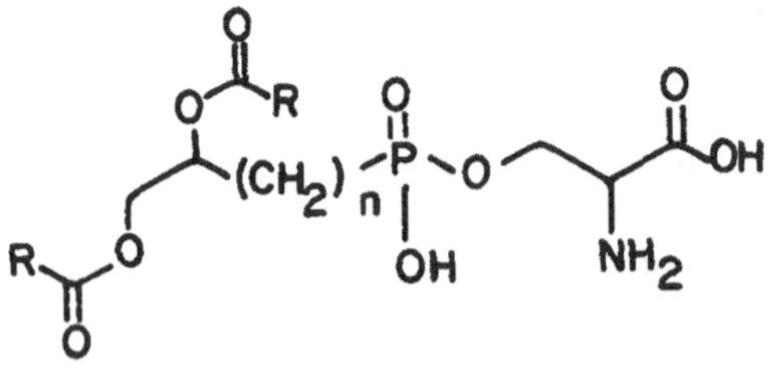

60

The preparation of the isosteric (Structure 61) and nonisosteric (Structure 62) analogues of CDP-diglyceride has also been reported.[187] These were prepared using the phosphotidic acids (29) and (30) in coupling to the monomorpholidate of CMP.

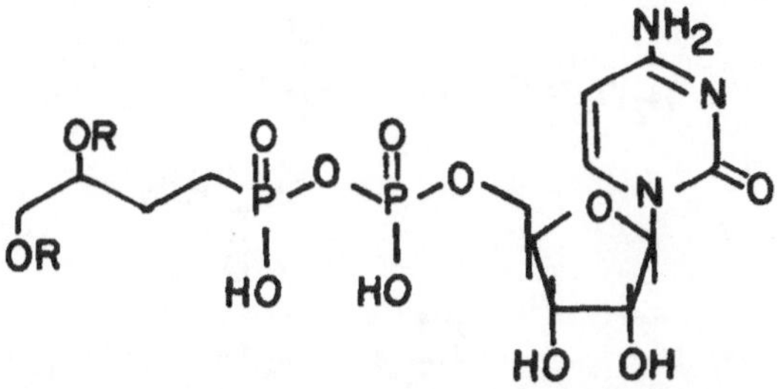

61

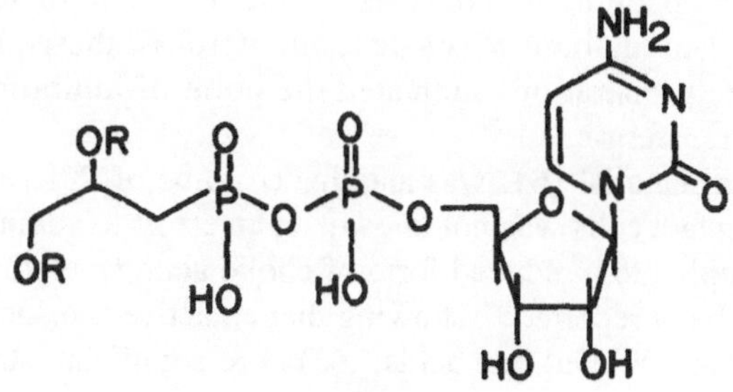

62

F. Compounds Related to Sterol Precursors

The series of phosphate and pyrophosphate species which serve as biochemical precursors to squalene, and thereby sterols,[188] provide an interesting target for inhibiting cholesterol biosynthesis in a systematic manner. Corey and Volante[189] reported the synthesis of a series of phosphophosphonates isosteric with the pyrophosphate species in the sequence leading to squalene beginning with the analogue (Structure 63) of isopentenylpyrophosphate.

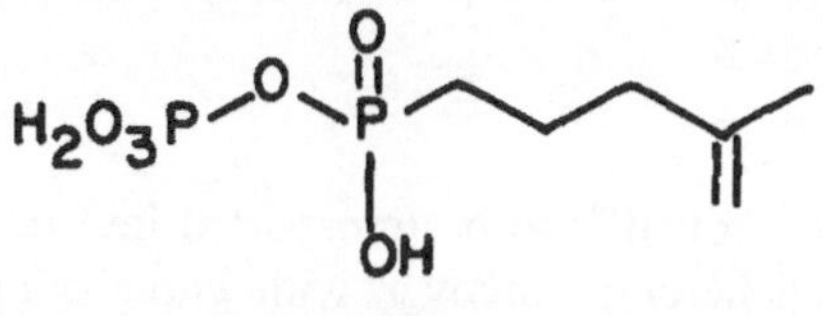

63

These were generated via a displacement reaction by dimethyl lithiomethylphosphonate on the appropriate bromides followed by cleavage of the ester linkages[190] and formation of the phosphophosphonate linkage by a standard method.[191] These species were studied[189] for antimetabolic activity toward squalene biosynthesis using squalene synthetase isolated from rat liver extracts. They are mildly inhibitory with the greatest activity being found for the species correlating with the intermediates of the latter stages of the sequence. Similar results were found[189] with kaurene synthetase from *Ricinus communis*.

Notable by its absence from this work was any analogue related to the first phosphorylated species in the biosynthesis of squalene, 5-phosphomevalonate. The isosteric phosphonic acid species (Structure 64) related to this intermediate was later reported by Sarin et al.[192] using the addition of the dianion of acetic acid to diethyl 4-oxopentyl-1-phosphonate followed by ester hydrolysis. This approach allows ready incorporation of 14-C into the material at a late stage of the synthesis.[193]

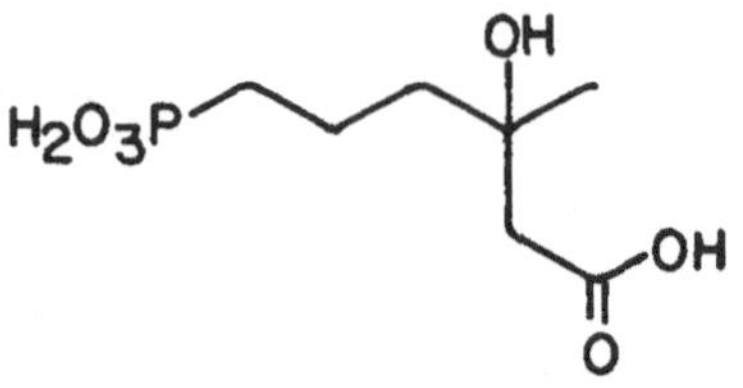

64

The antimetabolic nature of the material (64) has been investigated[194] using a preparation of homogenized rat liver containing all enzymes for the formation of squalene. It (64) has been found to be a significant inhibitor of squalene and sterol synthesis. Further studies with partially purified enzyme preparations indicated the point of inhibition to be at 5-phosphomevalonate phosphotransferase.

For in vivo activity the material (64) was anticipated to be of little value as transport of the natural material into intact cells was not known. In an effort to attain "illicit transport", the material (64) was coupled to a reduced form of cholic acid generating (Structure 65),[193] since previous work had been reported[195] showing that an active transport system existed in liver for taurocholic acid and related bile acids; (65) bore significant structural relationship to taurocholic acid.

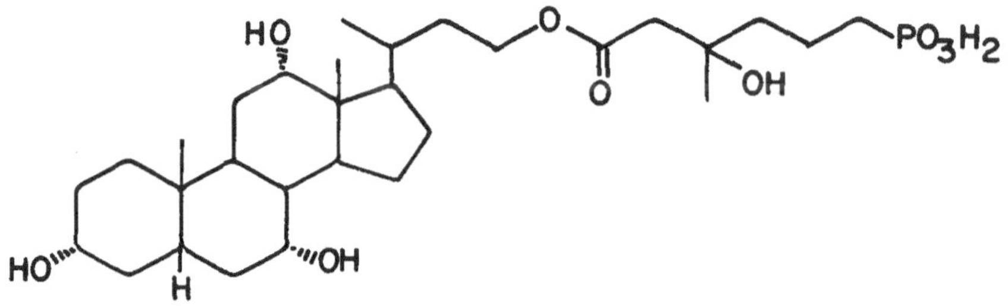

65

The coupled material (65) was found[193] to be transported into rat liver hepatocytes and to inhibit cholesterol biosynthesis therein; moreover, with whole animal injections it was found to be transported predominantly to the liver with only a much smaller amount going to intestine.

VI. ANTIMETABOLITES OF NUCLEOTIDE RELATED PROCESSES

As nucleotides are involved in a most extensive variety of biochemical processes, potential applications of related phosphonic acids as antimetabolites are manifold. It is the intent here, even more than in previous sections, to limit the view to the molecular systems most relevant to biochemical application rather than to be encyclopedic; i.e., applications or potential applications to biochemical processes of such molecular systems, rather than structural considerations, will serve as the basis for inclusion.

A. Protein Synthesis

The most noted phosphonic acid related to a natural phosphate for this process is the species (Structure 66), guanosine 5′-(beta,gamma-methylene)triphosphate, related to GTP. Hershey and Monro[196] prepared this material by coupling the monomorpholidate of GMP to (1) by the use of dicyclohexylcarbodiimide with pyridine and *o*-chlorophenol. This reaction system was developed for the synthesis of the corresponding analogue of ATP.[197,198]

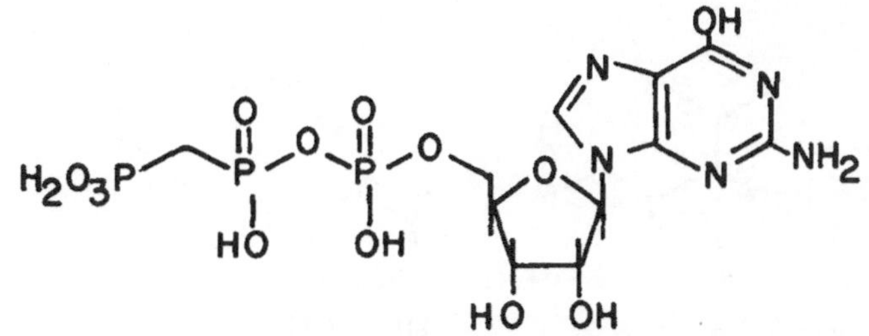

66

The action of (66) as an antimetabolite of protein synthesis has been well-documented[199-203] and its use in determination of the role of GTP therein has been thoroughly reviewed.[204-211] Fundamentally, the phosphonic acid species (66) has been o use in the determination of the stages in which cleavage of the normal beta-gamma pyrophosphate linkage of GTP occurred; since the species (66) is incapable of undergoing such a cleavage process it serves as an inhibitor at those points (translocation and termination) but does not inhibit initial binding. This inhibitor capability of (66) is quite interesting in light of two structural considerations; first, there are significant geometrical deviations of the methylenediphosphonate system from the natural pyrophosphate *(vide supra)*, and second, there is a loss of any binding capability normally present at that anhydride oxygen site. These deviations, the latter significant in simple pyrophosphate related systems, do not appear to hinder the interaction and inhibitory action of (66) in the protein synthesis scheme.

The related system (Structure 67), adenosine 5′-(alpha,beta-methylene)triphosphate, prepared[198,212,213] via dicyclohexylcarbodiimide mediated coupling of (1) to adenosine followed by phosphorylation, has been studied with tyrosyl-tRNA synthetase from *E. coli* and found to be a competitive inhibitor of ATP binding.[214] On the other hand, using (67) with Phe-tRNA synthetase from *E. coli* it was found[215] that (67) served as a noncompetitive inhibitor with respect to ATP. Presumably a specific alpha-beta pyrophosphate oxygen interaction is required in this system; all of this points to a myriad of complexities in the various protein synthesizing enzymes.

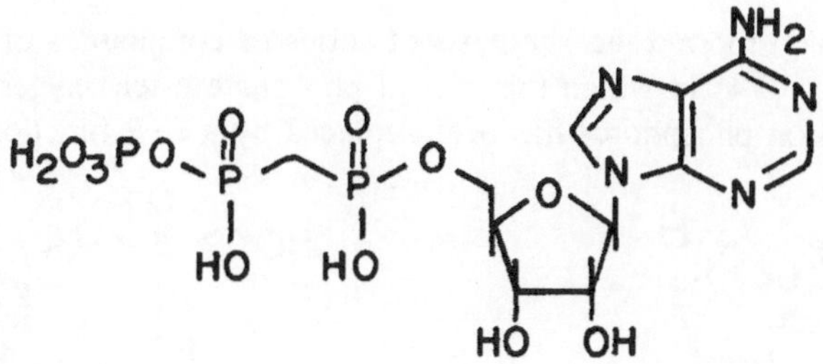

67

More recently, Southgate and Dixon[216] have investigated a series of species in which the anhydride oxygen of an aminoacyl-AMP has been replaced by a methylene function, such as (structure 68), related to valyl-AMP. The material (68) has been studied in binding to valyl-tRNA synthetase. It inhibits both ATP-PPi exchange and the valylation of tRNA, processes normally catalyzed by this enzyme.

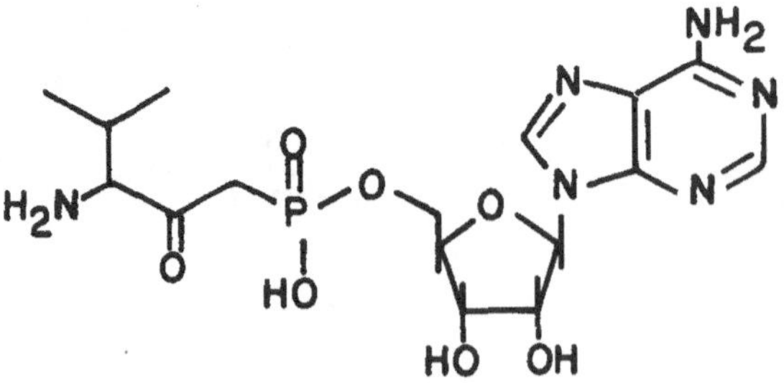

68

However, binding is not nearly as tight as with the natural material, presumably due to a loss of binding capability when the anhydride oxygen is replaced. Other related species[215] should prove equally useful in the investigation of processes of protein synthesis.

B. Nucleotidases

Several investigations of the effect of nucleotide analogues (phosphonates) on nucleotidase activity have been reported. The GTP analogue (66) serves as an inhibitor of RNase V isolated from *E. coli*,[217] and (Structure 69), the synthetic precursor to (67),[198,212,213] serves[218] as a noncompetitive inhibitor of 5′-nucleotidase from mouse liver plasma membranes, and exhibits "partially competitive" inhibition of 5′-nucleotidase from brain and smooth muscle of small intestine.[219,220] The role of the diphosphonic acid linkage here is not clear since (69) behaves much as does ADP.

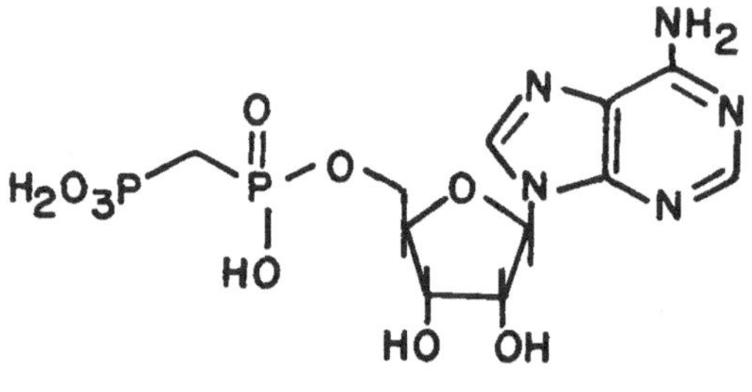

69

Holy et al.[221-223] have reported the synthesis of series of compounds of the general categories (Structures 70 to 74) in which the normal phosphate ester oxygen is still present, but an acidic hydroxyl on phosphorus has been replaced by a C–P function.

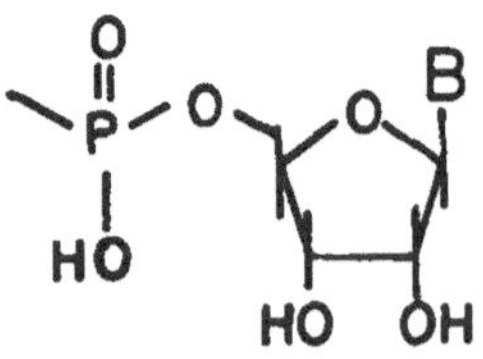

70

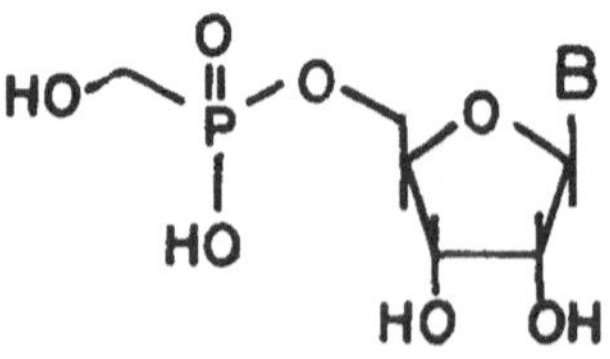

71

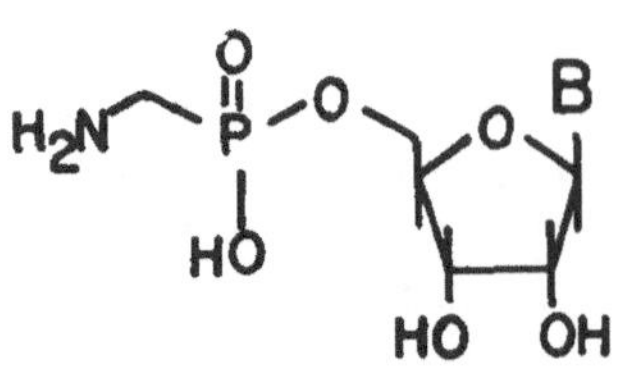

72

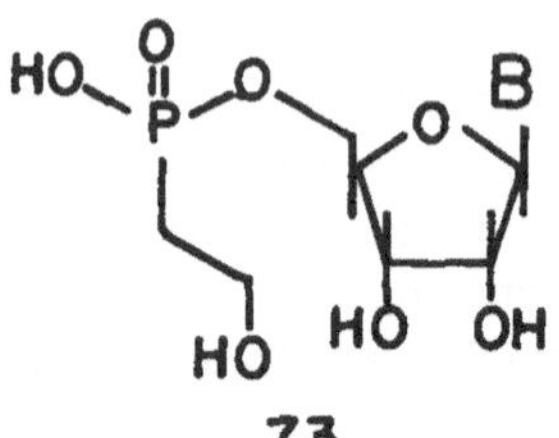

73

These materials appear to be resistant to cleavage by bacterial alkaline phosphatase,[221,224] but (71) and (72) are good substrates for snake venom 5′-nucleotidase.[221-224] In comparison, (70) and (73) are *not* good substrates for this enzyme system. A conclusion is made that hydrophilic character about the phosphorus site is necessary for activity.[221-224]

In a vaguely related vein, Jones et al.[225] reported the synthesis of dinucleoside phosphate analogues of the type (74) and found these not to be cleaved by spleen phosphodiesterase.

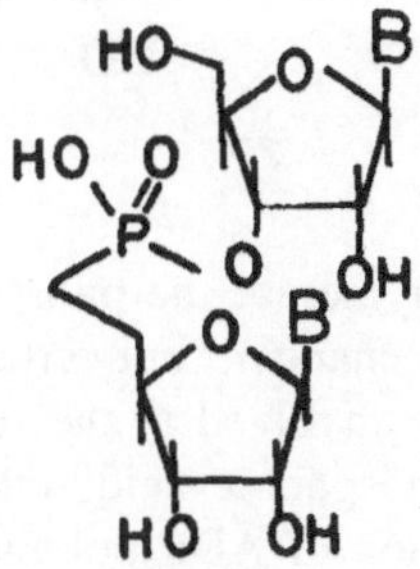

74

Physical chemical studies[226-228] on the material itself and on complexes with RNase indicate that (74) exhibits significant conformational differences from the natural materials, presumably due to a lack of favorable interactions with solvent at the site at which esteratic oxygen has been replaced by a methylene group. It might be of interest to perform these same experiments using species in which a hydroxymethylene function replaces the esteratic oxygen.

Work by Hampton et al.[229,230] has given data regarding the binding capabilities at critical sites. The two compounds (Structures 75 and 76) "isosteric" with AMP serve as inhibitors of snake venom 5′-nucleotidase. Some interesting stereochemically-dependent differentiations of epimers of these materials were also observed in related systems.

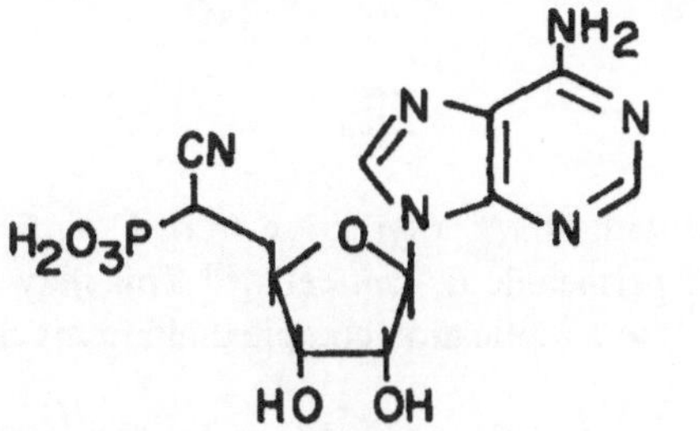

75

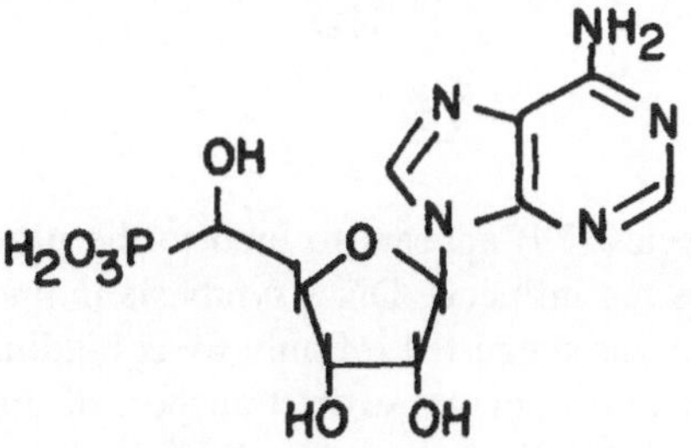

76

C. Polynucleotide Polymerases and Phosphorylases

A variety of systems have been investigated using phosphonic acid analogues of nucleotide species. The analogue of ATP, (Structure 77),[197,198] where the beta-gamma pyrophosphate oxygen is replaced by a methylene group[231] serves as a substitute for ATP with RNA polymerase and is a significantly better inhibitor than ATP of PPi incorporation into ADP by polynucleotide phosphorylase.[57]

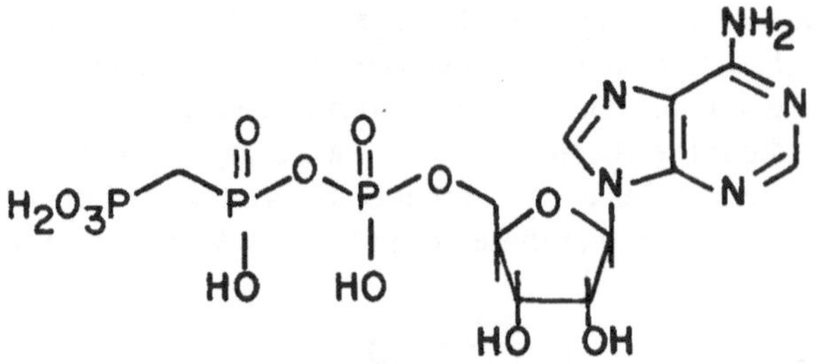

77

It is reasonable that with the RNA polymerase the phosphonic acid linkage simply affects the *rate* of reaction by an inductive mechanism, but certainly does not affect the *course* of the reaction as "normal" linkages are involved at the reaction sites. Moreover, with the analogue (69) of ADP, where the phosphonic acid linkage *is* in a critical position, no interaction with the enzyme or inhibition of ADP polymerization by polynucleotide polymerase is observed.[232] Nor was (69) capable of replacing ADP with polynucleotide phosphorylase.[233,234]

Englund et al.[235] synthesized a pair of compounds (Structures 78 and 79) related to dTDP and dTTP respectively, the former by coupling of (1) to protected deoxythymidine and the latter by a further coupling with phosphoric acid.

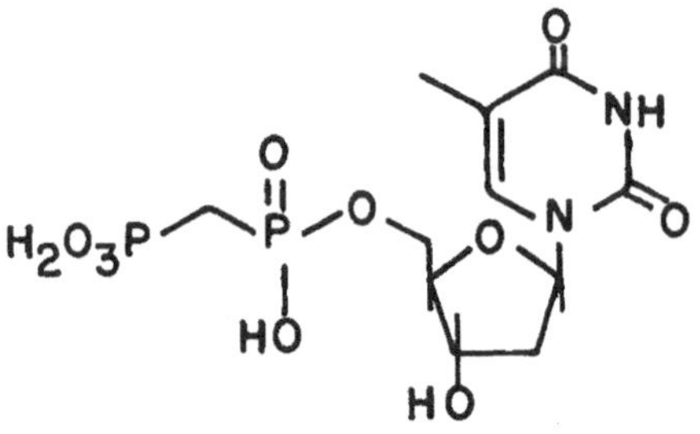

78

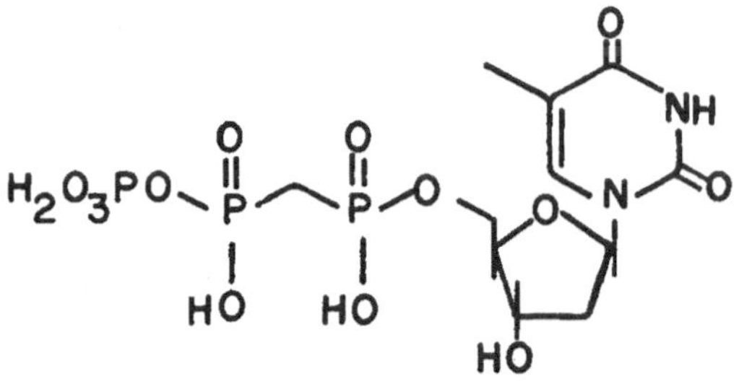

79

The species (79) appears to bind to the proper site with DNA polymerase from *E. coli*,[235] but does not influence DNA synthesis in nucleotide permeable *E. coli* cells.[236] This may be the result, as suggested, of only weak binding of (79) to a replication complex; alternatively, the substitution may prevent transport of the species.

In some relatively early work in this area Yengoyan and Rammler[237] prepared the nonisosteric analogues (Structure 80) of UMP, and (Structure 81) of UDP by coupling of (80) with phosphoric acid.

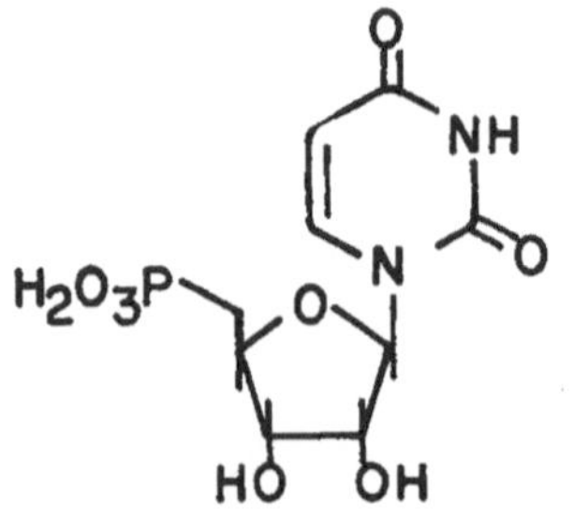

80

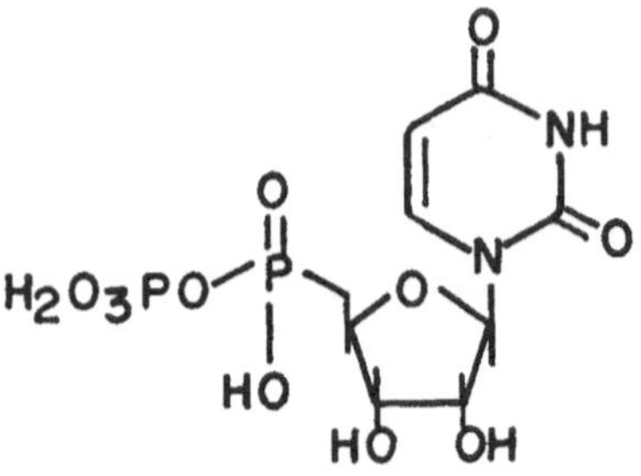

81

Upon investigation of the material (81) with polynucleotide phosphorylase from *Micrococcus lysodeikticus* it was found that it could not substitute for the natural substrate and was not polymerized. It would be interesting to test this system with the *isosteric* species (Structure 82). No report of the synthesis of (82) has yet appeared, but its preparation should proceed quite readily using the above noted coupling[237] on the isosteric UMP analogue (Structure 83) which was reported previously.[78,238-241]

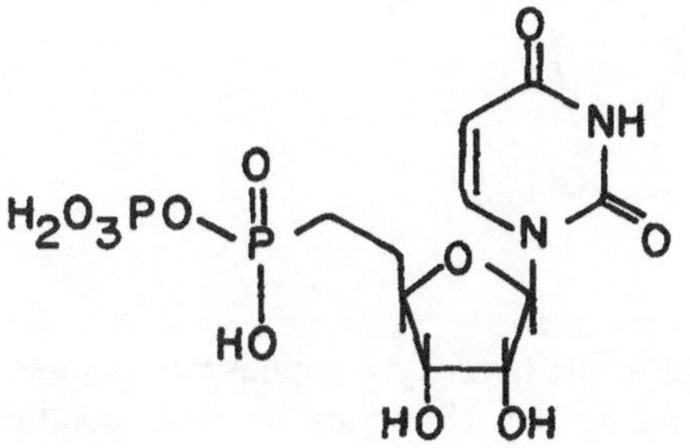

82

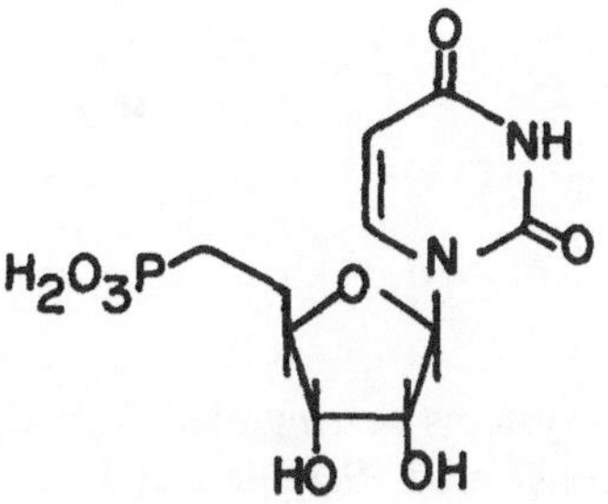

83

Finally, Huberman and Kornberg[242] studied the guanosine system (70)[221-223] wherein a methyl group replaces the acidic hydroxyl on phosphorus, and found it not to interact with DNA polymerase from *E. coli*. On the other hand, Lazarus et al.[243] have recently found the adenosine system (70)[221-223] to be capable of undergoing phosphoryl transfer with 5′-nucleotidase.

D. Kinases and Related Systems

The phosphonic acid analogues have been utilized as substitutes for natural materials (nucleotides) in a number of investigations on kinase mechanisms of phosphate transfer processes.

In a study by Berman and Cohn[244] of the role of ATP with phosphoenolpyruvate synthetase the ATP analogue (77) was utilized and found to be a reversible competitive inhibitor. However, no AMP was generated from (77) and it did not block the 14-C-pyruvate-phosphoenolpyruvate exchange reaction. This result has been rationalized by invoking differences in binding conformations of ATP and (77). Murray[245] has studied two ATP analogues (67) and (77) with adenosine kinase from Ehrlich-Ascites tumor cells and found neither to interact with the enzyme alone, but both caused significant inhibition in the presence of ATP.

The adenosine diphosphate (ADP) analogue (69) has been utilized in several studies. O'Sullivan et al.[246] found it to bind more strongly to creatine kinase than the natural material. Duee, et al.[247,248] studied (69) alone with (67) and (77) with several systems including adenylate kinase, pyruvate kinase, and hexokinase, and found none of them to interact. The materials served neither as substrates for the enzymes nor as inhibitors for the ADP or ATP reactions.

The analogues (75) and (76) are particularly interesting as they bear, on the carbon replacing the normal esteratic oxygen, a function capable of binding; that is, it has an unshared electron pair. Work with these materials illustrates a significant point of stereochemistry. Only one epimer of (75) serves as a substrate with pig muscle AMP kinase[230] and with muscle pyruvate kinase only one epimer of (76) could serve as a substrate.[229] Interestingly, both epimers of (76) could be phosphorylated by muscle adenylate kinase.[229] Again, with the species (Structure 84) related to ATP, only one epimer could serve as phosphoryl donor in the AMP kinase reaction; moreover, that epimer of (84) is *opposite in configuration* to that of (76) which is noted above as operating with muscle pyruvate kinase.[229]

Further structural analysis and coorelation of data will prove of value in better understanding of this enzyme process.

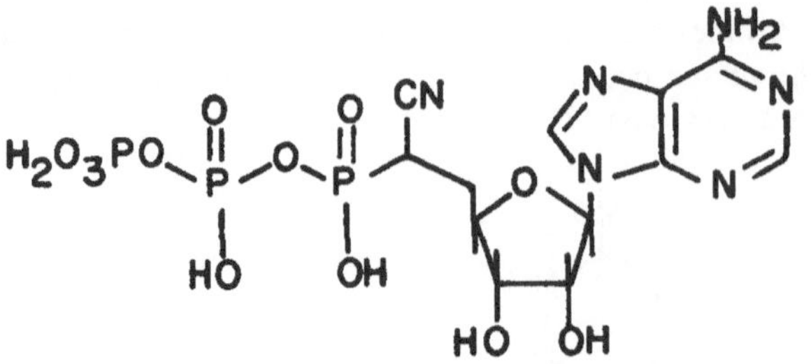

84

Several systems bearing methyl groups in place of acidic hydroxyls on phosphorus should also be considered. Wigler and Lozzio reported the system (Structure 85) incorporating variations from dTMP;[249] in addition to the methyl on phosphorus, a bromine has been introduced in replacement for the methyl on the pyrimidine ring.

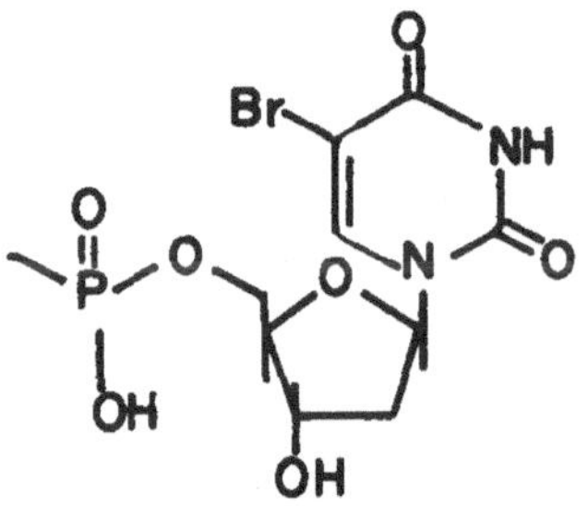

85

This material competes with dTMP as a substrate for the thymidylate kinase and in certain systems is transported and exhibits cytotoxicity. Lazarus et al.[243] found that the ATP analogue (Structure 86) is a competitive inhibitor of ATP with hexokinase and 3-phosphoglycerate kinase and of the utilization of ATP by phosphofructokinase.

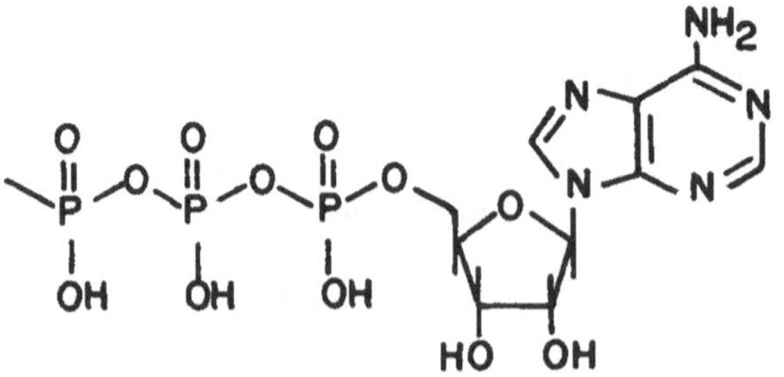

86

Finally, an interesting system (Structure 87) has been generated by Webster et al.[250] as an analogue of ADP. This incorporates both a phosphonic acid and an arsenic in place of phosphorus.

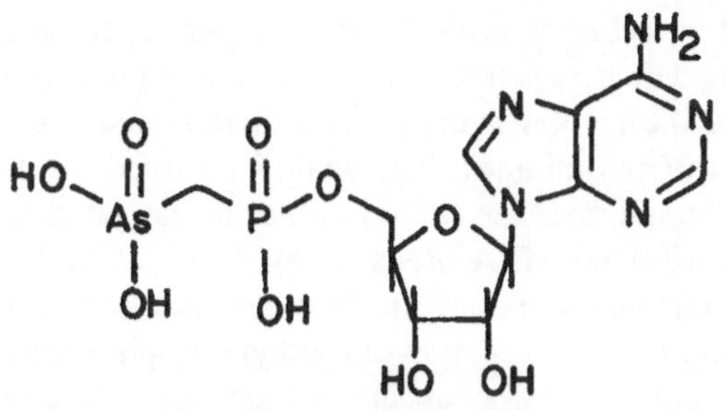

87

This material is only a poor substrate for 3-phosphoglycerate kinase and pyruvate kinase and has no substrate activity for adenylate kinase or creatine kinase, although (69) is a good substrate for these systems.[250-252]

E. Binding to Metal Ions and Macromolecules

With the introduction of a methylene group in place of an anhydride oxygen of a nucleotide polyphosphate, both geometrical and anhydride oxygen binding capabilities may be considered as being altered. Studies[254,255] using the ATP analogue (77) in binding to metal ions indicate the material to bind as well and in the same manner as ATP. Presumably such binding involves the phosphoryl oxygens and indicates binding of that type to be rather insensitive to the geometrical differences. With this as a "baseline", lack of binding of (77) comparable to ATP in other systems may be indicative of a requirement for interaction with the anhydride oxygen. Thus only phosphoryl oxygen type interaction is indicated in ATP binding to rat liver fructose-1,6-diphosphatase[256] and clostridial formyltetrahydrofolate synthetase.[257] However, with a variety of other macromolecular systems[197,258-262] it is incapable of binding; presumably ATP interaction involves reversible anhydride oxygen binding.

As nucleotides are involved in an extensive variety of enzyme processes, numerous phosphonic acid analogues have been used as possible antimetabolites in the study of their mechanisms of action. These are not as easily categorized by reaction type as those noted above and are not discussed here. The reader is referred to an earlier review[1] in which most of these were mentioned.

VII. PHOSPHONOACETATE AND PHOSPHONOFORMATE AS ANTIVIRAL AGENTS

In contrast to the previously discussed areas of phosphonic acid use and potential as antimetabolites, where very often knowledge of the metabolic processes involved led to the design and use of the phosphonic acid species, interest in phosphonoacetic acid (Structure 88) and phosphonoformic acid (Structure 89) resulted from routine drug-screening.

$H_2O_3PCH_2COOH$

88

H_2O_3PCOOH

89

Knowledge of the presence of antimetabolic activity was then followed with detailed investigations. I will not attempt here to give an encyclopedic review of the area of study, but rather to give an overview for further reading.

Phosphonoacetic acid (88) was generated[263] as an intermediate in a synthesis of an isosteric analogue of carbamoyl phosphate. The route used involved an Arbuzov reaction on methylbromoacetate followed by acidic hydrolysis. It is interesting to note that the analogue of carbamoyl phosphate was being prepared as a potential antitumor agent. The analogue was intended to substitute for the natural material (and thereby cause inhibition) in its role in the *de novo* biosynthesis of pyrimidines. The analogue turned out to be inactive and the biological properties of the intermediate (88) were not investigated until later.

Phosphonoformic acid (89) was first prepared by Nylen[264] by basic hydrolysis of the corresponding triethyl ester; acidic hydrolysis leads to decarboxylation. The triethyl ester had previously been generated[265] via reaction of triethyl phosphite with ethyl chloroformate.

As noted, (88) was found to serve as an in vivo antiviral agent by routine drug screen and was reported[266] to be an effective drug in the treatment of herpes simplex virus infections in mice and rabbits. Since the initial report, numerous studies have been performed to determine the activity of (88) against a variety of viruses. These studies have been selectively reviewed,[267-269] and only a portion of them will be noted here.

Studies of drug effectiveness against a variety of herpes viral infections have been reported,[270-278] variability of results being observed. The mechanism of action against herpes viruses has been the subject of a number of investigations. It is concluded that inhibition of replication occurs as a result of a specific inhibition by (88) of virus induced DNA polymerase, although normal cellular polymerases were relatively insensitive to its activity.[279-293] In addition, the genetic nature of resistance to (88) has been studied[294] along with the synergistic effect[295] of (88) in combination with other antiviral agents. Detailed comparisons of (88) with other antiviral agents against herpes viruses have been made.[277,296-298] Other agents related to (88) have been investigated for activity against herpes. There have been synthesized[299] a series of compounds wherein the P–C linkage has been maintained but with the phosphorus acid partially esterified or present as a phosphinic acid with several aromatic hydrocarbon groups. Activity was noted only for the partially esterified compounds and there it is probably the result of hydrolysis to (88). In another study[288] it was noted that phosphonopropionic acid (Structure 90) was without effect against herpes.

$$H_2O_3PCH_2CH_2COOH$$

90

The material (88) has also been observed to be of value in the inhibition of proliferation of Epstein-Barr virus.[300,301] Here the material has also been noted to interact with viral DNA polymerase.[302-312] Similarly, (88) has been investigated with a number of viruses and found to have variable activity against those bearing double-stranded DNA in the virion.[313-320]

As regards the species (89), phosphonoformic acid has been found to be effective in treatment of surface herpes infections[321] and Epstein-Barr viral infections.[306,322-324] For the latter system the mode of activity has been found to be similar to that for (88) and capsid antigen synthesis inhibition is found at subtoxic concentrations. This activity is found for both (89) and its triethyl ester, although the triethyl ester is less effective as an inhibitor.[322] It is entirely possible that the activity of the triethyl ester is the result of hydrolysis to (89).

VIII. OTHER PHOSPHONIC ACIDS AS ANTIMETABOLITES

In this "miscellaneous" category several classes of compounds will be considered. The first of these systems is pyridoxal phosphate (Structure 91), analogues of which are of interest for antimetabolic activity toward transaminase reactions for which (91) is a coenzyme.

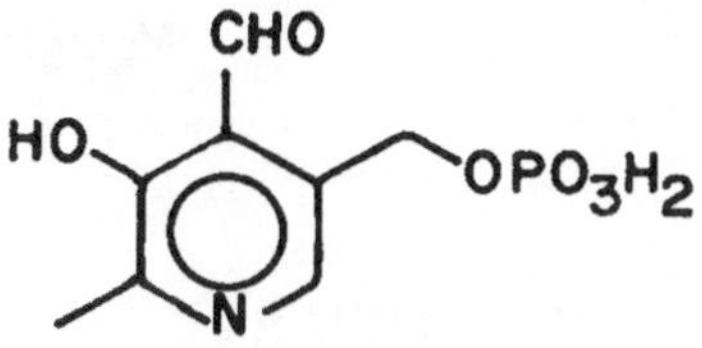

91

Several analogue species have been prepared including species (Structures 92 to 96).[325-330]

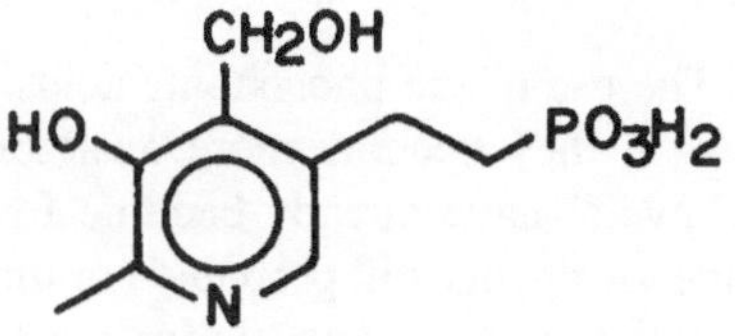

92

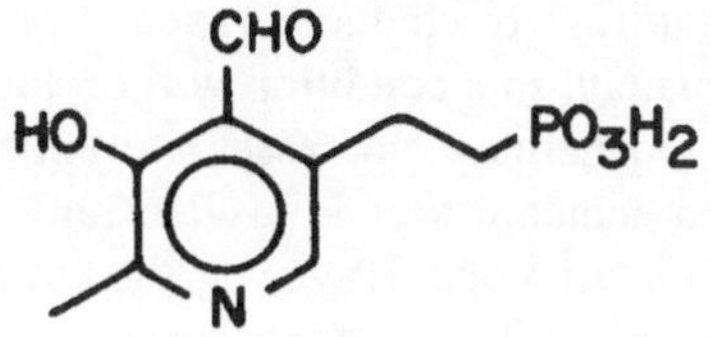

93

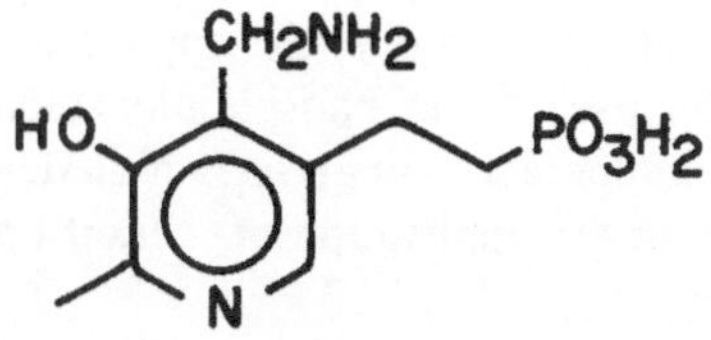

94

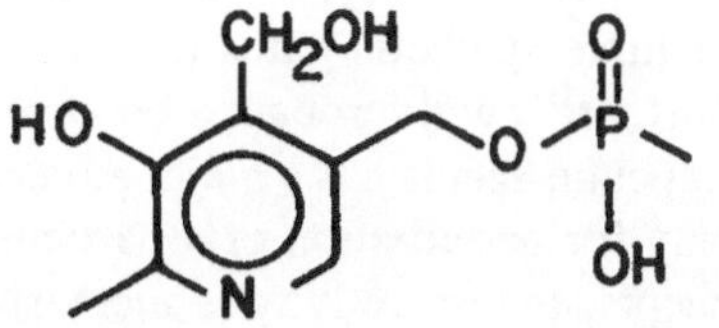

95

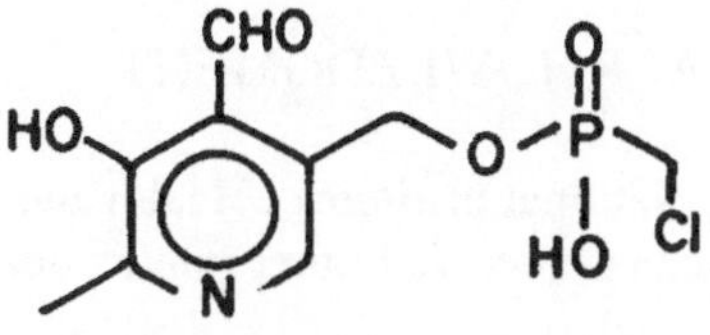

96

The species (92) has been found to have only minimal interaction with tyrosine decarboxylase,[325] but to be very similar to the natural system in its interaction with aspartate aminotransferase.[330] It has also been found that (93) and (94) are active for the reconstitution of rabbit muscle glycogen phosphorylase.[326-327] Species (95) and (96) have been inactive in all systems tested to date.[328]

There have also been recent efforts to effect an inhibition of human red cell acid phosphatase using phosphonic acids. As it is known that the enzyme is particularly effective for aromatic phosphate esters, a series of arylphosphonic and benzylphosphonic acids were prepared for testing of inhibitory capacity.[331] Several have been found to be reasonably potent inhibitors.

There has also been interest of late in the use of phosphonic acids as analogues of carboxylic acids. Preparations have been reported of species of the class (Structure 97)[332] related to alpha-amino acids and (Structure 98)[333] related to beta-amino acids. To date no biological data have been presented for them.

$H_2O_3PCHRNH_2$ **97** $H_2O_3PCHRCH_2NH_2$ **98**

IX. SUMMARY

In this review, a variety of biochemical systems in which phosphonic acids have been used as substitutes for natural phosphates have been considered in some detail. In most instances, the utility of these analogues of the natural phosphates has been for the elucidation of biochemical mechanisms. Only in a few instances have these phosphonic acids found significant utility for chemotherapeutic purposes.

Unfortunately, this condition will probably continue. The use of the phosphonic acids as probes of biochemical processes are expected to expand as they become more available, becoming a standard tool in a wide range of areas, as several have already become for a relatively limited scope. However, the use as chemotherapeutic agents will probably continue to lag. A major difficulty contributing to this is the complexity in the synthesis for most of these compounds. With certain exceptions, particularly in the categories of glycerol related materials and lipids, use outside the laboratory will be quite limited.

For those which do have a potential for commercial chemotherapeutic use, those with relatively simple synthetic pathways and a broad range of applicability, the application of "illicit transport" will probably become a factor. The use of the phosphonic acids as chemotherapeutic agents has a major advantage over other approaches in giving a high degree of selectivity for perturbation of a particular process. With the application of "illicit transport" concepts, this selectivity is increased even further.

Within the next few years, the mushrooming of applications of the phosphonic acids is anticipated and the continued synthesis of analogues for new systems with further carefully devised structural variations.

ACKNOWLEDGMENT

The author wishes to thank the National Institute of Health and DOW Chemical for support of work noted here which has not yet received other publication.

REFERENCES

1. **Engel, R.,** Phosponates as analogues of natural phosphates, *Chem. Rev.*, 77, 349, 1977.
2. **Rosenthal, A. F. and Geyer, R. P.,** Compounds related to alpha-glycerophosphoric acid, phosphorylcholine, and phosphorylethanolamine, *J. AM. Chem. Soc.*, 80, 5240, 1958.
3. **Baer, E. and Basu, H.,** Phosphonic acid analogues of carbohydrate metabolites. I. Synthesis of L- and D-dihydroxypropylphosphonic acid, *Can. J. Biochem.*, 47, 955, 1969.
4. **Baer, E., Nazir, D. J., and Basu, H.,** Phosphonic acid analogs of carbohydrate metabolites. Dihydroxypropylphosphonic acid. II. Substrate specificity of L-glycerol-3-phosphate:NAD oxidoreductase, *Can J. Biochem.*, 47, 992, 1969.
5. **Kabak, J., DeFilippe, L., Engel, R., and Tropp, B.,** Synthesis of the phosphonic acid isostere of glycerol-3-phosphate, *J. Med. Chem.*, 15, 1074, 1972.
6. **Adams, P. R., Harrison, R., and Inch, T. D.,** Dehydrogenation of a phosphonate substrate analog by glycerol-3-phosphate dehydrogenase, *Biochem. J.*, 141, 729, 1974.
7. **Tang, K.-C., Tropp, B. E., and Engel, R.,** The synthesis of phosphonic acid and phosphate analogues of glycerol-3-phosphate and related metabolites, *Tetrahedron,* 34, 2873, 1978.

8. **Goldstein, S. L., Braksmayer, D., Tropp, B. E., and Engel, R.,** Isosteres of natural phosphates. 2. Synthesis of the monosodium salt of 4-hydroxy-3-oxobutyl-1-phosphonic acid, an isostere of dihydroxyacetone phosphate, *J. Med. Chem.*, 17, 363, 1974.
9. **Lalinde, N., Tropp, B. E., and Engel, R.,** Unpublished data, 1981.
10. **Cheng, P.-J., Hickey, R., Engel, R., and Tropp, B. E.,** Rabbit muscle L-glycerol-3-phosphate dehydrogenase. Substrate activity of 3,4-dihydroxybutyl-1-phosphonate and 4-hydroxy-3-oxobutyl-1-phosphonate, *Biochim. Biophys. Acta*, 341, 85, 1974.
11. **Bax, P. C., Morris, F., and Rammler, D. H.,** Nucleoside phosphonic acids. III. A nucleoside phosphonic acid analog of UDP-glucose, *Biochim. Biophys. Acta*, 201, 416, 1970.
12. **LeMarechal, P., Froussios, C., Leval, M., and Azerad, R.,** The interaction of phosphate and homophosphonate analogues of 3-deoxy-D-arabinoheptulosante-7-phosphate with 3-dehydroquinate synthetase from *Escherichia coli, Biochem. Biophys. Res. Commun.*, 92, 1104, 1980.
13. **Kraut, J.,** Crystal structure of 2-aminoethanol phosphate, *Acta Crystallogr.*, 14, 146, 1961.
14. **Ferrier, W. G., Lindsay, A. R., and Young, D. W.,** Confirmation of the crystal structure of 2-aminoethanol phosphate, *Acta Crystallogr.*, 15, 616, 1962.
15. **Okaya, Y.,** Crystal structure of the stable modification of 2-aminoethylphosphonic acid, beta-ciliatine, *Acta Crystallogr.*, 20, 712, 1966.
16. **Abrahamsson, S. and Pascher, I.,** Crystal and molecular structure of L-alpha-glycerylphosphorylcholine, *Acta Crystallogr.*, 21, 79, 1966.
17. **Viswamitra, M. A., Reddy, B. S., Lin, G. H.-Y., and Sundaralingam, M.,** Stereochemistry of nucleic acids and their constituents. XVII. Crystal and molecular structure of deoxycytidine 5′-phosphate monohydrate. Possible puckering for the furanoside ring in B-deoxyribonucleic acid, *J. Am. Chem. Soc.*, 93, 4565, 1971.
18. **Larsen, M., Willett, R., and Yount, R. G.,** Imidophosphate and pyrophosphate: possible biological significance of similar structures, *Sciences*, 166, 1510, 1969.
19. **Crofts, P. C. and Kosolapoff, G. M.,** Preparation and determination of apparent dissociation constants of alkylphosphonic and dialkylphosphinic acids, *J. Am. Chem. Soc.*, 75, 3379, 1953.
20. **Rammler, D. H., Yengoyan, L., Paul, A. V., and Bax, P. C.,** Nucleoside phosphonic acids. II. Synthesis of 5′-deoxythymidine-5′-phosphonic acid and its pyrophosphate derivatives, *Biochemistry*, 6, 1828, 1967.
21. **VanWazer, J. R.,** *Phosphorus and Its Compounds*, Vol. 1, Wiley-Interscience, New York, 1958.
22. **Arbuzov, A. E. and Kushkova, N. P.,** Action of dihalohydrocarbons on ethyl phosphite and salts of diethyl phosphite, *Zh. Obshch. Khim.*, 6, 283, 1936.
23. **Ford-Moore, A. H. and Williams, J. H.,** Reaction between trialkyl phosphites and alkyl halides, *J. Chem. Soc.*, 1465, 1947.
24. **Schwarzenbach, G. and Zurc, J.,** Pyro- and hypophosphorous acids in comparison with organic diphosphonic acids, *Monatsh. Chem.*, 81, 202, 1950.
25. **Kosolapoff, G. M.,** The chemistry of aliphatic phosphonic acids. I. Alkylation of methanediphosphonic acid, *J. Am. Chem. Soc.*, 75, 1500, 1953.
26. **Kosolapoff, G. M.,** Some physical properties of aliphatic diphosphonates, Part I. Ethyl esters, *J. Chem. Soc.*, 3092, 1955.
27. **Cade, J. A.,** Methylenediphosphonates and related compounds, *J. Chem. Soc.*, 2266, 1959.
28. **Moedritzer, K. and Irani, R. R.,** Synthesis and properties of some mono- and poly(methylenediphosphonic acid) and esters, *J. Inorg. Nucl. Chem.*, 22, 297, 1961.
29. **Roy, C. H.,** Tetra-secondary alkyl methylenediphosphonates, U.S. Patent 3,251,907 (17 May 1966); *Chem. Abstr.*, 65, 3908, 1966.
30. **Hays, H. R. and Kosolapoff, G. M.,** gem-Bisphosphinoalkanes, U.S. Patent 3,445,522 (20 May 1969); *Chem. Abstr.*, 71, 81520k, 1969.
31. **Gupta, A., Sacks, K., Khan, S., Tropp, B. E., and Engel, R.,** An improved synthesis of vinylic phosphonates from ketones, *Syn. Commun.*, 10, 299, 1980.
32. **Fleisch, H., Russell, R. G. G., Bisaz, S., Casey, P. A., and Muhlbauer, R. C.,** The influence of pyrophosphate analogues (diphosphonates) on the precipitation and dissolution of calcium phosphate *in vitro* and *in vivo, Calcif. Tissue Res.*, 2 (Suppl. 10), 1968.
33. **Fleisch, H., Russell, R. G. G., and Francis, M. D.,** Diphosphonates inhibit hydroxyapatite dissolution *in vitro* and bone resorption in tissue culture and *in vivo, Science*, 165, 1262, 1969.
34. **Francis, M. D., Russell, R. G. G., and Fleisch, H.,** Diphosphonates inhibit formation of calcium phosphate crystals *in vitro* and pathological calcification *in vivo, Science*, 165, 1264, 1969.
35. **Francis, M. D.,** Inhibition of calcium hydroxyapatite crystal growth by polyphosphonates and polyphosphates, *Calcif. Tissue Res.*, 3, 151, 1969.
36. **Francis, M. D.,** Polyphosphonates to prevent abnormal mobilization and deposition of calcium phosphate in living tissues, German Offen. 1,813,659 (3 July 1969); *Chem. Abstr.*, 71, 89977k, 1969.

37. **Liggett, S. J. and Libby, R. A.**, Spectrotitration of ethane-1-hydroxy-1,1-diphosphonic acid with thorium diaminocyclohexanetetraacetate, *Talanta*, 17, 1135, 1970.
38. **Jethi, R. K., Inlow, C. W., and Wadkins, C. L.**, Mechanism of biological calcification. I. Kinetic properties of the *in vitro* calcification of collagen-containing matrix, *Calcif. Tissue Res.*, 6, 81, 1970.
39. **Russell, R. G. G., Muehlbauer, R. C., Bisaz, S., and Williams, D. A.**, Influence of pyrophosphate, condensed phosphates, phosphonates, and other phosphate compounds on the dissolution of hydroxyapatite *in vitro* and on bone resorption induced by parathyroid hormone in tissue culture and on thyroparathyroidectomized rats, *Calcif. Tissue Res.*, 6, 183, 1970.
40. **Fleisch, H., Russell, R. G. G., Bisaz, S., Muehlbauer, R. C., and Williams, D. A.**, The inhibitory effect of phosphonates on the formation of calcium phosphonate crystals *in vitro* and kidney calcification *in vivo*, *Eur. J. Clin. Invest.*, 1, 12, 1970.
41. **Jowsey, J., Holly, K. E., and Linman, J. W.**, Effect of etidronate in adult cats, *J. Lab. Clin. Med.*, 76, 126, 1970.
42. **Cooperman, B. S. and Mark, D. H.**, Aggregation in dilute solution of Mn(II) and inorganic pyrophosphate and its biochemical consequences, *Biochim. Biophys. Acta*, 252, 221, 1971.
43. **Jethi, R. K. and Wadkins, C. L.**, Mechanism of biological calcification. II. Evidence for a multistep mechanism of calcification by tendon matrix, *Calcif. Tissue Res.*, 7, 277, 1971.
44. **King, W. R., Francis, M. D., and Michael, W. R.**, Effect of disodium ethane-1-hydroxy-1,1-diphosphonate (disodium dihydrogen 1-hydroxyethylidenediphosphonate) on bone, *Clin. Orthop.*, 78, 251, 1971.
45. **Morgan, D. B., Bonjour, J.-P., Gasser, A. B., O'Brien, K., and Fleisch, H.**, Influence of a diphosphonate on the intestinal absorption of calcium, *Isr. Med. Sci.*, 7, 384, 1971.
46. **Gasser, A. B., Morgan, D. B., Fleisch, H. A., and Richell, L. J.**, Influence of two diphosphonates on calcium metabolism in the rat, *Clin. Sci.*, 43, 31, 1972.
47. **Hill, L. F., Mawer, E. B., Lumb, G. A., and Stanbury, S. W.**, Indirect inhibition of the biosynthesis of 1,25-dihydroxycholecalciferol in rats treated with a diphosphonate, *Clin. Sci.*, 42, 16P, 1972.
48. **Liggett, S. J.**, Determination of ethane-1-hydroxy-1,1-diphosphonic acid (EHDP) in human feces and urine, *Biochem. Med.*, 7, 68, 1973.
49. **Cooperman, B. S. and Chiu, N. Y.**, Yeast inorganic pyrophosphatase. II. Magnetic resonance and steady-state kinetic studies of metal ion and pyrophosphate analog binding, *Biochemistry*, 12, 1670, 1973.
50. **Schenk, R., Mera, R., Muehlbauer, W. A., Russell, R. G. G., and Fleisch, H.**, Effect of ethane-1-hydroxy-1,1-diphosphonate (EHDP) and dichloromethylene diphosphonate on the calcification and resorption of cartilege and bone in the tibial epiphysis and metaphysis of rats, *Calcif. Tissue Res.*, 11, 196, 1973.
51. **Russell, R. G. G., Kislig, A.-M., Casey, P. A., Fleisch, H., Thornton, J., Schenk, R., and Williams, D. A.**, Effect of diphosphonates and calcitonin on the chemistry and quantitative histology of rat bone, *Calcif. Tissue Res.*, 11, 179, 1973.
52. **Bonjour, J.-P., DeLuca, H. F., Fleisch, H. A., Trechel, U., Matjowec, L. A., and Omdahl, J. L.**, Reversal of the EHDP inhibition of calcium absorption by 1,25-dihydroxycholecalciferol, *Eur. J. Clin. Invest.*, 3, 44, 1973.
53. **Francis, M. D. and Centner, R. L.**, The development of diphosphonates as significant health care products, *J. Chem. Ed.*, 55, 760, 1978.
54. **Mundy, G. R. and Raisz, L. G.**, Drugs for disorders of bone. Pharmacological and clinical considerations, *Drugs*, 8, 250, 1974.
55. **Zamecnik, P. C. and Stephenson, M. L.**, Possible regulatory site located at the gateway to protein synthesis, *Regul. Mech. Protein Sym. Mamm., Kettering Symp., 3rd*, Pietro, A. S., Lamborg, M. R., and Kenney, F. T., Eds., Academic Press, New York, 1968, 3.
56. **Zamecnik, P. C. and Stephenson, M. L.**, Nucleoside pyrophosphate compounds related to the first step in protein synthesis, in *The Role of Nucleotides for the Function and Conformation of Enzymes*, Kalckar, H. M., Kenow, H., Munch-Peterson, A., Otteson, M., and Tayson, Y. H., Eds., Academic Press, New York, 1969, 276.
57. **Simon, L. N., and Myers, T. C.**, Effect of some phosphonic acid analogues of Adenosine-5′-diphosphate on polynucleotide phosphorylase, *Biochim. Biophys. Acta*, 51, 178, 1961.
58. **Kelly, S. J, Feldman, F., Sperow, J. W., and Butler, L. G.**, Kinetic effects of inorganic pyrophosphate analogs on several inorganic pyrophosphate, *Biochemistry*, 12, 3338, 1973.
59. **Kelly, S. J., Sperow, J. W., and Butler, L. G.**, Stimulation of alkaline phosphatase by analogs of inorganic pyrophosphate, *Biochemistry*, 13, 3503, 1974.
60. **Kjellstrom, W. A. and Bishop, S. H.**, Methylene diphosphonic acid does not inhibit inorganic pyrophosphatase, *Life Sci.*, 9, 943, 1970.
61. **Sperow, J. W., Moe, O. A., Ridlington, J. W., and Butler, L. G.**, Yeast inorganic pyrophosphatase. VI. Specificity and mechanism, *J. Biol. Chem.*, 248, 2062, 1973.
62. **Cooperman, B. S. and Chiu, N. Y.**, Yeast inorganic pyrophosphatase. III. Active-site mapping by electrophilic reagents and binding measurements, *Biochemistry*, 12, 1676, 1973.

63. **Klement, R. and Biberacher, G.,** Preparation of imidodiophosphate and nitrodiotriphosphate, *Z. Anorg. Allg. Chem.*, 283, 246, 1956.
64. **Quimby, O. T., Prentice, J. B., and Nicholson, D. A.,** Tetrasodium carbonyldiphosphonate. Synthesis, reactions and spectral properties, *J. Org. Chem.*, 32, 4111, 1967.
65. **Griffin, B. S. and Burger, A.,** D-glucopyranose-6-deoxy-6-phosphonic acid, *J. Am. Chem. Soc.*, 78, 2336, 1956.
66. **Jung, A. K. and Engel, R.,** Catalytic hydrogenolysis-reduction of aryl phosphate esters, *J. Org. Chem.*, 40, 244, 1975.
67. **Harvey, R. G. and DeSombre, E. R.,** The Michelis-Arbuzov and related reactions, in *Topics in Phosphorus Chemistry*, Vol. 1, Grayson, M. and Griffith, E. J., Eds., Wiley-Intescience, New York, 1964, 57.
68. **Rabinowitz, R.,** Reactions of phosphonic acid esters with acid chlorides — mild hydrolytic route, *J. Org. Chem.*, 28, 2975, 1963.
69. **Blackburn, G. M. and Ingleson, D.,** Specific dealkylation of phosphonate esters using iodotrimethylsilane, *Chem. Commun.*, 870, 1978.
70. **Morita, T., Okamoto, Y., and Sakurai, H.,** The preparation of phosphonic acids having labile functional groups, *Bull. Chem. Soc. Jpn.*, 51, 2169, 1978.
71. **Blackburn, G. M. and Ingleson, D.,** The dealkylation of phosphate and phosphonate esters by iodotrimethylsilane: a mild and selective procedure, *J. Chem. Soc. Perkin I*,, 1150, 1980.
72. **Moffatt, J. G. and Jones, G. H.,** Didealkylation of phosphate esters, U.S. Patent 3,524,846 (18 August 1970); *Chem. Abstr.*, 74, 31940p, 1971.
73. **Orlov, N. F., Kaufman, B. L., Sukhi, L., Slesar, L. N., and Sudakova, E. V.,** Synthesis of triorganosilyl derivatives of phosphorous acid, *Khim. Prakt. Primen. Kremniiorg. Soldin., Tr. Sovesch.*, 111, 1966; *Chem. Abstr.*, 72, 21738y, 1970.
74. **Belokrinitskii, M. A. and Orlov, N. F.,** Reaction of alkyl haides with triorganosilyl derivatives of phosphorous acid, *Kremniiorg. Mater.*, 145, 1971; *Chem. Abstr.*, 78, 29929f, 1973.
75. **Rosenthal, A. F., Vargas, L. A., Isaacson, Y. A., and Bittman, R.,** A simple synthesis of phosphonate-containing lipids, introduction of the phosphonic acid moiety into hydrolytically labile compounds, *Tetrahedron Lett.*, 977, 1975.
76. **Deroo, P. W., Rosenthal, A. F., Isaacson, Y. A., Vargas, L. A., and Bittman, R.,** Synthesis of DL-2,3-diacyloxypropylphosphonylcholines from DL-2,3-diacyloxyiodopropanes, *Chem. Phys. Lipids*, 16, 60, 1976.
77. **LeMarchal, P., Froussios, C., Leval, M., and Azerad, R.,** Enzymic properties of phosphonic analogues of D-erythrose-4-phosphate, *Biochem. Biophys. Res. Commun.*, 92, 1097, 1980.
78. **Jones, G. H. and Moffatt, J. G.,** Hexose and heptose phosphonates, phosphonic acids, phosphonic acid salts, and their derivatives, U.S. Patent 2,583,974 (8 June 1971); *Chem. Abstr.*, 75, 130091q, 1971.
79. **Adams, P. R., Harrison, R., Inch, T. D., and Rich, P.,** Dehydrogenation of the phosphonate analog of glucose-6-phosphate by glucose-6-phosphate dehydrogenase, *Biochem. J.*, 155, 1, 1976.
80. **Jones, G. H., Hamamura, E. K., and Moffatt, J. G.,** A new stable Wittig reagent for the synthesis of alpha,beta-unsaturated phosphonates, *Tetrahedron Lett.*, 5731, 1968.
81. Syntex Corp., Phosphorylated phosphonium ylides, British Patent 1,234,213 (18 August 1971); *Chem. Abstr.*, 75, 130083p, 1971.
82. **Gotlinsky, B., Mildener, B., Tropp, B. E., and Engel, R.,** Unpublished data, 1981.
83. **Paulsen, H., Bartsch, W., and Thieme, J.,** Phosphorus-containing carbohydrates. III. Preparation of sugar phosphonates by Wittig and Horner reaction, *Chem. Ber.*, 104, 2545, 1971.
84. **Horner, L., Hoffmann, H., and Wippel, H. G.,** Phosphine oxides as olefin forming reagents, *Chem. Ber.*, 91, 61, 1958.
85. **BeMiller, J. N., Chaklos, M. S., and Yee, Y. S.,** Synthesis of a methylene analog of alpha-D-glucopyranose, Paper #3, Division of Carbohydrate Chemistry, 2nd Chem. Congr. N. Am. Cont., August 1980, Las Vegas.
86. **BeMiller, J. N.,** private communication, 1981.
87. **Paulsen, H. and Bartsch, W.,** Reaction of carbonylsugars with lithio methanephosphonic acid esters. Preparation of dimethyl 1-deoxy-D-fructose-1-phosphonic acid, *Chem. Ber.*, 108, 1229, 1975.
88. **Tang, J.-C., Tropp, B. E., and Engel, R.,** Isosteres of natural phosphates. 8. The preparation and biological properties of 1-deoxy-1-dihydroxyphosphinylmethylfructose, an analogue of fructose-1-phosphate, *Tetrahedron Lett.*, 723, 1978.
89. **Stribling, D.,** Properties of the phosphonomethyl isosteres of two phosphate ester glycolytic intermediates, *Biochem. J.*, 141, 725, 1979.
90. **Hennon, G. F. and Kupiecki, F. P.,** Some reactions of 2-butyne-1,4-diol, *J. Org. Chem.*, 18, 1601, 1953.
91. **Dixon, H. B. F. and Sparkes, M. J.,** Phosphonomethyl analogues of phosphate ester glycolytic intermediates, *Biochem J.*, 141, 715, 1974.

92. **Sheikh, M., Tropp, B. E., and Engel, R.**, Unpublished data, 1981.
93. **Gilvarg, C.**, Peptide transport in bacteria, in *Peptide Transport in Bacteria and Mammalian Gut*, Ciba Foundation Symposium, Excerpta Medica, North Holland Associate Scientific, Amsterdam, 1972, 11.
94. **Payne, J. W.**, Mechanisms of bacterial peptide transport, in *Peptide Transport in Bacteria and Mammalian Gut*, Ciba Foundation Symposium, Excepta Media, North Holland Associate Scientific, Amsterdam, 1972, 17.
95. **Matthews, D. W.**, Rates of peptide uptake by small intestine, in *Peptide Transport in Bacteria and Mammalian Gut*, Ciba Foundation Symposium, Excerpta Medica, North Holland Associate Scientific, Amsterdam, 1972, 71.
96. **Payne, J. W. and Gilvarg, C.**, Peptide transport, *Adv. Enzymol.*, 35, 187, 1971.
97. **Payne, J. W.**, Peptides and micro-organisms, *Adv. Microb. Physiol.*, 13, 55, 1976.
98. **Barak, Z. and Gilvarg, C.**, Triornithine-resistant strains of *Escherichia coli*, *J. Biol. Chem.*, 249, 143, 1974.
99. **Naider, F. and Becker, J.**, Multiplicity of transport systems in *Escherichia coli*, *J. Bacteriol.*, 122, 1208, 1975.
100. **Barak, A. and Gilvarg, C.**, Specialized peptide transport system in *Escherichia coli*, *J. Bacteriol.*, 122, 1200, 1975.
101. **Naider, F., Becker, J. M., and Kazir-Katchalski, E.**, Utilization of methionine-containing peptides and their derivatives by a methionine-requiring auxotroph of *Saccharomyces cervisiae*, *J. Biol. Chem.*, 249, 9, 1974.
102. **Wolfinbarger, L. and Marzluf, G.**, Specificity and regulation of peptide transport in *Neurospora crassa*, *Arch. Biochem. Biophys.*, 171, 637, 1975.
103. **Baer, E. and Robinson, R.**, Phosphonic acids analogs of carbohydrate metabolites. IV. Synthesis of DL-glyceraldehyde-3-phosphonic acid, *Can. J. Chem.*, 51, 104, 1973.
104. **Goldstein, S. L., Pulcrano, M., Tropp, B. E., and Engel, R.**, Isosteres of natural phosphates. 3. Synthesis of the dilithium salt of 4,4-diethoxy-3-hydroxybutyl-1-phosphonic acid, an isostere of glyceraldehyde-3-phosphate, *J. Med. Chem.*, 17, 1115, 1974.
105. **Connolly, M., Tang, C.-T., Engel, R., and Tropp, B. E.**, Unpublished data, 1981.
106. **Tang, C.-T., Engel, R., and Tropp, B. E.**, Glycerol-3-phosphate analogues as metabolic inhibitors in *Escherichia coli:* 3-hydroxy-4-oxobutyl-1-phosphonate, a durg that interferes with normal phosphoglyceride metabolism, *Biochim. Biophys. Acta*, 572, 472, 1979.
107. **Baer, E. and Robinson, R.**, Synthesis of 2,3-diphosphono-DL-glyceric acid, the cofactor of both phosphoglyceric acid mutase and hemoglobin, *Can. J. Biochem.*, 49, 300, 1971.
108. **Pfeiffer, F. R., Mier, J. D., and Weisbach, J. A.**, Synthesis of phosphonic acid isosteres of 2-phospho-, and 2,3-diphosphoglyceric acid, *J. Med. Chem.*, 17, 112, 1974.
109. **Benesch, R. E., Benesch, R., Yung, S., Baer, E., and Robinson, R.**, A comparison of diphosphonoglyceric acid and diphosphoglyceric acid as allosteric effectors for hemoglobin, *Can. J. Biochem.*, 51, 1120, 1973.
110. **Orr, G. A. and Knowles, J. R.**, The interaction of the phosphonate analogue of 3-phospho-D-glycerate with phosphoglycerate kinase, *Biochem. J.*, 141, 721, 1974.
111. **Stubbe, J. A. and Kenyon, G. L.**, Analogs of phosphoenolpyruvate. Substrate specificities of enolase and pyruvate kinase from rabbit muscle, *Biochemistry*, 11, 338, 1972.
112. **Reed, G. H. and Cohn, M.**, Electron paramagnetic resonance studies of manganese (II)-pyruvate kinase-substrate complexes, *J. Biol. Chem.*, 248, 6436, 1973.
113. **James, T. L. and Cohn, M.**, Structural aspects of manganese-pyruvate kinase substrate and inhibitor complexes deduced from proton magnetic relaxation rates of pyruvate and a phosphoenolpyruvate analog, *J. Biol. Chem.*, 249, 3519, 1974.
114. **Nowak, T., Mildvan, A. S., and Kenyon G. L.**, Nuclear relaxation and kinetic studies of the role of Mn(II) in the mechanism of enolase, *Biochemistry*, 12, 1690, 1973.
115. **Lane, R. H. and Hurst, J. K.**, Intermediates in enolase-catalyzed reactions, *Biochemistry*, 13, 3292, 1974.
116. **Paulsen, H. and Bartsch, W.**, Rearrangement of sugar-1-enolphosphonates to sugar-2-enolphosphonates, *Chem. Ber.*, 108, 1745, 1975.
117. **Paulsen, H. and Bartsch, W.**, Preparation of olefinic sugar phosphonates and allenic sugar phosphonates, *Chem. Ber.*, 108, 1732, 1975.
118. **Leifer, Z., Engel, R., and Tropp, B. E.**, Transport of 3,4-dihydroxybutyl-1-phosphonate, an analogue of *sn*-glycerol-3-phosphate, *J. Bacteriol.*, 130, 968, 1977.
119. **Guth, A., Engel, R., and Tropp, B. E.**, Uptake of glycerol-3-phosphate and some of its analogues by the hexose phosphate transport system of *Escherichia coli*, *J. Bacteriol.*, 143, 538, 1980.
120. **Shopsis, C. S., Engel, R., and Tropp, B. E.**, Effects of phosphonic acid analogues of glycerol-3-phosphate on the growth of *Escherichia coli*, *J. Bacteriol.*, 112, 408, 1972.

121. **Shopsis, C. S., Nunn, W. D., Engel, R., and Tropp, B. E.,** Effects of phosphonic acid analogues of glycerol-3-phosphate on the growth of *Escherichia coli:* phospholipid metabolism, *Antimicrob. Agents Chemother.*, 4, 467, 1973.
122. **Shopsis, C. S., Engel, R., and Tropp, B. E.,** The inhibition of phosphatidylglycerol synthesis in *Escherichia coli* by 3,4-dihydroxybutyl-1-phosphonate, *J. Biol. Chem.*, 249, 2473, 1974.
123. **Cheng, P.-J., Nunn, W. D., Tyhach, R. J., Goldstein, S. L., Engel, R., and Tropp, B. E.,** Investigations concerning the mode of action of 3,4-dihydroxybutyl-1-phosphonate on *Escherichia coli, J. Biol. Chem.*, 250, 1633, 1975.
124. **Tyhach, R. J., Engel, R., and Tropp, B. E.,** Metabolic fate of 3,4-dihydroxybutyl-1-phosphonate in *Escherichia coli, J. Biol. Chem.*, 251, 6716, 1976.
125. **Cattopadhyay, P. K., Engel, R., Tropp, B. E., and Wu, H. C.,** Biosynthesis of murein lipoprotein in *Escherichia coli:* effects of 3,4-dihydroxybutyl-1-phosphonate, *J. Bacteriol.*, 138, 944, 1979.
126. **Klein, D. A., Engel, R., and Tropp, B. E.,** Incorporation of 3,4-dihydroxybutyl-1-phosphonate, a glycerol-3-phosphate analogue, into the cell wall of *Bacillus subtilis, J. Bacteriol.*, 129, 550, 1977.
127. **Deutsch, R. M., Engel, R., and Tropp, B. E.,** Effect of 3,4-dihydroxybutyl-1-phosphonate on phosphoglyceride and lipoteichoic acid synthesis in *Bacillus subtilis, J. Biol. Chem.*, 255, 1521, 1980.
128. **Tang, K.-C., Mildener, B., Eisikowitz, L., Tropp, B. E., and Engel, R.,** Inhibition of growth of *Escherichia coli* by 1,3,4-trihydroxybutyl-1-phosphonic acid, Paper #54, Division of Biological Chemistry, American Chemical Society/Chemical Society of Japan Joint Meeting, April 1979, Honolulu.
129. **Mildener, B., Engel, R., and Tropp, B. E.,** Unpublished data, 1981.
130. **Horiguchi, M. and Kandatsu, M.,** Isolation of 2-aminoethane phosphonic acid from rumen protozoa, *Nature (London)*, 184, 901, 1959.
131. **Kittredge, J., Roberts, E., and Simonsen, D. G.,** The occurrence of free 2-aminoethylphosphonic acid in the sea anemone, *Anthopleura elegantissima, Biochemistry*, 1, 624, 1962.
132. **Kittredge, J. S. and Hughes, R. R.,** The occurrance of alpha-amino-beta-phosphonopropionic acid in the zoanthid *Zoanthus sociatus*, and the ciliate *Tetrahymena pyriformis, Biochemistry*, 3, 991, 1964.
133. **Hori, T., Itasaka, O., and Inoue, H.,** Biochemistry of shellfish lipids. III. Purification and elemental analysis of ceramide aminoethyl phosphonate from corbicula complex lipid mixtures, *J. Biochem. (Tokyo)*, 59, 509, 1966.
134. **Liang, C. R. and Rosenberg, H.,** The biosynthesis of the phosphonic analogue of cephalin tetrahymena, *Biochim. Biophys. Acta*, 125, 548, 1966.
135. **Simon, G. and Rouser, G.,** Phospholipids of the sea anemone: quantitative distribution; absence of carbon-phosphorus linkages in glycerol phospholipids; structural elucidation of ceramide aminoethylphosphonate, *Lipids*, 2, 55, 1967.
136. **Kittredge, J. S. and Roberts, E.,** A carbon-phosphorus bond in nature, *Science*, 164, 37, 1969.
137. **Smith, J. D., Snyder, W. R., and Law, J. H.,** Phosphonolipids in *Tetrahymena* cilia, *Biochem. Biophys. Res. Commun.*, 39, 1163, 1970.
138. **Rosenthal, A. F., Kosolapoff, G. M., and Geyer, R. P.,** Nonhydrolyzable synthetic analogs of phosphatidic acid, *Recl. Trav. Chim. Pays-Bas*, 83, 1273, 1964.
139. **Rosenthal, A. F.,** An isosteric non-hydrolyzable phosphatidic acid analogue, *J. Chem. Soc.*, 7345, 1965.
140. **Rosenthal, A. F.,** New, partially hydrolyzable synthetic analogues of lecithin, phosphatidyl ethanolamine, and phosphatidic acid, *J. Lipid Res.*, 7, 779, 1966.
141. **Rosenthal, A. F. and Pousada, M.,** Inhibition of particulate phosphatidate phosphohydrolase by an analogue of its substrate, *Biochim. Biophys. Acta*, 125, 265, 1966.
142. **Baer, E. and Basu, H.,** Synthesis of alpha-phosphotidic acids, *Can. J. Biochem.*, 48, 1010, 1970.
143. **Bonsen, P. P. M., Burbach-Westerhuis, G. S., DeHaas, G. H., and VanDeenen, L. L. M.,** Chemical synthesis of some lecithin analogues potential inhibitors of phospholipase A, *Chem. Phys. Lipids*, 8, 199, 1972.
144. **Tang, J.-C., Tropp, B. E., and Engel, R.,** Isosteres of natural phosphates. 4. The synthesis of phosphonic acid analogues of phosphatidic acid and acyldihydroxyacetone phosphate, *Chem. Phys. Lipids*, 17, 169, 1976.
145. **Spencer, R. P.,** Private communication, 1978.
146. **Waters, P., Tropp, B. E., and Engel, R.,** Unpublished data, 1981.
147. **Tang, J.-C., Tang, C.-T., Tropp, B. E., and Engel, R.,** The synthesis and enzymatic properties of *lyso*-phosphotidic acid, paper #164, Division of Biological Chemistry, 172nd Meeting of the American Chemical Society, September 1976, San Francisco.
148. **Tang, J.-C., Tang, C.-T., Tropp, B. E., and Engel, R.,** Isosteres of natural phosphates. 6. The preparation and properties of *lyso*-phosphotidic acid, *Chem. Phys. Lipids*, 19, 99, 1977.
149. **Rosenthal, A. F. and Pousada, M.,** Synthesis of 2-aminoethylphosphonate-containing cephalin analogs, *Proc. Chem. Soc.*, 358, 1964.

150. **Baer, E. and Stanacev, N. Z.,** Phosphonolipids. I. Synthesis of a phosphonic acid analogue of cephalin, *J. Biol. Chem.*, 239, 3209, 1964.
151. **Rosenthal, A. F. and Pousada, M.,** Phosphonate analogs of cephalin, *Recl. Trav. Chim. Pays-Bas*, 84, 833, 1965.
152. **Baer, E. and Sarma, G. R.,** Phosphonolipids. IV. Synthesis of phosphonic acid analogs of L-alpha-cephalins, *Can. J. Biochem.*, 43, 1353, 1965.
153. **Baer, E., Basu, H., and Pal, B. C.,** Phosphonolipids. X. A new synthesis of phosphonic acid analogues of L-alpha-lecithins, *Can. J. Biochem.*, 45, 1467, 1967.
154. **Baer, E. and Basu, H.,** Phosphonolipids. XVI. An alternative method for the synthesis of phosphonic acid analogues of L-alpha-cephalins, *Can. J. Biochem.*, 46, 351, 1968.
155. **Turner, D. L., Silver, M. J., Holburn, R. R., Baczynski, E., and Brown, A. B.,** An unsaturated phosphonic acid analog of phosphatidylethanolamine and its activity in blood-clotting systems, *Lipids*, 3, 234, 1968.
156. **Baer, E.,** Phosphonolipids. XXVI. Mixed-acid phosphonocephalins: synthesis of alpha'-stearoyl-beta-L-alpha-glyceryl-(2-aminoethyl) phosphonate, *Can. J. Biochem.*, 52, 570, 1974.
157. **Pfeiffer, F. R., Miao, C. K., Hoke, S. C., and Weisbach, J. A.,** Potential renin inhibitors. 2. Ethanolamine and ethylamine derivatives of phospholipids, *J. Med. Chem.*, 15, 58, 1972.
158. **Baer, E. and Stanacev, N. Z.,** Phosphonolipids. II. Synthesis of dialkyl L-alpha-glyceryl-(2-aminoethyl)phosphonates, *J. Biol. Chem.*, 240, 44, 1965.
159. **Chacko, G. K. and Hanahan, D. J.,** Chemical synthesis of phosphonic acid monoether analogues of phosphatidylethanolamine and phosphatidylcholine, *Biochim. Biophys. Acta*; 176, 190, 1969.
160. **Baer, E. and Pal, B. C.,** Phosphonolipids. XV. Synthesis of a dihydroceramide aminoethylphosphonate, *Can. J. Physiol. Pharmacol.*, 46, 525, 1968.
161. **Baer, E. and Pal, B. C.,** Synthesis of L-alpha-phosphatidyl-(alpha-methyl)cholines, *Can. J. Physiol. Pharmacol.*, 45, 1478, 1967.
162. **Baer, E. and Sarma, G. R.,** Phosphonolipids. XX. Total synthesis of a naturally occurring ceramide aminoethylphosphonate and of its enantiomer, *Can. J. Biochem.*, 47, 603, 1969.
163. **Rouser, G., Kritchevsky, G., Heller, D., and Lieber, E.,** Lipid components of beef brain, beef liver, and the sea anemone — fractionation of complex lipid mixtures, *J. Am. Oil Chem. Soc.*, 40, 425, 1963.
164. **Baer, E. and Basu, H.,** Phosphonolipids. XVIII. Synthesis of ethylamine glycophosphonic acid analogues of cephalin, *Can. J. Biochem.*, 46, 1279, 1968.
165. **Braksmayer, D., Tropp, B. E., and Engel, R.,** Isosteres of natural phosphates. 5. The prepration of phosphotidylcholine, phosphotidylethanolamine, and phosphotidylglycerol, *Chem. Phys. Lipids*, 19, 93, 1977.
166. **Baer, E. and Stanacev, N. Z.,** Phosphonolipids. III. Synthesis of a phosphonic acid analog of L-alpha-(distearoyl)lecithin, *J. Am. Chem. Soc.*, 87, 679, 1965.
167. **Baer, E. and Stanacev, N. Z.,** Phosphonolipids. V. Synthesis of phosphonic acid analogues of L-alpha-lecithins, *J. Biol. Chem.*, 240, 3754, 1965.
168. **Baer, E. and Rao, K. V. J.,** Phosphonolipids. VI. Synthesis of phosphonic acid analogs of diether L-alpha-lecithins, *Lipids*, 1, 291, 1966.
169. **Baer, E. and Rao, K. V. J.,** Phosphonolipids. VIII. Synthesis of phosphonic acid analogues of L-alpha-cephalins, *Can. J. Biochem.*, 45, 317, 1967.
170. **Baer, E. and Robinson, R.,** Phosphonolipids XI. Synthesis of phosphonolipid metabolites. L-alpha-glyceryl(2-trimethylammoniumethyl)phosphonate, *Can. J. Biochem.*, 45, 1747, 1967.
171. **Baer, E. and Stanacev, N. Z.,** Phosphonolipids. VI. Chemical and enzymatic degradation for study of structure, *Can. J. Biochem.*, 44, 893, 1966.
172. **Rosenthal, A. F. and Pousada, M.,** Inhibition of phospholipase C by phosphonate analogs of glycerophosphatides, *Biochim. Biophys. Acta*, 164, 226, 1968.
173. **Rosenthal, A. F. and Pousada, M.,** Two synthetic phosphonate analogs of lecithin, *Lipids*, 4, 37, 1969.
174. **Rosenthal, A. F. and Chodsky, S. V.,** A completely non-hydrolyzable synthetic analog of lecithin, *Chem. Commun.*, 1504, 1968.
175. **Bjerve, K. S.,** Lecithin biosynthesis in the rat studied with phosphonate analogs of phosphorylcholine, *Biochim. Biophys. Acta*, 270, 348, 1972.
176. **Rosenthal, A. F. and Chodsky, S. V.,** A completely non-hydrolyzable synthetic analog of lecithin, *Chem. Commun.*, 1504, 1968.
177. **Rosenthal, A. F. and Chodsky, S. V.,** New synthetic analogs of lecithin, *J. Lipid Res.*, 12, 277, 1971.
178. **Rosenthal, A. F. and Chodsky, S. V.,** Completely isosteric nonhydrolyzable analog of lecithin, *Biochim. Biophys. Acta*, 239, 248, 1971.
179. **Rosenthal, A. F. and Han, A. C.-H.,** Phospholipase A inhibition by glycerophosphatide analogs in various systems, *Biochim. Biophys. Acta*, 218, 213, 1970.

180. **Hendrickson, H. S., Rustard, D. G., Scattergood, E. M., and Engle, D. E.,** Action of phospholipase C and lipase on black film bilayer membranes, *Chem. Phys. Lipids,* 13, 63, 1974.
181. **Bittman, R. and Blau, L.,** The phospholipid-cholesterol interaction. Kinetics of water permeability in liposomes, *Biochemistry,* 11, 4831, 1972.
182. **DeKruyff, B., Demel, R. A., Slotboom, A. J., VanDeenen, L. L. M., and Rosenthal, A. F.,** Effect of the polar head group on the lipid-cholesterol interaction. Monolayer and differential scanning calorimetry study, *Biochim. Biophsy. Acta,* 307, 1, 1973.
183. **Clejan, S., Bittman, R., Deroo, P. W., Isaacson, Y. A., and Rosenthal, A. F.,** Permeability properties of sterol-containing liposomes from analogues of phosphatidylcholine lacking acyl groups, *Biochemistry,* 18, 2118, 1979.
184. **Kamp, H. H., Wirtz, K. W. A., Baer, P. R., Slotboom, A. J., Rosenthal, A. F., Paltauf, F., and VanDeenen, L. L. M.,** Specificity of the phosphatidylcholine exchange protein from bovine liver, *Biochemistry,* 16, 1310, 1977.
185. **Isaacson, Y. A., Deroo, P. W., Rosenthal, A. F., Bittman, R., McIntyre, J. O., Bock, H.-G., Gazzotti, P., and Fleischer, S.,** The structural specificity of lecithin for activation of purified D-beta-hydroxybutyrate apodehydrogenase, *J. Biol. Chem.,* 254, 117, 1979.
186. **Doerr, I. L., Tang, J.-C., Rosenthal, A. F., Engel, R., and Tropp, B. E.,** Synthesis of phosphonate and ether analogs of rac-phosphatidyl-L-serine, *Chem. Phys. Lipids,* 19, 185, 1977.
187. **Tyhach, R. J., Rosenthal, A. F., and Tropp, B. E.,** Substrate activity of phosphonic acid analogues of CDP-diglyceride in the synthesis of phosphoglycerides in *Escherichia coli, Biochim. Biophys. Acta,* 388, 29, 1975.
188. **Bloch, K.,** The biosynthesis of cholesterol, *Science,* 150, 19, 1965.
189. **Corey, E. J. and Volante, R. P.,** Application of unreactive analogs of terpenoid pyrophosphates to studies of multistep biosynthesis. Demonstration that "presqualene pyrophosphate" is an essential intermediate on the path to squalene, *J. Am. Chem. Soc.,* 98, 1291, 1976.
190. **Zervas, L. and Dilaris, I.,** Dealkylation and debenzylation of triesters of phosphoric acid — phosphorylation of hydroxy and amino compounds, *J. Am. Chem. Soc.,* 77, 5354, 1955.
191. **Cornforth, R. H. and Popjak, G.,** Chemical synthesis of isopentenyl (3-methyl-3-en-1-yl) pyrophosphate, *Methods Enzymol.,* 15, 382, 1969.
192. **Sarin, V. K., Tropp, B. E., and Engel, R.,** Isosteres of natural phosphates. 7. The preparation of 5-carboxy-4-hydroxy-4-methylpentyl-1-phosphonic acid, *Tetrahedron Lett.,* 351, 1977.
193. **Parker, T., Gotlinsky, B., Sarin, V. K., Tropp, B. E., and Engel, R.,** *In vivo* inhibition of sterol biosynthesis, Paper #118, Division of Biological Chemistry, Second Chemical Congress of the North American Continent, August 1980, Las Vegas.
194. **Popjak, G., Parker, T. S., Sarin, V. K., Tropp, B. E., and Engel, R.,** Inhibition of 5-phosphomevalonate kinase by an isosteric analogue of 5-phosphomevalonate, *J. Am. Chem. Soc.,* 100, 8014, 1978.
195. **Schwarz, L. R., Burr, R., Schwenk, M., Pfaff, E., and Grein, H.,** Uptake of taurocholic acid into isolated rat liver cells, *Eur. J. Biochem.,* 55, 617, 1975.
196. **Hershey, J. W. B. and Monro, R. E.,** Competitive inhibitor of the guanosine triphosphate reaction in protein synthesis, *J. Mol. Biol.,* 18, 68, 1966.
197. **Myers, T. C., Nakamura, K., and Flesher, J. W.,** Phosphonic acid analogues of nucleoside phosphates. (I) Synthesis of 5′-adenylyl methylenediphosphonate, a phosphonic acid analogue of adenosine triphosphate, *J. Am. Chem. Soc.,* 85, 3292, 1963.
198. **Myers, T. C.,** Phosphonic acid analogs of nucleoside phosphates, U.S. Patent 3,238,191 (1 March 1966); *Chem. Abstr.,* 64, 15972h, 1966.
199. **Hershey, J. W. B. and Thach, R. E.,** Role of guanosine 5′-triphosphate in the initiation of protein synthesis. I. Synthesis of formylmethionyl-puromycin, *Proc. Natl. Acad. Sci. U.S.A.,* 57, 759, 1967.
200. **Thach, R. E., Dewey, K. F., and Mydolajewycz, N.,** Role of guanosine 5′-triphosphate in the initiation of protein synthesis. II. Synthesis of dipeptides, *Proc. Natl. Acad. Sci. U.S.A.,* 57, 1103, 1967.
201. **Ohta, T., Sarkar, S., and Thach, R. E.,** Role of guanosine 5′-triphosphate in the initiation of protein synthesis. III. Binding of formylmethionyl-transfer RNA to ribosomes, *Proc. Natl. Acad. Sci. U.S.A.,* 58, 1638, 1967.
202. **Ravel, J. M., Shorey, R. L., Froehner, S., and Shive, W.,** The enzymic transfer of aminoacyl-RNA to *Escherichia coli* ribosomes, *Arch. Biochem. Biophys.,* 125, 514, 1968.
203. **Brot, N., Ertel, R., and Weissbach, H.,** Effect of a soluble transfer factor on the reaction of aminoacyl-tRNA with puromycin, *Biochem. Biophys. Res. Commun.,* 31, 563, 1968.
204. **Lippmann, F.,** Polypeptide chain elongation in protein biosynthesis, *Science,* 163, 1024, 1969.
205. **Haselkorn, R. and Rothman-Denes, L. B.,** Protein synthesis, *Annu. Rev. Biochem.,* 42, 397, 1973.
206. **Lucas-Leonard, J. and Lippmann, F.,** Protein biosynthesis, *Annu. Rev. Biochem.,* 40, 409, 1971.
207. **Pestka, S.,** Inhibitors of ribosome functions, *Annu. Rev. Biochem.,* 40, 697, 1971.
208. **Kurland, C. G.,** Structure and function of the bacterial ribosome, *Annu. Rev. Biochem.,* 41, 377, 1972.

209. **Kozak, M. and Nathans, D.**, Translation of a ribonucleic acid bacteriophage genome, *Bacteriol. Rev.*, 36, 109, 1972.
210. **Lengyel, P. and Soll, D.**, Mechanism of protein biosynthesis, *Bacteriol. Rev.*, 33, 264, 1969.
211. **Frisch, L., and Hershey, H., Eds.**, *The Mechanism of Protein Synthesis*, Cold Spring Harbor Symp. Quant. Biol. Vol. XXXIV, Cold Spring Harbor Laboratories, Cold Spring Harbor, N.Y., 1969.
212. **Myers, T. C., Nakamura, K., and Danielzadeh, A. B.**, Phosphonic acid analogues of nucleoside phosphates. III. The synthesis of adenosine-5′-methylenediphosphonate, a phosphonic acid analog of adenosine-5′-diphosphate, *J. Org. Chem.*, 30, 1517, 1965.
213. **Myers, T. C.**, Analogues of nucleoside phosphates, U.S. Patent 3,569,478 (2 February 1971); *Chem. Abstr.*, 75, 20921c, 1971.
214. **Santi, D. V. and Pena, V. A.**, Tyrosyl transfer ribonucleic acid synthetase from *Escherichia coli B*. Analysis of tryosine and adenosine 5′-triphosphate binding sites, *J. Med. Chem.*, 16, 273, 1973.
215. **Santi, D. V., Danenberg, P. V., and Montgomery, K. A.**, Phenylalanyl transfer ribonucleic acid synthetase from *Escherichia coli*. Analysis of adenosine triphosphate binding site, *Biochemistry*, 10, 4821, 1971.
216. **Southgate, C. C. D. and Dixon, H. B. F.**, Phosphonate analogues of aminoacyl adenylates, *Biochem. J.*, 175, 461, 1978.
217. **Kuwano, M., Kwan, C. N., Apirion, D., and Schlessinger, D.**, Ribonuclease V of *Escherichia coli*. I. Dependence on ribosomes and translocating, *Proc. Natl. Acad. Sci. U.S.A.*, 64, 693, 1969.
218. **Evans, W. H. and Gurd, J. W.**, Properties of a 5′-nucleotidase purified from mouse liver plasma membranes, *Biochem. J.*, 133, 189, 1973.
219. **Burger, R. M. and Lowenstein, J. M.**, Preparation and properties of 5′-nucleotidase from smooth muscle of small intestine, *J. Biol. Chem.*, 245, 6274, 1970.
220. **Burger, R. M. and Lowenstein, J. M.**, 5′-Nucleotidase from smooth muscle of small intestine and from brain. Inhibition by nucleotides, *Biochemistry*, 14, 2362, 1975.
221. **Holy, A. and Hong, N. D.**, Nucleic acid components and their analogues. CL. Preparation and properties of some nucleoside hydroxyalkanephosphonates, *Collect. Czech. Chem. Commun.*, 37, 2066, 1972.
222. **Holy, A. and Hong, N. D.**, Synthesis of ribonucleoside 5′-*O*-hydroxymethanephosphonates, *Collect. Czech. Chem. Commun.*, 36, 316, 1971.
223. **Gulyaev, N. N. and Holy, A.**, Ribonucleoside 5′-aminomethanephosphonates: synthesis and affinity toward some phosphomonoesterases, *FEBS Lett.*, 22, 294, 1972.
224. **Zemlicka, J. and Chaldek, N.**, Aminoacyl derivatives of nucleosides, nucleotides, and polynucleotides. VI. Phosphonic analogues of 2′(3′)-*O*-aminoacyl ribonucleosides: synthesis of 2′(3′)-*O*-(R,S)-(1-aminoethanephosphonyl)- and 2′(3′)-*O*-(R,S)-(1-amino-2-phenylethane-1-phosphonyl) adenosine, *Collect. Czech. Chem. Commun.*, 34, 1007, 1969.
225. **Jones, D. H., Albrecht, H. P., Damodaran, N. P., and Moffatt, J. G.**, Synthesis of isosteric phosphonate analogs of some biologically important phosphodiesters, *J. Am. Chem. Soc.*, 92, 5510, 1970.
226. **Richards, F. M., Wyckoff, H. W., Carlson, W. D., Allewell, N. M., Lee, B., and Mitsui, Y.**, Protein structure, ribonuclease-S, and nucleotide interactions, *Cold Spring Harbor Symp. Quant. Biol.*, 36, 35, 1972.
227. **Griffin, J. H., Schechter, A. N., and Cohen, J. S.**, Nuclear magnetic resonance studies of a ribonuclease-dinucleoside phosphonate complex and their implications for the mechanism of the enzyme, *Ann. N.Y. Acad. Sci.*, 222, 693, 1973.
228. **Johnson, N. P. and Schleich, T.**, Circular dichroism studies of the conformational stability of dinucleoside phosphates and related compounds in aqueous neutral salt solutions, *Biochemistry*, 13, 981, 1974.
229. **Hampton, A., Perini, F., and Harper, P. J.**, Synthesis of homoadenosine-6′-phosphonic acid and studies of its substrate and inhibitor properties with adenosine monophosphate utilizing enzymes, *Biochemistry*, 12, 1730, 1973.
230. **Hampton, A., Sasaki, T., and Paul, B.**, Synthesis of 6′-cyano-6′-deoxyhomoadenosine-6′-phosphonic acid and its phosphoryl and pyrophosphoryl anhydrides and studies of their interactions with adenine nucleotide utilizing enzymes, *J. Am. Chem. Soc.*, 95, 4404, 1973.
231. **Simon, L., Myers, T., and Mednieks, M.**, Enzymic studies of polyribonucleotides using 5′-adenylylmethylenediphosphonate — phosphonic acid analog of adenosine triphosphate, *Biochim. Biophys. Acta*, 103, 189, 1965.
232. **Chou, J. Y. and Singer, M. F.**, Synthesis of a copolymer containing adenylic and deoxyadenylic acid residues with polynucleotide phosphorylase, *J. Biol. Chem.*, 246, 7497, 1971.
233. **Chou, J. Y. and Singer, M. F.**, Kinetic analysis of the phosphorolysis of oligonucleotides by polynucleotide phosphorylase, *J. Biol. Chem.*, 245, 995, 1970.
234. **Raue, H. A. and Cashel, M.**, Regulation of RNA synthesis in *Escherichia coli*. II. Polynucleotide phosphorylase activity in cold-shocked cells, *Biochim. Biophys. Acta*, 340, 40, 1974.

235. **Englund, P. T., Huberman, J. A., Jovin, T. M., and Kornberg, A.**, Enzymatic synthesis of deoxyribonucleic acid. XXX. Binding of triphosphates to deoxyribonucleic acid polymerase, *J. Biol. Chem.*, 244, 3038, 1969.
236. **Geider, K.**, DNA synthesis in nucleotide permeable *Escherichia coli* cells, *Eur. J. Biochem.*, 27, 554, 1972.
237. **Yengoyan, L. and Rammler, D.H.**, Nucleoside phosphonic acids. I. The synthesis of 5′-deoxyuridine 5′-phosphonic acids and derivatives, *Biochemistry*, 5, 3629, 1966.
238. **Pfitzner, K. E. and Moffatt, J. G.**, A new and selective oxidation of alcohols, *J. Am. Chem. Soc.*, 85, 3027, 1963.
239. **Jones, G. H. and Moffatt, J. G.**, The synthesis of 6′-deoxyhomonucleoside-6′-phosphonic acids, *J. Am. Chem. Soc.*, 90, 5337, 1968.
240. **Jones, G. H. and Moffatt, J. G.**, Physiologically active nucleoside phosphonates and phosphonic acids, German Offen. 2,009,834 (17 September 1970); *Chem. Abstr.*, 74, 54150v, 1971.
241. **Syntex Corp.**, Nucleoside phosphonic acids and the corresponding phosphonates, British Patent 3,238,214 (18 August 1971); *Chem. Abstr.*, 75, 118548m, 1971.
242. **Huberman, J. A. and Kornberg, A.**, Enzymatic synthesis of deoxyribonucleic acid, *J. Biol. Chem.*, 245, 5326, 1970.
243. **Lazarus, R. A., Benkovic, P. A., and Benkovic, S. J.**, The synthesis and properties in enzymic reactions of substrate analogs containing the methylphosphonyl group, *Arch. Biochem. Biophys.*, 197, 218, 1979.
244. **Berman, K. M. and Cohn, M.**, Phosphoenolpyruvate synthetase, *J. Biol. Chem.*, 245, 5319, 1970.
245. **Murray, A. W.**, Some properties of adenosine kinase from Ehrlich ascites tumour cells, *Biochem. J.*, 106, 549, 1968.
246. **O'Sullivan, A. W., Reed, G. H., Marsden, K. H., Gough, G. R., and Lee, C. S.**, The binding of manganese-nucleoside diphosphates to creatine kinase as determined by proton relaxation rate measurements, *J. Biol. Chem.*, 247, 7839, 1972.
247. **Duee, E. D., Vignais, P. V., and Moreau, M.**, Atractyloside sensitive translocation of phosphonic acid analogues of adenine nucleotides in mitochondria, *Biochem. Biophys. Res. Commun.*, 30, 420, 1968.
248. **Duee, E. D., Vignais, P. V., and Moreau, M.**, Atractyloside sensitive translocation of phosphonic acid analogues of adenine nucleotides in mitochondria, *Biochem. Biophys. Res. Commun.*, 30, 546, 1968.
249. **Wigler, P. W. and Lozzio, C. B.**, Cytotoxic nucleotide analogs as potential anticancer agents, 5-bromodeoxyuridine 5′-methylphosphonate, *J. Med. Chem.*, 15, 1020, 1972.
250. **Webster, D., Sparkes, M. J., and Dixon, H. B. F.**, An arsenical analogue of adenosine diphosphate, *Biochem. J.*, 169, 239, 1978.
251. **Lagunas, R. and Sols, A.**, Arsenate-induced activity of certain enzymes on their dephosphorylated substrates, *FEBS Lett.*, 1, 32, 1968.
252. **Long, J. W. and Ray, W. J.**, Kinetics and thermodynamics of the formation of glucose arsenate. Reaction of glucose arsenate with phosphoglucomutase, *Biochemistry*, 12, 3932, 1973.
253. **Yount, R. G., Babcock, D., Ballantyne, W., and Ojala, D.**, Adenylyl imidodiphosphate, an adenosine triphosphate analog containing a P-N-P linkage, *Biochemistry*, 10, 2484, 1971.
254. **Yount, R. G., Ojala, D., and Babcock, D.**, Interaction of P-N-P and P-C-P analogs of adenosine triphosphate with heavy meromyosin, myosin, and actomyosin, *Biochemistry*, 10, 2490, 1971.
255. **Grisham, C. M. and Mildvan, A. S.**, Magnetic resonance and kinetic studies of the mechanism of sodium and potassium ion-activated adenosine triphosphatase, *J. Biol. Chem.*, 249, 3187, 1974.
256. **Taketa, K., Sarngadharan, M. G., Watanabe, A., Aoe, H., and Pogell, B. M.**, Reversible inactivation of rabbit liver fructose 1,6-diphosphatase by adenosine triphosphate and adenosine diphosphate, *J. Biol. Chem.*, 246, 5676, 1971.
257. **Curthoys, N. P. and Rabinowitz, J. C.**, Formyltetrahydrofolate synthetase, *J. Biol. Chem.*, 246, 6942, 1971.
258. **Moss, C., Alpert, N. R., and Myers, T. C.**, Effects of a phosphonic acid analog of adenosine triphosphate on actomyosin system, *Arch. Biochem. Biophys.*, 88, 183, 1960.
259. **Seidel, J. C.**, The effects of ionic conditions, temperature, and chemical modification on the fluorescence of myosin during the steady state of ATP hydrolysis, *J. Biol. Chem.*, 250, 5681, 1975.
260. **Koretz, J. F. and Taylor, E. W.**, Transient state kinetic studies of proton liberation by myosin and subfragment 1, *J. Biol. Chem.*, 250, 6344, 1975.
261. **Werber, M. M., Szent-Gyrogyi, A. G., and Fasman, G. D.**, Fluorescence studies on heavy meromyosin-substrate interaction, *Biochemistry*, 2872, 1972.
262. **Cooke, R. and Murdoch, L.**, Interaction of actin with analogs of adenosine triphosphate, *Biochemistry*, 12, 3927, 1973.
263. **Basinger, R. W., Jones, D. J., and Montgomery, J. A.**, Synthesis of potential anticancer agents. XVIII. Analogs of carbamoyl phosphate, *J. Org. Chem.*, 24, 434, 1959.

264. **Nylen, P.,** Organophosphorus compounds. II. beta-Phosphonopropionic acid and gamma-phosphonobutyric acid, *Chem. Ber.*, 57B, 1023, 1924.
265. **Arbuzov, A. E. and Dunin, A. A.,** Action of HI derivatives of aliphatic esters on alkyl phosphites, *J. Chem. Soc.*, 653, 1914.
266. **Shipkowitz, N. L., Bower, R. R., Appell, R. N., Nordeen, C. W., Overby, L. R., Roderick, W. R., Sleicher, J. B., and vonEsch, A. M.,** Suppression of herpes simplex virus infection by phosphonoacetic acid, *Appl. Microbiol.*, 26, 264, 1973.
267. **Overby, L. R., Duff, R. G., and Mao, J. C.-H.,** Antiviral properties of phosphonoacetic acid, *Ann. N.Y. Acad. Sci.*, 284, 310, 1977.
268. **Hay, J., Brown, J. M., Jamieson, A. T., Rixon, F. J., Moss, H., and Dargan, D. A.,** The effect of phosphonoacetic acid on herpes virus, *J. Antimicrob. Chemother.*, 3 (Suppl. A), 63, 1977.
269. **Castaner, J. and Hopkins, S. J.,** Phosphonoacetic acid, *Drugs Future*, 2, 677, 1978.
270. **Hwang, E. S.,** Human cytomegalovirus. IV. Specific inhibition of virus-induced DNA polymerase activity and viral DNA replication by phosphonoacetic acid, *J. Virol.*, 16, 1560, 1975.
271. **Barahona, H., Daniel, M. D., Bekesi, J. G., Fraser, C. E. O., King, N. W., Hunt, R. D., Ingalls, J. K., and Jones, T. C.,** *In vitro* suppression of herpesvirus Saimiri replication by phosphonoacetic acid, *Proc. Soc. Exp. Biol. Med.*, 154, 431, 1977.
272. **Meyer, R. F., Vanell, E. D., and Kaufmann, H. E.,** Phosphonoacetic acid in the treatment of experimental ocular herpes simplex infections, *Antimicrob. Agents Chemother.*, 9, 308, 1976.
273. **Descamps, J., DeClerq, E., Barr, P. J., Jones, A. S., Walker, R. T., Torrence, P. F., and Shugar, D.,** Relative potencies of different anti-herpes agents in the topical treatment of cutaneous herpes simplex virus infections of athymic nude mice, *Antimicrob. Agents Chemother.*, 16, 680, 1979.
274. **Gordon, Y. J., Lahaw, M., Photiou, S., and Becker, Y.,** Effect of phosphonoacetic acid in the treatment of experimental herpes simplex keratitis, *Br. J. Ophthalmol.*, 61, 506, 1978.
275. **Koment, R. W. and Haines, H.,** Decreased antiviral effect of phosphonoacetic acid on the poikilothermic herpesvirus of channel catfish disease, *Proc. Soc. Exp. Biol. Med.*, 159, 21, 1978.
276. **Felsenfeld, A. D., Abee, C. R., Gerone, P. J., Soike, K. F., and Williams, S. R.,** Phosphonoacetic acid in the treatment of simian Varicella, *Antimicrob. Agents Chemother.*, 14, 331, 1978.
277. **Kern, E. R., Richards, J. T., Overall, J. C., Jr., and Glasgow, L. A.,** Alteration of mortality and pathogenesis of 3 experimental herpesvirus Hominis infections of mice with adenine arabinosidemonophosphate, adenine arabinoside, and phosphonoacetic acid, *Antimicrob. Agents Chemother.*, 13, 53, 1978.
278. **Kern, E. R., Richards, J. T., Overall, J. C., Jr., and Glasgow, L. A.,** Genital herpesvirus hominis infection in mice. Part 2. Treatment with phosphonoacetic acid, adenine arabinoside, and adenine arabinoside 5′-monophosphae, *J. Infect. Dis.*, 135, 557, 1977.
279. **Overby, L. R., Robishaw, E. E., Schleicher, J. B., Reuter, A., Shipkowitz, N. L., and Mao, J.-C.,** Inhibition of herpes simplex virus replication by phosphonoacetic acid, *Antimicrob. Agents Chemother.*, 6, 360, 1974.
280. **Mao, J. C.-H., Robishaw, E. E., and Overby, L. R.,** Inhibition of DNA polymerase from herpes simplex virus infected WI-38 cells by phosphonoacetic acid, *J. Virol.*, 15, 1281, 1975.
281. **Mao, J. C.-H. and Robishaw, E. E.,** Mode of inhibition of herpes simplex virus DNA polymerase by phosphonoacetic acid, *Biochemistry*, 14, 5475, 1975.
282. **Bolden, A. J., Aucker, J., and Weissbach, A.,** Synthesis of herpes simplex virus, vaccinia virus, and adenovirus DNA in isolated Hela cell nuclei. I. Effect of viral-specific antisera and phosphonoacetic acid, *J. Virol.*, 16, 1584, 1975.
283. **Gerstein, D. D., Dawson, C. R., and Oh, J. O.,** Phosphonoacetic acid in the treatment of experimental herpes simplex keritit is, *Antimicrob. Agents Chemother.*, 7, 285, 1975.
284. **Klein, R. J. and Friedman-Kien, A. E.,** Phosphonoacetic acid-resistant herpes simplex virus infection in hairless mice, *Antimicrob. Agents Chemother.*, 7, 289, 1975.
285. **Huang, E. S., Huang, C. H., Huang, M. S., and Selgrade, M. J.,** Preferential inhibition of herpes-group viruses by phosphonoacetic acid: effect on virus DNA synthesis and virus-induced DNA polymerase activity, *Yale J. Biol. Med.*, 49, 93, 1976.
286. **Honess, R. W. and Watson, D. H.,** Herpes simplex virus resistance and sensitivity to phosphonoacetic acid, *J. Virol.*, 21, 584, 1977.
287. **Hay, J. and Subak-Sharpe, J. A.,** Mutants of herpes simplex virus types 1 and 2 that are resistant to phosphonoacetic acid induced altered DNA polymerase activities in infected cells, *J. Gen. Virol.*, 31, 145, 1976.
288. **Becker, Y., Asher, Y., Cohe, Y., Weinberg-Zahlering, E., and Shlomai, J.,** Phosphonoacetic acid-resistant mutants of herpes simplex virus: effect of phosphonoacetic acid on virus replication and *in vitro* deoxynucleic acid synthesis in isolated nuclei, *Antimicrob. Agents Chemother.*, 11, 919, 1977.
289. **Knopf, K. W.,** Properties of herpes simplex virus DNA polymerase EC-2.7.7.7 and characterization of its associated exo-nuclease activity, *Eur. J. Biochem.*, 98, 231, 1979.

290. **Friedlander, B., Chejanovsky, N., and Becker, Y.,** DNA polymerase of wild type and phosphonoacetic acid resistant mutant of herpes simplex virus, *Antimicrob. Agents Chemother.*, 13, 124, 1978.
291. **King, N. W., Jr., Barahona, H., Daiel, M. D., Bekesi, J. G., and Jones, T. C.,** The effect of phosphonoacetic acid on the ultrastructure of cell cultures infected with simian oncogenic herpesvirus, *Lab. Invest.*, 38, 181, 1978.
292. **Burke, J. F., Duff, P. M., and Pearson, C. K.,** Effect of drugs on DNA synthesis in isolated mammalian cell nuclei — comparison with partially purified DNA polymerases, *Biochem. J.*, 178, 621, 1979.
293. **Allaudeen, H. S. and Bertion, J. R.,** Inhibition of activities of DNA polymerase alpha beta gamma and reverse transcriptase of L-1210 cells by phosphonoacetic acid, *Biochim. Biophys. Acta*, 520, 490, 1978.
294. **Jofre, J. T., Schaffer, P. A., and Parris, D. S.,** Genetics of resistance to phosphonoacetic acid in strain KOS of herpes simplex virus type 1, *J. Virol.*, 23, 83, 1977.
295. **Ayisi, N. K., Gupta, V. S., Meldrum, J. B., Taneja, A. K., and Babiuk, L. A.,** Combination chemotherapy interaction of 5-methoxymethyldeoxyuridine with adenine arabinoside 5-ethyldeoxyuridine, 5-iododeoxyuridine and phosphonoacetic acid against herpes simplex virus types 1 and 2, *Antimicrob. Agents Chemother.*, 17, 558, 1980.
296. **Alenius, S. and Oberg, B.,** Comparison of the therapeutic effects of 5 anti-viral agents on cutaneous herpesvirus infection in Guinea pigs, *Arch. Virol.*, 58, 277, 1978.
297. **Harris, S. R. B. and Boyd, M. R.,** The activity of iododeoxyuridine, adenine arabinoside, cytosine arabinoside, ribavirin and phosphonoacetic acid against herpes virus in the hairless mouse model, *J. Antimicrob. Chemother.*, 3 (Suppl. A), 91, 1977.
298. **Pearson, G. R. and Beneke, J. S.,** The inhibition of herpesvirus Saimiri replication by phosphonoacetic acid benzo(a)pyrene, and methylcholanthrene, *Cancer Res.*, 37, 42, 1977.
299. **Herrin, T. R., Fairgrieve, J. S., Bower, R. R., Shipkowitz, N. L., and Mao, J. C.-H.,** Synthesis and anti-herpes activity of analogs of phosphonoacetic acid, *J. Med. Chem.*, 20, 660, 1977.
300. **Thorley-Lawson, D. and Strominger, J. L.,** Transformation of human lymphocytes by Epstein-Barr virus is inhibited by phosphonoacetic acid, *Nature (London)*, 263, 332, 1976.
301. **Thorley-Lawson, D. A. and Strominger, J. L.,** Reversible inhibition by phosphonoacetic acid of human B lymphocyte transformation by Epstein-Barr virus, *Virology*, 86, 423, 1978.
302. **Miller, R. L., Glaser, R., and Rapp, F.,** Studies of an Epstein-Barr Virus induced DNA polymerase, *Virology*, 76, 494, 1977.
303. **Summers, W. C. and Klein, G.,** Inhibition of Epstein-Barr Virus DNA synthesis and late gene expression by phosphonoacetic acid, *J. Virol.*, 18, 151, 1976.
304. **Patel, P. and Mendzes, J.,** Differential effect of phosphonoacetic acid on early antigen synthesis in 2 Epstein-Barr Virus producer cell lines, *J. Virol.*, 92, 336, 1979.
305. **Lemon, S. M., Hut, L. M., and Pagano, J. S.,** Phosphonoacetic acid inhibits transformation of human cord lymphocytes by Epstein-Barr Virus, *Clin. Res.*, 25, 490A, 1979.
306. **Manor, D. and Margalith, M.,** Phosphonoacetic acid inhibition of human cord blood lymphocytes by Epstein-Barr Virus, *Cancer Biochem. Biophys.*, 3, 157, 1979.
307. **Lemon, S. M., Hutt, L. M., and Pagano, J. S.,** Cytofluorometry of lymphocytes infected with Epstein-Barr Virus: effect of phosphonoacetic acid on nucleic acid, *J. Virol.*, 25, 138, 1978.
308. **Yajima, Y., Tanaka, A., and Nonoyama, M.,** Inhibition of productive replication of Epstein-Barr Virus DNA by phosphonoacetic acid, *Virology*, 71, 352, 1976.
309. **Seebeck, T., Shaw, J. E., and Pagano, J. S.,** Synthesis of Epstein-Barr Virus DNA *in vitro:* effect of phosphonoacetic acid, *N*-Ethylmaleimide, and ATP, *J. Virol.*, 21, 435, 1977.
310. **Strnad, B. C., Neubauer, R. H., Rabin, H., and Mazur, R. A.,** Correlation between Epstein-Barr Virus membrane antigen and 3 large cell surface glyco proteins, *J. Virol.*, 32, 885, 1979.
311. **Gussander, E. and Adams, A.,** Intra-cellular state of Epstein-Barr Virus DNA in producer cell lines, *J. Gen. Virol.*, 45, 331, 1979.
312. **Thorley-Lawson, D. A.,** Characterization of cross reacting antigens on the Epstein-Barr Virus envelope and plasma membranes of producer cells, *Cell*, 16, 33, 1978.
313. **Leinbach, S. S., Reno, J. M., Lee, L. F., Isbell, A. F., and Bozzi, J. A.,** Mechanism of phosphonoacetate inhibition of herpesvirus-induced DNA polymerase, *Biochemistry*, 15, 426, 1976.
314. **Nazerian, K.,** Selective inhibition by phosphonoacetic acid of MDV DNA replication in a lymphoblastoid cell line, *Virology*, 74, 188, 1976.
315. **Elliott, R. M., Bateson, A., and Kelly, D. C.,** Phosphonoacetic acid inhibition of frog virus 3 replication, *J. Virol.*, 33, 539, 1980.
316. **Andzhaparidze, O. G., Bektemiorov, T. A., Burgasova, M. P., and Chekunova, E. V.,** Effect of phosphonoacetic acid on experimental vaccina infection, *Vopr. Virusol.*, 56, 1978.
317. **Nishibe, Y. and Inoue, Y. K.,** Effects of phosphonoacetic acid on subacute myeloopticoneuropathy virus *in vitro* and *in vivo*, *J. Med. Virol.*, 2, 225, 1978.

318. **Young, J. F. and Mayor, H. D.,** Effect of phosphonoacetic acid on the replication of adeno-associated virus, *Replication Mamm. Parvoviruses,* 269, 1978.
319. **Moreno, M. A., Carrascosa, A. L., Ortin, J., and Vinuela, J.,** Inhibition of African Swine Fever (ASF) virus replication by phosphonoacetic acid, *J. Gen. Virol.,* 39, 253, 1978.
320. **May, D. C., Miller, R. L., and Rapp, F.,** The effect of phosphonoacetic acid on the *in vitro* replication of Varicella-Zoster Virus, *Intervirology,* 8, 83, 1977.
321. **Alenius, S., Dinter, Z., and Oberg, B.,** Therapeutic effect of trisodium phosphonoformate on cutaneous herpesvirus infection in Guinea pigs, *Antimicrob. Agents Chemother.,* 14, 408, 1978.
322. **Margalith, M., Manor, D., Usieli, V., and Goldblum, N.,** Phosphonoformate inhibits synthesis of Epstein-Barr Virus capsid antigen and transformation of human cord blood lymphocytes, *Virology,* 102, 226, 1980.
323. **Menezes, J., Patel, P., Durrault, H., and Bourkas, A. E.,** Comparative studies on the induction of virus-associated nuclear antigens and early antigen by lymphocyte-transforming (B95-8) and non-transforming (P3HR-1) strains of Epstein-Barr Virus, *Intervirology,* 9, 86, 1978.
324. **Nyormoi, O., Thorley-Lawson, D. A., Elkinson, J., and Strominger, J. L.,** Differential effect of phosphonoacetic acid on the expression of Epstein-Barr Viral antigens and virus production, *Proc. Natl. Acad. Sci. U.S.A.,* 73, 1745, 1976.
325. **Hullar, T. L.,** Vinyl phosphonates: a convenient route to phosphonic acid analogues of phosphate monoesters, *Tetrahedron Lett.,* 4921, 1967.
326. **Vidgoff, J. M., Pocker, A., Hullar, T. L., and Fischer, E. H.,** Interaction of muscle glycogen phosphorylase with pyridoxal 5′-methylenephosphonate, *Biochem. Biophys. Res. Commun.,* 57, 1166, 1974.
327. **Feldman, K., Gaugler, B. J. M., Winkler, H., and Helmreich, E. J. M.,** Conformational transitions in glycogen phosphorylase reported by covalently bound pyridoxamine derivatives, *Biochemistry,* 13, 2222, 1974.
328. **Korytnyk, W., Lachman, B., and Angelino, N.,** Phosphorylated analogs of vitamin B-6 modified in the 5′-position and on the phosphate group: synthesis and interaction with pyridoxine phosphate oxidase and certain apoenzymes, *Biochemistry,* 11, 722, 1972.
329. **Gulyaev, N. N., Agalarova, M. B., Gnuchev, N. V., Chukanov, V. V., Khonutov, R. M., and Severin, E. S.,** Synthesis of coenzyme inhibitors, *Biokhimiya,* 39, 862, 1974.
330. **Hullar, T. L.,** Pyridoxal phosphate. I. Phosphonic acid analogs of pyridoxal phosphate, synthesis *via* Wittig reactions and enzymic evaluation, *J. Med. Chem.,* 12, 58, 1969.
331. **Dissing, J., Dahl, O., and Svensmark, O.,** Phosphonic and arsenic acids as inhibitors of human red cell acid phosphatase and their use in affinity chromatography, *Biochim. Biophys. Acta,* 569, 159, 1979.
332. **Varlet, J.-M., Collignon, N., and Savignac, P.,** A new route to 2-aminoalkanephosphonic acids, *Syn. Commun.,* 8, 335, 1978.
333. **Brigot, D., Collignon, N., and Savignac, P.,** Preparation of aminoalkylphosphonic acids *via* phosphorylation of omega-haloalkylamines, *Tetrahedron,* 35, 1345, 1979.

Chapter 6

THE EFFECTS OF SYNTHETIC PHOSPHONATES ON LIVING SYSTEMS

Richard L. Hilderbrand*

TABLE OF CONTENTS

I. INTRODUCTION

The biochemical processes involving normal metabolism of phosphorus are limited to compounds containing phosphorus at the +V level of oxidation, although phosphorus itself can exist in a variety of oxidation states. In addition, the metabolic processes usually involve phosphorus esters with alkoxy groups although, in some cases, the metabolic processes may extend to the carbon-phosphorus (C–P) bond of the phosphonates (see Chapter 3). The diversity of the normal biochemical reactions involving phosphorus and the specificity of those reactions allow many possibilities for phosphorus containing xenobiotics to influence phosphorus metabolism. The metabolic alterations may be intended for therapeutic purposes or as a biological insult to prohibit or inhibit life processes. In addition, the exposure may be intentional or unintentional, as in the case of a nontarget organism which lives in the area of pesticidal treatment of a target organism.

The specific metabolic effect produced by an organophosphorus compound may result from one or more of a variety of structural factors in the xenobiotic. In the phosphonate class of organic derivatives of phosphorus, the C–P bond may be responsible for inhibiting metabolic activity. Alternatively, the chemical characteristics of the C–P bond may enhance the reactivity of the molecule so that a specific metabolic effect is obtained, or the C–P bond may be inconsequential to the metabolic activity of a compound. The metabolic effects of phosphonoates are, in most cases, dependent on the chemical properties of the phosphonate moiety and the other substituents rather than on a direct inhibition of a biological process by the C–P bond. This is evident with the chemicals discussed in this chapter. For a discussion of the effects of structural factors on biochemical activity, see Chapter 5.

Although pesticides are often thought of as being only insecticides; the broad category of pesticides includes insecticides, herbicides, fungicides, rodenticides, and a number of other biocidal materials useful in the control of biological species considered to be pests. The insecticides and herbicides are of primary interest in this chapter since they contain the phosphonate compounds most significant as pesticides.

The many industrial applications of organophosphorus chemicals and their economic importance are discussed by Fisher and Van Wazer.[1] In that volume, Chadwick and Watt[2] note that, due to the difficulty of synthesis, the industrial applications for phosphonates have been slower to develop than for the corresponding phosphates. Thus, the commercial uses of phosphonate compounds have developed relatively recently. The industrial materials are selected for their chemical properties and on the basis of performance in a particular process; however, toxicological activity is now being considered prior to extensive development of new organophosphonates for industrial processes. The term "industrial chemicals" is used to denote specific chemicals used in industrial processes and not designed for intentional application to living systems. A discussion of these chemicals is beyond the scope of this work.

The utilization of phosphonates as insecticides requires diesters of alkyl, aryl, or mixed alkyl/aryl phosphonates. The effects of all of these chemicals on biochemical systems would be expected to be qualitatively similar (regardless of the alkyl or aryl nature), although not necessarily quantitatively similar. This quantitative difference is due to the variety of substituents and subsequent differences in biological reactivity. In addition, the chemical warfare agents exhibit effects qualitatively similar to those of the insecticides although they have greater biochemical reactivity. Representative phosphonates of the respective types and their primary effects on living systems (the inhibition of acetylcholinesterase [AChE][3] and the production of delayed peripheral neuropathy [OPIDN]) will be presented in this chapter. Miscellaneous biological effects of these chemicals will also be presented.

Information on the synthetic phosphonate antibiotics, which in a broad sense may be categorized as pesticides, and herbicides will also be presented. Both antibiotics and herbicides are usually phosphonic acids or monoesters of phosphonic acid. As a result, they do not have the same biochemical effects as do the diesters of phosphonic acids. The naturally occurring phosphonate antibiotics are presented in Chapter 2 and the factors involved in antibacterial activity are discussed in Chapter 5.

The herbicides and plant growth regulators, while their mechanisms of action are not well-defined, will also be presented in this chapter. The distinctive feature of the compounds presented in this chapter is that they are exclusively "synthetic" and are not normally anabolized by biological systems, although they may undergo catabolism.

II. EFFECTS ON LIFE PROCESSES IN ANIMALS

A. Representative Agents

O'Brien[4] has presented detailed information on the historical aspects of the development of toxic phosphorus esters. The toxic nature of the phosphate esters was recognized sometime in the 1930s and intensively studied, primarily in Germany and England, in the 1940s. The work in Germany was headed by G. Schrader and was directed toward development of insecticides. At that same time, Sarin, Soman, and Tabun, compounds of extreme toxicity to mammals, were developed as potential chemical warfare agents.[4] The work in England was directed by B. C. Saunders and was primarily involved with diisopropyl fluorophosphate and other compounds containing fluorine. The materials (both from Germany and England) were commonly applied as mists and came to be known as "nerve gases". This work provided the foundation for the chemistry of the organophosphate (including phosphonate) insecticides which are currently in use.[4]

1. Insecticides

The use of organophosphates as insecticides has developed significantly in the last decade because of their relatively low chronic toxicity to mammals and limited environmental persistence compared to the organochlorine insecticides.[5] Organophosphorus insecticides of the phosphonate class are represented by the general structure

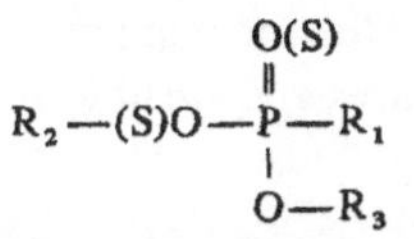

where R_n is an alkyl, aryl, or substituted alkyl or aryl group. Sulfur (S) can substitute for oxygen in thiono- or thiolophosphonates. Representative insecticides of the phosphonate class are shown in Figure 1. Other phosphonates with insecticidal activity are considered by Eto[6] in his comprehensive discussion of the organophosphonate insecticides. These materials are unusual among chemicals since, by definition, they must exhibit some deleterious biological effect in order to be beneficial as insecticides.

The efficacy with which an organophosphonate insecticide acts is related to several factors, including chemical reactivity, metabolic characteristics, ability to penetrate cuticular and membrane barriers, and ability to be transported to and inhibit the target enzyme, AChE. These factors must be balanced to provide an insecticide which is toxic to the target species without presenting unreasonable hazard to other organisms. Few pesticides are of sufficient specificity to act on a target organism without presenting some hazard to nontarget species, especially since application is often made around, or directly to, food sources for other species or where nontarget species cohabit. The degree of hazard presented to nontarget organisms by a particular pesticide is dependent on the quantities used, the method of application, the degree of translocation, and the chemical, physical, and biological char-

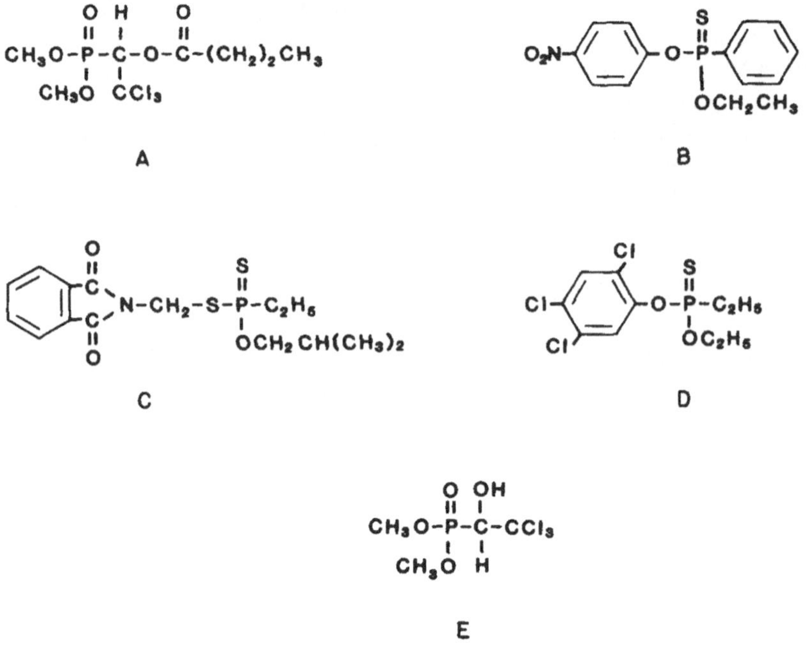

FIGURE 1. Representative insecticides containing a C–P bond. (A) Butonate; *O,O*-dimethyl 2,2,2-trichloro-1-*n*-butyryl-oxyethylphosphonate; (B) EPN; *O*-ethyl *O*-*p*-nitrophenyl phenylphosphonothioate; (C) *N*-(mercaptomethyl)phthalimide *S*-(*O*)-isobutylethylphosphonodithioate); (D) trichloronate; *O*-ethyl *O*-2,4,5-trichlorophenyl ethylphosphonothioate; (E) trichlorfon; *O,O*-dimethyl (1-hydroxy-2,2,2-trichloroethyl) phosphonate. Serial numbers of other phosphonate pesticides are 324, 333, 400, 515, 516, 518, 596, 762, 763, 776, 778, 782, 899, 1040, 1042, 1043, 1050, and 1246. (Data from Frear, D. E. H., *Pesticide Index*, 4th ed., College Science Publishers, State College, Pa., 1969.)

acteristics of the pesticide. The hazard can be minimized and specificity of action maximized by an understanding of the mechanism (or mechanisms) producing the pesticidal and toxicological effect and the structural features responsible for that effect.

2. Chemical Warfare Agents

An extensive review of the early work on chemical warfare agents and relationships between structure and toxicity was completed by Saunders.[7] The three compounds of primary interest are Soman, Sarin, and Agent VX (Figure 2).[8] The most significant effect of the chemical warfare agents is their extreme acute toxicity. The LC_{50} for Sarin (for humans) is estimated to be 70 mg/m^3.[9] Although LD_{50} values are dependent on the route of administration and on the species used to determine them, they are in the mg/kg range for the most species. For example, the oral LD_{50} in rats is 500 mg/kg and the subcutaneous LD_{50} in the guinea pig is 38 mg/kg.[9] The range of toxicity is very similar for Soman.[9] Toxicity may be seen with a dose as low as 1 mg/kg.[4]

These chemical warfare agents are extremely effective inhibitors of AChE and inhibit the enzyme in the same manner (Figure 2) as do the organophosphonate and carbamate insecticides. A comparison of the structure of acetylcholine, the normal substrate for AChE, with that of the inhibitors shows similarities in structure and in ionic charge placement (Figure 2). In cases where fluorine is the leaving group it is not surprising that the inhibitors have such a high affinity for the AChE, due to the high electronegativity of the fluorine to facilitate nucleophilic attack of the active site hydroxyl on the phosphorus.[30]

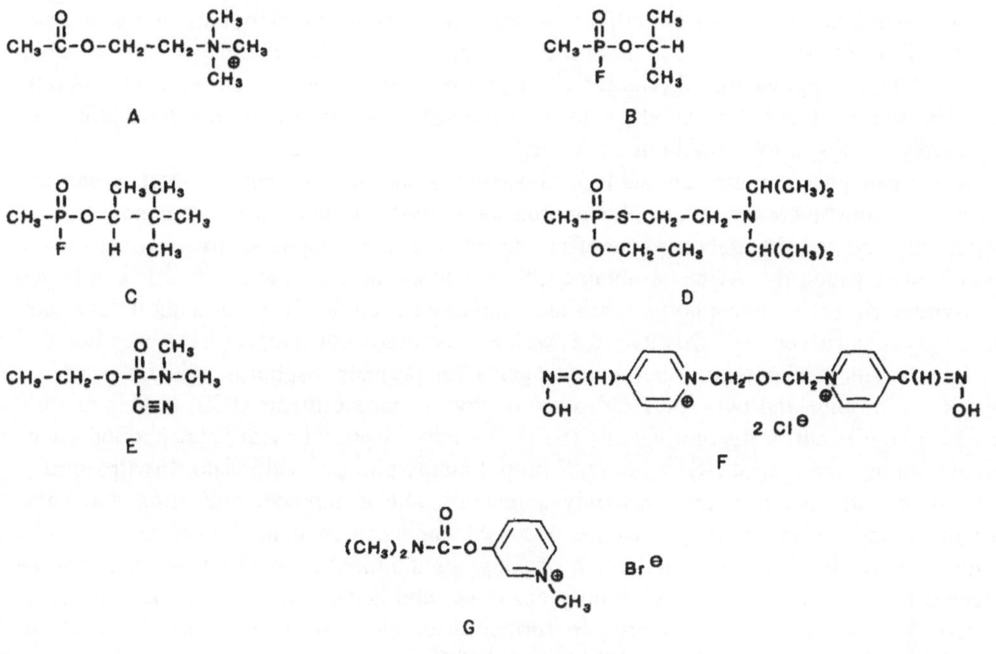

FIGURE 2. Representative compounds which bind to acetylcholinesterase (AChE). (A) Acetylcholine; the normal substrate; (B) Sarin, Agent GB, *O*-isopropyl methylphosphonofluoridate; (C) Soman, Agent GD, *O*-1,2,2-trimethylpropyl methylphosphonofluoridate; (D) Agent VX, *O*-ethyl *S*-2-(diisopropylamino)ethyl methyl-phosphonothiolate; (E) Tabun, *O*-ethyl *N,N*-dimethylphosphonoamidocyanidate; (F) obidoxime chloride, oxybis-(4-hydroxyiminomethylpyridinium-1-methyl) dichloride; a reactivator of inhibited AChE; (G) pyridostigmine bromide, 3-hydroxy-1-methylpyridinium bromide dimethylcarbamate; a protector of uninhibited AChE.

B. Biological Activity

1. Inhibition of Acetylcholinesterase

a.Structure-Activity Studies

The inhibition of AChE requires diesters of phosphonic acid and is primarily dependent on the reactivity of the compound toward alkaline hydrolysis.[10] The reaction with AChE that results in inhibition is a nucleophilic displacement reaction, which is facilitated by the presence of substituents on the phosphonate which are electron withdrawing and results in a monoesters of an alkylphosphonate esterified to the hydroxyl of the active site. Other factors, besides the reactivity, which influence anticholinesterase activity are steric factors, stereoisomerism, and pseudorotation.

Phosphonates are, in general, more reactive than the corresponding phosphate esters because the carbon has no unpaired electrons to contribute to allow a $p\pi$ - $d\pi$ contribution to the C–P bond.[11,12] This makes the phosphorus atom of phosphonates more electrophilic than the phosphorus atom of the corresponding phosphate ester. The esters of phosphonic acids are thus more susceptible to alkaline hydrolysis than the phosphate esters and more stable to acid hydrolysis than the phosphate esters.[12] The C–P bond is usually stable to hydrolytic procedures; however, α-hydroxy alkylphosphonates, in the presence of alkali, will rearrange to give phosphate with cleavage of the C–P bond. This is the mechanism of conversion of trichlorfon (or the antischistosomal pharmaceutical of trichlorfon, Metrifonate) into dichlorvos, which is an active inhibitor of AChE.[11,13] In addition, Tabun (Figure 2) is an unusual compound in the manner in which it is hydrolyzed. Although the C–P bond of phosphonates is normally not cleaved, Tabun is hydrolyzed at the phosphorus to nitrile bond.[14]

Steric factors are very influential in the inhibition of AChE by phosphonates.[10] If the alkyl group is increased in length from ethyl to *n*-hexyl in ethyl *p*-nitrophenyl alkylphosphonate, the inhibition of AChE is less than would be expected from the change in hydrolyzability.[11,15,16] This suggests that the larger alkyl groups prevent interaction with the esteratic site. The steric effect of the alkyl group is of greater importance than the steric effects of the alkoxy groups in the inhibition of AChE.[11]

Most organophosphonates are achiral; however, if one or more chiral center is present, either in a substituent or at the phosphorus atom itself, it may affect the inhibition of AChE.[10,11] The actual stereochemical effect depends on the compound involved and on the species from which the AChE is obtained.[10] Of interest here is that the AChE is affected by asymmetric organophosphorus inhibitors, although acetylcholine, its natural substrate, has no asymmetric center.[11] Qualitative as well as quantitative differences have been noted.[17] An in vivo study of potential therapeutic agents for organophosphorus poisonings used a mixture of atropine and pyridine-2-aldoxime methylmethanesulfonate (P2S).[18] This mixture provided significant protection against (S) (−) - ethyl *S*-propyl methylphosphonothionate but did not protect against (R)(+)-ethyl *S*-propyl methylphosphonothionate. In vitro studies showed this stereodependence was only a reaction rate difference, indicating that some metabolic modification of the (+) isomer occurred which caused an inhibition not responsive to the therapy. In vitro, the reactivation of enzymes inhibited by both enantiomers can be affected by oximes whereas in vivo only enzymes inhibited by the (−) isomers are reactivated.[18] A variety of sources provide further discussion on stereochemical aspects of phosphorus chemistry and of the inhibition of AChE.[17,19-25]

A nucleophilic attack on phosphate-phosphorus, would be expected to result in inversion of configuration. Pseudorotation is a reaction whereby a tetracoordinate organophosphorus compound undergoes nucleophilic attack at the phosphorus to form a pentacoordinate trigonal bipyramidal intermediate followed by a permutational isomerization and the departure of the leaving group with retention (rather than inversion) of configuration.[18] The factors which influence the occurrence of pseudorotation are not well-understood. However, the C–P bond provides a restraint on pseudorotational process with a resultant inversion of configuration upon hydrolysis. This may, in fact, facilitate rates of reaction of phosphonates with AChE and have other biological implications as well.[12,18]

A greater variety of structures for insecticides can be expected in the phosphonate class than in the phosphate class.[6] This is a result of the alkyl and alkoxy groups having less effect on the biological activity of the phosphonates than the alkoxy groups in the phosphates and thus allowing a greater number of substituents to be used.[11] It is apparent that the many factors influencing AChE inhibiting activity have complex interactions. The chemical reactivity and structure of a compound will be influenced by a dynamic relationship between activation and detoxification by metabolic processes and the availability of all of the materials to the AChE. Thus, the prediction of structure activity relationships is difficult.[27]

b. Metabolism

Metabolic processes can have a great effect on the in vivo toxicity of organophosphonates; if a mechanism exists for the rapid metabolic degradation of the compound, toxicity may never be exhibited. The metabolic process may be the hydrolysis of an ester with a consequent increase in the hydrophilic nature of the phosphonate. The metabolites may then be excreted directly or undergo further degradation for ultimate elimination from the body. The enzymatic hydrolysis of one of the esters is normally a detoxification and prevents subsequent inhibition of AChE by the organophosphorus compound. DuBois et al.[28] have shown that inhibition of hydrolytic detoxification by one organophosphorus compound may potentiate the toxicity of another organophosphorus compound or of other drugs and chemicals. Although the phosphonates may be metabolically and chemically more reactive than the corresponding phosphate esters, there is no evidence that the C–P bond is hydrolyzed enzymatically by

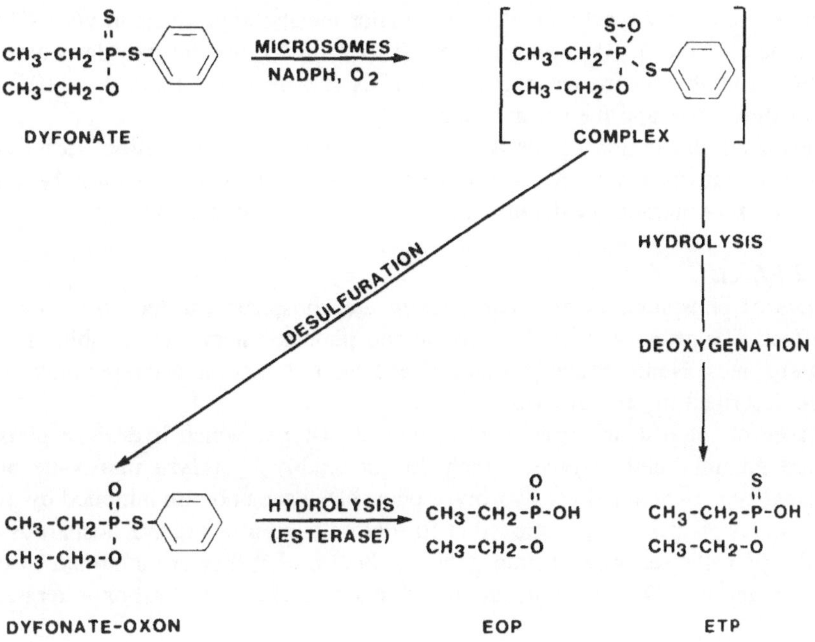

FIGURE 3. Metabolic pathway for the activation and inactivation of dyfonate (via a postulated oxygenated intermediate). (From McBain, J. B., Yamamoto, I., and Casida, J. E., *Life Sci.*, 10, 947, 1971. With permission.)

mammalian systems. Thus, as shown by Hollingworth et al.,[29] the phosphonates offer fewer metabolic possibilities than do phosphates. Other reactions of importance to the metabolism of organophosphonates are the oxidation of thiono (=S) or thiolo(–S–) groups, the hydroxylation of substituent alkyl or aryl groups, and oxidative dealkylation.

Many phosphonate insecticides are either thioates or dithioates (Figure 1). The thiono compounds are very poor inhibitors of AChE in vitro and must be metabolized to oxoanalogs to exhibit biological activity.[27] The microsomal mixed function oxidase system is responsible for the oxidative desulfuration of the thiono to the oxon in mammals and requires NADPH and molecular oxygen for the reaction to occur.[11] The poor electron-withdrawing capability of the sulfur may be the main reason for the negligible activity against AChE although other factors such as the lack of hydrogen bonding between the sulfur atom and the AChE active site should be taken into account.[27] In contrast to the minimal biological activity of the thiono group, the thiolo-esters generally have greater biological activity than the corresponding oxygen analogs.[27] This thiolo effect can be accounted for by the chemical reactivity of the thiolo analogs and is especially pronounced if the oxygen analog is a poor AChE inhibitor.[30]

Menn et al.[31] have reported on the metabolism of organophosphonates in rats by using dyfonate, a phosphonodithioate, as a model compound. The dyfonate was proposed to undergo oxidative desulfuration through an intermediate complex (see Figure 3). This complex can then proceed on to the dyfonate-oxon or undergo hydrolysis and deoxygenation to allow the oxidation of the thioether to *O*-ethyl ethyl phosphonothioic acid (ETP). The complex can undergo desulfuration and hydrolysis to give *O*-ethyl ethyl phosphonic acid (EOP). ETP accounts for over 50% of the urinary excretion of dyfonate. Of interest here is that when ETP (=S) is administered to rats there is no EOP (=O) found in the urine. Thus, if oxidative desulfuration of the dyfonate does not occur prior to hydrolysis of the thiolo linkage, the desulfuration is not likely to occur to a detectable level.[31] The oxon and other oxidized intermediates are relatively reactive and are not major metabolic products in vivo. The oral LD_{50} for dyfonate in rats is 15 mg/kg compared to 3 mg/kg for the oxo derivative. The relative ability to inhibit bovine erythrocyte AChE is shown by the I_{50} of 2×10^{-5} and 3×10^{-8} *M* for dyphonate and the oxon, respectively.[31]

A comprehensive discussion of the effects of metabolism on activation, deactivation, excretion, and biotransformation of organophosphorus compounds in general is beyond the scope of this work. A number of detailed reviews are readily available.[4,11,32,33]

c. *Biological Effects*

The diesters of phosphonates are analogous to the phosphate triesters in structure and biological effect. The primary toxic effect of the phosphonate esters is the inhibition of ester hydrolyzing enzymes. Neurological processes[34] and the role that the esterases play in these functions are described in previous works.[34-36]

The esterases of interest are broadly classified as A-type, which hydrolyze phosphate esters and are not inhibited; B-type, which do not readily hydrolyze phosphate but are inhibited; and C-type, which neither hydrolyze phosphate esters nor are inhibited by them.[36] The A-type esterase is primarily responsible for detoxification of organophosphorus pesticides. The B-type esterases are of primary interest here in that they are inhibited by organophosphorus materials. The esterases in this category include the carboxyesterases and cholinesterases.[36]

The cholinesterases are subdivided functionally into two types, acetylcholinesterase and butyrylcholinesterase. Acetylcholinesterase (EC 3.1.1.7, AChE or true cholinesterase) is found in the central nervous system, ganglia, motor end-plates, and erythrocytes. AChE preferentially hydrolyzes the neurotransmitter acetylcholine, which is produced at cholinergic synapses by the activity of choline acetylase with choline and acetyl coenzyme A. Following release acetylcholine migrates to a receptor in the postsynaptic membrane of another neuron and produces a change in cation conductance, which is followed by depolarization of the membrane. Following this synaptic transmission, the acetylcholine is released from the neuroeffector site and is available for hydrolysis by AChE while the neuronal membrane returns to a resting potential. Inhibited AChE is unable to hydrolyze acetylcholine and allows accumulation of acetylcholine at the synapse. Some organophosphates will interfere with binding of acetylcholine to the receptor; however, under normal conditions this is not a significant effect since the concentration required to produce this effect is 100 to 1000 times the concentration required to inhibit cholinesterase.[36]

The second type of cholinesterase is butyrylcholinesterase (EC 3.1.1.8, BuChE, pseudocholinesterase) which is present in various types of glial or satellite cells, and in the plasma, liver, and other organs.[34] BuChE is most effective with butyryl and propionyl esters of choline but can hydrolyze acetylcholine and other carboxylate esters. However, its physiological function is unknown and inhibition of BuChE produces no known functional abnormalities. Although BuChE is inhibited by many organophosphonate agents and is more sensitive than AChE to inhibition by some agents, the significant physiological effects of organophosphonates are due to the accumulation of endogenous acetylcholine.[34]

d. *Mechanism*

The normal hydrolysis of acetylcholine by AChE is depicted in Figure 4. The base, B_2, is proposed to be the imidazole group of histidine and responsible for the activation of the serine hydroxyl of the active site. A catalytically active acid group (H–A), which is likely to be the phenolic hydroxyl of tyrosine, is present near the esteratic oxygen of acetylcholine. The quaternary nitrogen of the acetylcholine interacts with the anionic site of the AChE and positions the substrate for nucleophilic attack by the activated hydroxyl of the active site. The acetylated serine is then quickly deacetylated (hydrolyzed) as a result of a conformational change of the enzyme and the activation of a water molecule by an imidazol base, B_1.[36,37]

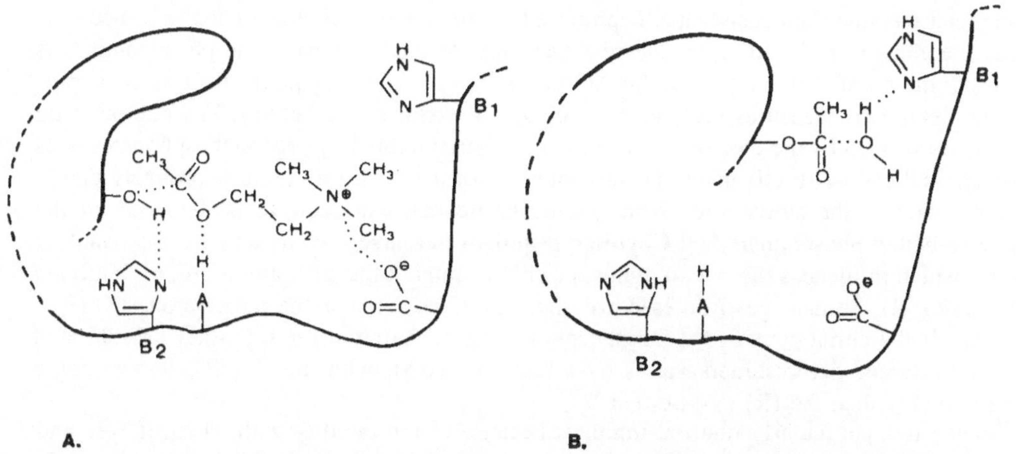

FIGURE 4. Schematic mechanism for AChE. (A) AChE complex with acetylcholine; (B) deacetylation of acetyl-AChE. (From Eto, M., *Organophosphorus Pesticides: Organic and Biological Chemistry*, CRC Press, Boca Raton, Fla., 1979, 133.)

The phosphonylation (or phosphorylation) of AChE proceeds by the same basic mechanism. An enzyme-ester complex is formed and stabilized. The hydroxyl group of the active site is activated and phosphonylated. The result is a phosphonylated serine which can either undergo "aging" and be essentially irreversibly phosphonylated or undergo dephosphonylation. If the phosphonate has steric and charge similarities to acetylcholine the inhibition will be facilitated, as illustrated by organophosphonates used as chemical warfare agents and the therapeutic compounds used in treatment of AChE inhibition (see Figure 2). Spontaneous reactivation (dephosphonylation) occurs by the hydrolysis of the enzyme-phosphonate ester linkage to produce the active enzyme and a free phosphonate.

The active site of the enzyme must have a suitable steric configuration to allow the alignment of catalytic groups in the active site with the substrate (or inhibitor). The AChE is proposed to undergo conformational change, to allow the appropriate sites to align. Investigators have recently shown that ligands such as quaternary nitrogen inhibitors which bind with the anionic site alter the active site conformation.[38,39] Although occupation of the anionic site is rapid, the active site conformation change may be relatively slow. In addition, there are apparently different active site conformation effects depending on the ligand which has associated with the peripheral site.[38]

Following the covalent linkage of an organophosphonate inhibitor to the active site of AChE, the phosphonate will have one alkoxy group as a substituent, which is necessary to allow "aging" to occur. Aging is a process whereby a phosphonylated (or phosphorylated) AChE which is stored for a given period can no longer be reactivated by oximes. Aging is due to the dealkylation of the phosphonate at an alkoxy group, as shown below, and is proportional to the electron donating capacity of the alkyl group; thus, alkyl groups of high alkylating ability age rapidly.[40,41,61]

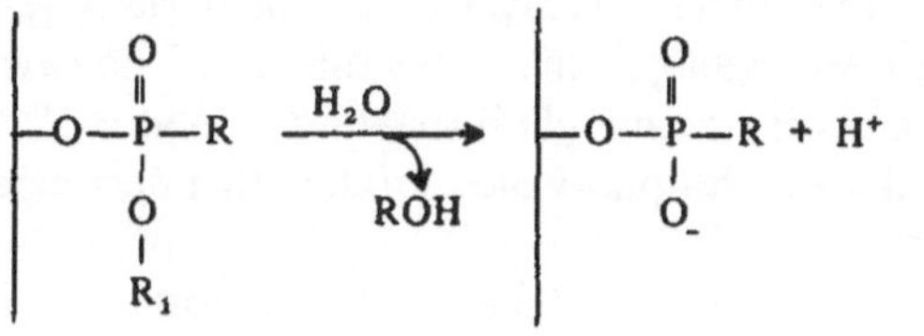

The aged enzyme then resists nucleophilic attack of oximes because of the presence of a negative charge on the phosphonate residue bound to AChE.[36] Until the phosphonate has "aged" the possibility exists that the AChE can be reactivated spontaneously or by appropriate therapeutic measures such as the nucleophilic oximes (see below). The rate of aging is also influenced by the enzyme, since chymotrypsin inhibited by organophosphonates does not age and can be reactivated. The susceptibility of the AChE to aging is probably due to the presence at the active site of an anionic group that can catalyze dealkylation of the enzyme-bound phosphonate.[40,41] Chymotrypsin does not have such an anionic site. Another factor which influences the rate of aging is a chiral center at the phosphorus. AChE inhibited by the $(R)_c(R)_p$-Soman ages 7 times slower than the stereoisomer at the carbon atom, $(S)_c(R)_p$-Soman. If the chiral center is at the phosphorus atom, the rate of aging of AChE inhibited by the different stereoisomers varies by a factor of 1000, with the $(R)_c(R)_p$-Soman aging more quickly than the $(R)_c(S)_p$-Soman.[36]

Soman is a particularly unusual inhibitor because of the rapidity with which it ages and the concomitant loss of potential for reactivation.[41] In addition, the *O*-alkyl group is 1,2,2-trimethylpropanol, a large substituent, which makes the reactivation of the AChE by oximes more difficult due to steric factors. These factors in combination with the effect of Soman on the central nervous system and the possibility of a storage depot for Soman, make therepeutic treatment of Soman poisoning unusually difficult.[42,43]

e. Therapeutic Approach

The symptoms of acute exposure to organophosphonate AChE inhibitors are due primarily to the accumulation of acetylcholine at the synapse. They initially include headache, miosis, weakness, and pallor. As the intoxication progresses, other symptoms, including salivation, lacrimation, fasciculations, and convulsions will appear.[44] The lethal effect is asphyxia which is due to the compromise of the respiratory system at several levels: (1) broncho constriction and excessive tracheobronchial secretions, (2) paralysis of the muscles involved in respiration, (3) lowered blood pressure, and (4) paralysis of the respiratory center.[45,46] These effects may be followed by a secondary cardiovascular collapse.[47]

If therapy is to be effective it must be instituted immediately upon exposure. The therapeutic approach is to treat the effects of excess acetylcholine, reactivate AChE and treat other symptoms as required. Atropine given every 10 to 15 min until signs of atropinization (dry, flushed skin, tachycardia, and dilation of the pupils) occur is used to counter the acetylcholine accumulation. The atropine will compete with acetylcholine at muscarinic receptor sites. A mild degree of atropinization should be maintained as long as symptoms are in evidence.[44]

An oxime, such as pyridine-2-aldoxime or obidoxime chloride (Figure 2), can be used to reactivate the inhibited enzyme. The proposed mechanism of this reactivation is that the positive charge on the oxime is attracted to the anionic site of the AChE. An inhibited enzyme-reactivator complex is formed and the nucleophilic group of the reactivator is in close proximity to the active site. The reactivation is then essentially a nucleophilic displacement reaction on the phosphorus atom to regenerate the hydroxyl of the enzyme. This mechanism and the chemical parameters affecting it are discussed extensively by Eto.[36]

The introduction of a second cationic center into a reactivator molecule (as in the obidoxime) makes a much more effective reactivator than the pyrid ne-2-aldoxime. In addition, an oxime can react with free organophosphonates directly,[12] however, the phosphorylated products of the oximes are very active cholinesterase inhibitors.[36,43,48] Thus, the oximes selected for use must produce a phosphorylated product that decomposes rapidly to prevent inhibition of the AChE.[36]

Soman toxicity is difficult to treat therapeutically; in addition to inhibiting AChE and being an agent that ages rapidly, Soman is believed to cause respiratory paralysis by affecting the respiratory center in the central nervous system.[49] Recent work indicates that, even if Soman inhibited AChE in the central nervous system is reactivated, the reactivation might not be therapeutic.[49] This is based on the development of an oxime which regenerates Soman inhibited central nervous system AChE but does not protect against poisoning by Soman. In addition, other therapeutic compounds, which contained no oxime group and do not reactivate central nervous system AChE, were protective when combined with atropine. The efficacy of an oxime found to be protective against Soman is attributed to the reactivation of AChE of respiratory musculature of the diaphragm.[49] The same compound apparently has antinicotinic and antimuscarinic activity as well as being a ganglionic blocking agent.[49] Other oximes and therapeutic regimes for treatment of Soman poisoning are being developed and investigated.[50-52,62]

In the cases where an exposure to organophosphonates can be predicted there is another approach to therapy which can be taken. This is not so much a prophylactic measure as it is a pretreatment to facilitate subsequent therapy. This pretreatment involves the "protection" of 30 to 40% of the AChE activity with a carbamate.[53-55] Carbamates (including the carbamate insecticides) exert their toxic effect by inhibiting AChE. However, as opposed to the phosphonylated enzyme, the carbamoylated enzyme regenerates rapidly, a factor which allows selected carbamates to act as protective agents. Pyridostigmine effectively protects AChE from phosphonylation while the organophosphonate is being detoxified by spontaneous hydrolysis, binding to aliesterases (which are not inhibited by carbamates),[47] metabolism, and excretion. The AChE then regenerates quickly from the carbamate inhibition.[53]

Although the phosphonylation of AChE is the primary toxic effect of most organophosphonates, there are a variety of indirect or other biochemical effects which can occur.[56-59] One effect is a neuromuscular dysfunction induced by AChE inhibition.[56,57] The mechanism is apparently independent of acetylcholinesterase inhibition and may result from effects on receptor and receptor-channel complexes. Other effects include behavioral changes and electroencephalographic changes even following relatively long recovery periods.[59,60]

2. Delayed Neurotoxicity

a. Introduction

In 1930 the widespread occurrence in the U.S. of a peripheral neuropathy was reported and attributed to the ingestion of tri-*o*-cresyl phosphate (TOCP).[63] The TOCP (Figure 5) was used to replace castor oil in a drink reputed to be "Fluid Extract of Ginger, U.S.P." to reduce costs. This was done without consideration of possible toxicity and the result was the paralysis of many thousands of persons. That extract was actually intended for use as a beverage during prohibition since true Fluid Extract of Ginger, U.S.P., was not potable due to the taste.[64] Further work demonstrated that TOCP is metabolized in vitro and in vivo to produce a cyclic phosphate metabolite (Figure 5)[65,66] which produced neurotoxicity in the chicken.[67] The neurotoxic metabolite is a potent inhibitor of esterases and is a saligenen cyclic *O*-tolyl phosphate.[67,68]

Since that time a number of poisonings by organophosphates followed by a delay period of several days with few untoward signs and then the rapid onset of paralysis have been reported[69-73] and reviewed.[74-77] Despite the efforts at prevention, these incidents continue to occur, although with less frequency and on a smaller scale than previously. In many of the cases the material responsible for the toxicity was identified as TOCP. However, the occurrence of organophosphate induced delayed peripheral neuropathy (OPIDN) as a result of exposure to other compounds such as the insecticide mipafox (*N,N*′-diisopropylphosphorodiamidic fluoride) and the phosphonate insecticide leptophos (*O*-[2,5-dichloro-4-bromophenyl]-*O*-methyl phenylphosphonothioate, Figure 5) emphasizes that a variety of materials can produce the effect.[69,78]

OPIDN is characterized by symptoms which appear 5 days to 4 weeks postexposure.

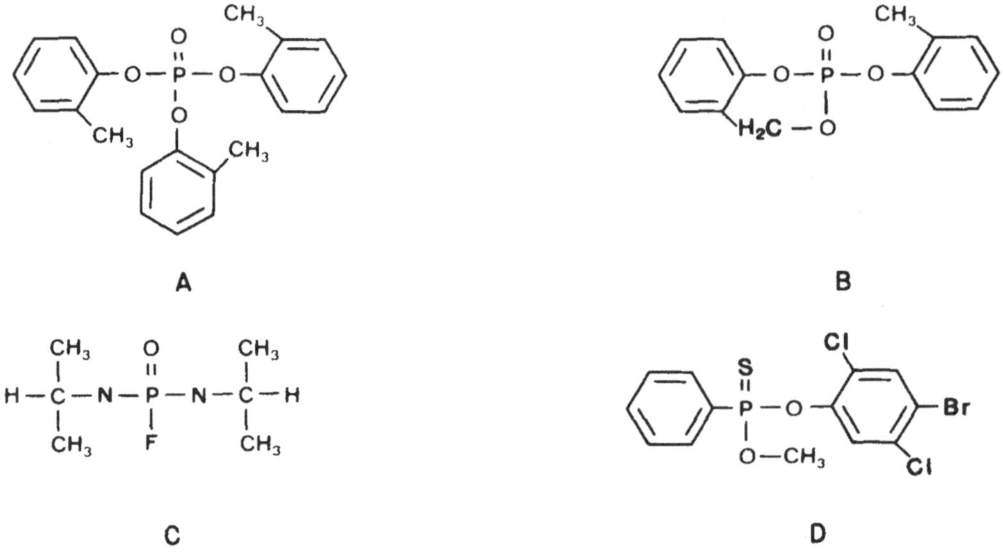

FIGURE 5. Representative compounds which produce OPIDN. (A) TOCP, tri-ortho-cresyl phosphate; (B) the toxic metabolite of TOCP, a cyclic saligenin phosphate; (C) Mipafox, *N,N'*-diisopropylphosphorodiamidofluoridate; (D) leptophos, *O*-(4-bromo-2,5-dichlorophenyl) *O*-methyl phenylphosphonothioate.

Initially, soreness of leg muscles occurs and is followed by ataxia and weakness progressing to paralysis of lower extremities. About 1 week later, similar symptoms occur in the upper extremities and, in severe cases, there is paralysis to the elbows. The lower extremities are the most severly affected and any improvement will be seen first in the upper extremities. Sensory symptoms soon disappear but the symmetrical paraplegia or quadriplegia remains. Children and young animals are not as sensitive to toxicity as are adults, in whom recovery is slow and seldom complete.[69,79,80] Delayed neurotoxicity has been reviewed recently by several authors.[74-77,81]

b. Proposed Mechanisms

The mechanism producing delayed neuropathy is not entirely known but must account for a number of divergent experimental findings.[77]

1. The first and most severe effect is on the lower extremities while the upper extremities are not so severely affected and may in fact recover.
2. Young chicks (<60 to 70 days) are resistent but chick embryos treated on incubation day 14 are susceptible to OPIDN.[82] Adult rodents are resistent to a single dose but may be affected by repeated dosing. Adults of species such as the cat or the chicken are susceptible to a single dose.
3. The effect is not observed until 2 to 3 weeks following exposure and is not reversed by AChE reactivating agents.
4. Certain nerves and areas of the central nervous system show cellular damage while others do not. The lesion is primarily dying back, rather than demyelination, of long axons and large diameter peripheral nerves.
5. Some AChE inhibitors produce OPIDN while others do not. Most compounds giving a delayed neurotoxicity inhibit AChE to some degree.
6. Some materials, such as carbamates, which can inhibit an esterase identified as neurotoxic esterase are protective against the delayed neurotoxicity of other compounds.

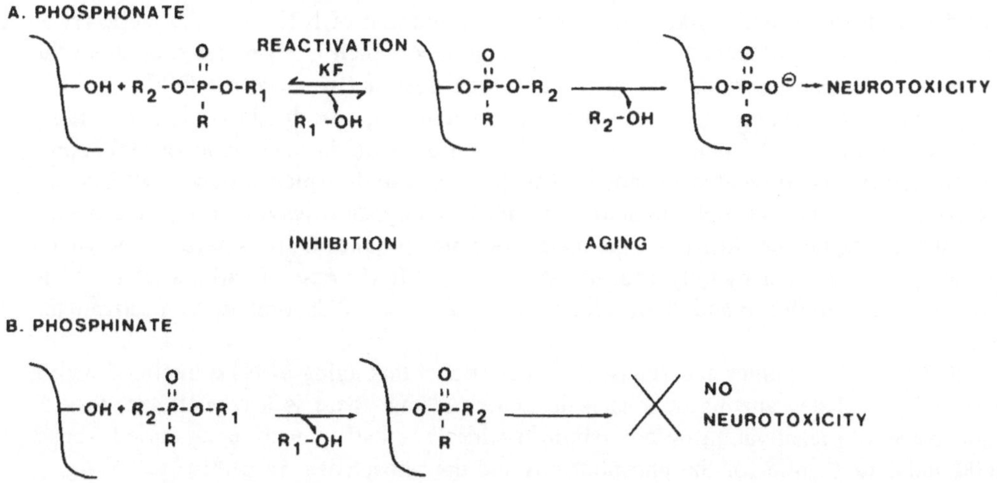

FIGURE 6. (A) Schematic mechanism for the production of OPIDN by a phosphonate. The esterase is inhibited and produces neurotoxicity following "aging"; (B) the esterase is inhibited by a phosphinate but OPIDN does not result.

A compelling proposal, based on the binding of labeled neurotoxic phosphate esters to brain esterases, has been made by Johnson[83] and is consistent with much of the available data (Figure 6). From the expected behavior of organophosphates in biological systems, it was likely that the delayed neurotoxicity was a result of the phosphorylation of a target protein. Johnson then found that under given conditions, 700 pmol of ^{32}P would bind to 1 g of hen brain following a neurotoxic dose of labeled DFP. If a nonneurotoxic inhibitor, TEPP (tetraethylpyrophosphate), was preincubated with the tissue and the ^{32}P-DFP then added, the binding of ^{32}P was reduced to 215 pmol/g. The further incorporation of the neurotoxic material, mipafox, with the nonneurotoxic inhibitor for a preincubation followed by ^{32}P-DFP led to a further reduction in binding to 178 pmol/g tissue. The 37 pmol reduction by the binding of neurotoxic mipafox was assumed to represent the esterase specifically responsible for OPIDN. This esterase was designated as neurotoxic esterase (NTE).[75,83] Further studies indicated that >80% of NTE needed to be inhibited for OPIDN to occur. All of the neurotoxic compounds inhibited at least 75% of the NTE and no compound that produced more than 80% inhibition of NTE failed to produce OPIDN (dichlorvos required a second dose).[84,85]

Johnson[87] has found that NTE is inhibited by a variety of compounds and that the esterase hydrolyzes both phenylphenylacetate and phenyl valerate. The latter substrate is hydrolyzed with greater specificity than the phenylphenylacetate. Excellent correlation has been found between the inhibitiion of this esterase, the labeling of the brain homogenate and the production of a delayed neurotoxic effect.

The proposed mechanism (Figure 6) is that the active site of the esterase is phosphorylated specifically by organophosphonates (or phosphates) which produce the delayed neurotoxicity. The effective compounds then must undergo "aging" (or the hydrolysis of the remaining phosphate ester bond) to produce a negative charge on the phosphorylated esterase.[88] This negative charge then interferes with normal biochemical or physiological processes in an undefined manner, but may interfere with control mechanisms or other physiological processes.[75]

c. *Aging and Protection*

The aging is a requirement since the actual inhibition of the esterase does not produce delayed neurotoxicity. This is known because the inhibition of NTE by either sulfonyl or phosphinyl moieties produces a stable inhibited enzyme which is regenerated at about the same rate as the phosphorylated enzyme but which does not produce the OPIDN.[87,88] In fact, these compounds and carbamates provide protection against the delayed neurotoxicity of other compounds.[75,89,90] One significant difference between those compounds which produce the effect and those that do not is that the compounds which produce OPIDN can undergo a dealkylation to yield an acidic phosphate group, thus leaving a negative charge on the phosphorylated esterase (i.e., the materials which produce OPIDN have a bond which can undergo ''aging'' or hydrolysis at the active site).[87] In the case of carbamates the C–N bond would be hydrolyzed and the resultant intermediate would decompose to reactivate the enzyme.[75]

Further work by Clothier and Johnson[91,92] has shown that aging of NTE inhibited with a variety of phosphates and phosphonates is rapid and that there is a concomitant loss of responsiveness to reactivation by potassium fluoride. The half-lives of aging varied from 3 to 600 and 1 to 10 min for the phosphonates and the phosphates, respectively.[91,92] Aging occurred rapidly with aryl and straight chain alkyl substituents on oxygen and slowly with highly branched alkoxy substituents. The mechanism of aging is suggested to be different than that for AChE because of the effects of structure on the aging process being different in the two cases and because the aging of NTE is somewhat insensitive to pH.[92]

These studies also indicate that the isopropyl group of DFP is not released in a free state upon aging and is probably transferred on the protein.[91] The involvement of macromolecular structures in the aging process opens the possibility that either the transfer of the substituent to another site can result in toxicity or that the active site can be cross-linked to another site on the protein and produce the toxicity.[92] In either case, the effect is still delayed and may be either metabolically mediated or the result of an abnormality in a control mechanism.

d. Structure-Activity Relationships

Many alkyl and alkyl aryl esters of phosphonates, as well as fluorophosphonates, have been shown to be neurotoxic. Although, in some cases, the OPIDN is observed only after significant therapeutic measures have been taken to allow the test animal to survive the acute toxicity resulting from the inhibition of AChE. This is especially true of the alkylfluorophosphonates.[84] In evaluating structure-activity relationships in the production of OPIDN, the alkylphosphonates, alkyl/arylphosphonates, and the arylphosphonates are treated separately. In the alkyl ester group (Table 1) the phosphonates and the fluorophosphonates are more toxic than the corresponding phosphates (as is seen with the inhibition of AChE).[77] The thiono analogs were not neurotoxic based on the results from the few compounds tested.[77] In the alkylfluorophosphonates all the compounds tested were neurotoxic and the toxicity depended on the alkoxy rather than the alkyl group. Increasing length (or hydrophobicity) increased neurotoxicity up to an iso-propyl or iso-butyl substituent.[84] This effect is not seen with AChE activity; therefore, the neurotoxicity/lethality ratio increases up a homologous series to about *n*-butyl.[84] Compounds having 2,2-dichlorovinyl as a leaving group were also neurotoxic but the dialkyl pyrophosphonates did not produce neurotoxicity.[93]

In the mixed ester group (Table 2) the majority of phosphonate and phosphonothionates produce OPIDN, although in the thio analogs the neurotoxicity is likely due to the oxo metabolite.[84] The alkyl phenylphosphonates are more potent than the corresponding thiono analogs. In the ω-phenylalkylphosphonates the neurotoxicity depended on the length of the aliphatic moiety and increased up to two methylenes and decreased to nontoxic as the methylenes increased to four.[84] These results are similar to those of the alkylphosphonates and have led Abou-Donia[77] to propose that there are hydrophobic areas of limited size on the active surface of the ''neurotoxicity protein'' which can only accomodate certain hydrocarbon moieties. This is supported by his findings with phenylphosphonothioates and by other data as well.[77]

Table 1
ALIPHATIC PHOSPHORUS ESTERS TESTED FOR DELAYED NEUROTOXICITY IN CHICKENS

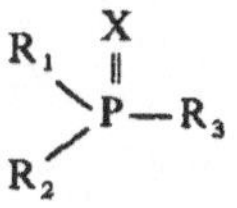

Substituents							
R_1	R_2	R_3	X	Dose (mg/kg)	Route of administration[a]	Delayed neurotoxicity	Ref.
			Phosphonates				
CH_3	CH_3O	tri-ClC–CH(OH)O	O	200 + 100[b]	s.c.	+	87
C_2H_5	ClC_2H_4O	di-ClC=CHO	O	2.5	s.c.	+	84
CH_3	CH_3O	tri-ClC–CH(OH)	O	200 + 100[b]	s.c.	+	87
			Phosphonothioates				
C_2H_5	C_2H_5O	C(CN)=NO	S	75	p.o.	–	84
C_2H_5	ClC_2H_4O	ClC_2H_4O	S	40	p.o.	–	156
			Phosphonofluoridates				
iso-C_3H_7	CH_3O	F	O	5	i.m.	+	93
CH_3	C_2H_5O	F	O	3	i.m.	+	93
CH_3	iso-C_3H_7O	F	O	1	i.m.	+	93
C_2H_5	iso-C_3H_7O	F	O	1	i.m.	+	93
C_2H_5	iso-C_4H_9O	F	O	1	i.m.	+	93
CH_3	iso-C_4H_9O	F	O	3	i.m.	+	93

[a] Abbreviations: s.c., subcutaneous; p.o., oral, i.p., intraperitoneal injection; i.m., intramuscular injection.
[b] Interval between successive doses is 1 day.

e. General

The mechanism responsible for the delayed neurotoxicity has not been identified unequivocally. Davies et al.[93] suggested that the fluorine ion released by the reaction of the phosphonofluoridate with a hydroxyl was responsible for the toxicity; however, this did not account for the involvement of TOCP or many other chemicals in OPION. Butyrylcholinesterase was proposed as the target protein in the production of OPIDN; however, tri-4-ethyl phenylphosphate was neurotoxic and did not inhibit butyrylcholinesterase.[94] Other compounds inhibited butyrylcholinesterase but were not neurotoxic,[75] indicating that the inhibition of that enzyme was not responsible for OPIDN. Another proposal concerning the neurotoxic potential of two brain esterases, which hydrolyzed phenyl phenylacetate and phenyl 3-phenylpropionate, respectively, was made.[95] However, further study showed that the production of OPIDN did not correlate with inhibition of these esterases.[96] Other possibilities including altered lipid metabolism[97] and autoimmune mechanisms[98] have been proposed and discounted. Abou-Donia[77] has proposed that the toxicity results from the phosphorylation of a target protein in an axon with a consequent breakdown in energy production and axoplasmic transport. This would result in accumulation of mitochondria

Table 2
ALIPHATIC AROMATIC PHOSPHORUS ESTERS TESTED FOR DELAYED NEUROTOXICITY IN CHICKENS

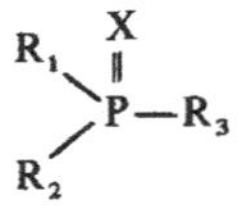

Substituents							
R_1	R_2	R_3	X	Dose (mg/kg)	Route of administration[a]	Delayed neurotoxicity	Ref.
			Phosphonates				
C_2H_5O	C_6H_5	$4\text{-}NO_2\text{–}C_6H_4O$	O	10	s.c.	+	156,157
C_2H_5O	$C_6H_5\text{–}CH_2$	$4\text{-}NO_2\text{–}C_6H_4O$	O	10	s.c.	+	156
C_2H_5O	$C_6H_5(CH_2)_2$	$4\text{-}NO_2\text{–}C_6H_4O$	O	8	s.c.	+	156
C_2H_5O	$C_6H_5(CH_2)_3$	$4\text{-}NO_2\text{–}C_6H_4O$	O	8 × 4[b]	i.p.	+	156
C_2H_5O	$C_6H_5(CH_2)_4$	$4\text{-}NO_2\text{–}C_6H_4O$	O	8 × 4[b]	i.p.	–	156
CH_3O	C_6H_5	2,5-di-Cl-4-BrC_6H_2O	O	50	p.o.	+	77,158
CH_3O	C_6H_5	2,5-di-ClC_6H_3O	O	10	p.o.	+	77
C_2H_5O	C_6H_5	2,4-di-ClC_6H_3O	O	100	p.o.	+	77
C_2H_5O	C_6H_5	4-CN–C_6H_4O	O	5	p.o.	+	77
CH_3O	C_6H_5	$CH_3\text{–}C_6H_4O$	O	500	p.o.	–	159
2,5-di-Cl-4-BrC_6H_2O	C_6H_5	2,5-di-Cl-4-BrC_6H_2O	O	500	p.o.	–	159
C_2H_5O	$n\text{-}C_5H_{11}$	$4\text{-}NO_2\text{–}C_6H_4O$	O	12 × 2[b]	i.p.	+	156
C_2H_5O	$n\text{-}C_{10}H_{21}$	$4\text{-}NO_2\text{–}C_6H_4O$	O	16	s.c.	+	84
$n\text{-}C_5H_{11}O$	$n\text{-}C_5H_{11}$	$4\text{-}NO_2\text{–}C_6H_4O$	O	6	s.c.	+	84
CH_3O	CH_3	2,5-di-ClC_6H_3O	O	50	s.c.	+	84
C_2H_5O	C_2H_5	2,4,5-tri-ClC_6H_2O	O	30	s.c.	+	84

Phosphonothioates							
CH_3O	C_6H_5	H	S	500	p.o.	–	159
CH_3O	C_6H_5	2-ClC_6H_4O	S	300	p.o.	+	159
CH_3O	C_6H_5	3-ClC_6H_4O	S	275	p.o.	+	159
CH_3O	C_6H_5	4-ClC_6H_4O	S	285	p.o.	+	159
CH_3O	C_6H_5	2,3-di-ClC_6H_3O	S	100	p.o.	+	159
CH_3O	C_6H_5	2,4-di-ClC_6H_3O	S	160	p.o.	+	159
CH_3O	C_6H_5	2,5-di-ClC_6H_3O	S	50	p.o.	+	158—161
CH_3O	C_6H_5	2,6-di-ClC_6H_3O	S	40	p.o.	+	159
CH_3O	C_6H_5	3,4-di-ClC_6H_3O	S	100	p.o.	+	159
CH_3O	C_6H_5	3,5-di-ClC_6H_3O	S	175	p.o.	+	159
CH_3O	C_6H_5	2,5-di-Cl-4-BrC_6H_2O	S	100	p.o.	+	78,159,160
C_2H_5O	C_6H_5	2,4-di-ClC_6H_3O	S	800	p.o.	+	160,162,163
C_2H_5O	C_6H_5	2,5-di-ClC_6H_3O	S	1000	p.o.	–	159
C_3H_7O	C_6H_5	2,5-di-ClC_6H_3O	S	500	p.o.	–	159
C_4H_9O	C_6H_5	2,5-di-ClC_6H_3O	S	333	p.o.	–	159
C_2H_5O	C_6H_5	2,5-di-Cl-3-BrC_6H_2O	S	1000	p.o.	–	159
C_2H_5O	C_6H_5	2,4,5-tri-ClC_6H_2O	S	300	p.o.	–,+	84,159
2,5-di-Cl-4-BrC_6H_2O	C_6H_5	2,5-di-Cl-4-BrC_6H_2O	S	500	p.o.	–	84,159
C_2H_5O	C_6H_5	4-NO_2–C_6H_4O	S	25	p.o.	+	25,71,157,160
C_2H_5O	C_6H_5	4-CN–C_6H_4O	S	10	p.o.	+	160,164

[a] Abbreviations: s.c., subcutaneous; p.o., oral, i.p., intraperitoneal injection; i.m., intramuscular injection.
[b] Interval between successive doses is 1 day.

followed by release of calcium into the axoplasm and a focal degeneration which spreads distally in the axon. This process can be exacerbated by further exposure or can reverse itself to some limited degree.[77]

The event currently known which best correlates with OPIDN is the inhibition of NTE, although the possibility still exists that the actual target has not been identified. The inhibition of NTE has not been clearly shown to produce a direct biochemical or physiological effect resulting in OPIDN. The time required for the effect to occur decreases the likelihood of the toxic effect being direct. The modification of a control mechanism seems to be a real possibility except that if the NTE has a half-life of about 4 days and the effect occurs at 14 days, there would apparently be time for recovery to start prior to the occurrence of irreversible damage. The continuation of the toxic effect beyond recovery of NTE would seem to indicate that a product of the inhibited protein is remaining *in situ*, following turnover, to perpetuate toxicity. The occurrence of a cross-linked, alkylated, or otherwise "unnatural" protein or protein catabolite might well produce an inhibited enzyme or transport mechanism at a second site, with the toxic material being transported to the distal portion of the axon by axoplasmic flow.

Despite what the mechanism ultimately proves to be, inhibition of NTE correlates well with neurotoxicity and allows for in vitro testing for OPIDN and for the gradiation of responses to neurotoxic materials.[99] Perhaps more importantly, the hypothesis has stimulated a great deal of research into possible mechansms.

The two primary neurotoxic events following exposure to organophosphonates, the inhibition of AChE and the production of delayed peripheral neuropathy, are very different in both mechanism and effect. The loss of enzymatic activity of AChE is directly responsible for the observed effects which are dose related. These effects, which are acute, can be treated by rational therapeutic measures and if the organism survives the acute effects, complete recovery is likely. "Aging" occurs and makes the phosphorylation of the enzyme irreversible but does not modify the actual acute effect.

The production of the OPIDN can result from a single exposure to an appropriate material but is delayed in effect. OPIDN can also result from repeated administration of subneurotoxic doses of organophosphonate compounds.[69,99] There is no known therapy (including the oximes)[100] and only limited recovery occurs, although prophylactic measures can be taken in experimental situations. In addition, there is an apparent threshold below which effects are not seen, or are minimal, and above which increasing the dose can exacerbate effects, but not markedly alter the outcome. The one element in common is that both effects are produced by alkyl, aryl, or mixed esters of phosphonic acid. The absence of correlation between the inhibition of AChE and the inhibition of NTE indicates that unique factors, whether they be chemical or metabolic, are producing OPIDN.

For many inhibitors of AChE, the dose required for lethality is greater than the dose required to produce OPIDN.[100] Lotti and Johnson[101] investigated the relative propensity of a number of organophosphonates and organophosphates to produce either AChE inhibition or delayed neuropathy in both hens and humans. By the use of in vitro NTE and AChE assays the concentrations required to provide a 50% inhibition of each enzyme by a particular compound was calculated (Table 3). The values c and f are ratios which represent the fold-difference in concentration required to inhibit NTE as compared to AChE. (If the ratio is greater than 1, the NTE requires a higher concentration to cause inhibition.) The means of the values for the four phosphonates are 13.3 and 16.4 for the hen and human, respectively. These values are influenced by the ratios for the ethyl 4-nitrophenylpropylphosphonate but, in general, are within a few fold difference of the concentration required for AChE inhibition. This is of significance because delayed peripheral neuropathy may become more of a problem as therapeutic measures are improved and heroic measures are taken to sustain life in cases of inhibition of AChE by organophosphonates.

Table 3

Compound	Hen a NTE I_{50} (μM)	Hen b AChE I_{50} (μM)	Hen c (a/b)	Human d NTE I_{50} (μM)	Human e AChE I_{50} (μM)	Human f (d/e)
Methyl 2,5-dichloro-4-bromophenyl phenylphosphonate	0.18	0.034	5.3	0.31	0.12	2.6
Ethyl 4-nitrophenyl 3-phenyl propylphosphonate	0.9	0.025	36.0	2.2	0.038	57.9
2-Chloroethyl 2,2-dichlorovinyl ethylphosphonate	0.045	0.022	2.0	0.038	0.069	0.6
Ethyl 2,4,5-trichlorophenyl ethylphosphonate	1.5	0.15	10.0	2.6	0.6	4.3
Diisopropyl fluorophosphate	0.7	0.62	1.1	1.2	0.83	1.4
Phenyl benzylcarbamate	4	N[a]	∞	5	N[a]	∞

[a] N = No significant inhibition at 100 μM.

Based on data from M. Lotti and M. K. Johnson, *Arch. Toxicol.*, 41, 215, 1978.

These data also indicate that the hen and human have the same magnitude of sensitivity to materials producing delayed neuropathy and that the hen is an appropriate model by which to establish the safety of organophosphonate compounds. Lotti and Johnson[101] concluded that their in vitro tests could be used to improve prediction of in vivo effects of compounds producing OPIDN. Dudek et al.[102] proposed that the inhibition of NTE in lymphocytes could be used to measure exposure and to predict potential neurotoxic hazard.

3. *Miscellaneous Effects on Nervous Tissue*

Amino acids have been found to play a variety of roles in the central nervous system.[103] Glycine and γ-aminobutyric acid are generally accepted as being important inhibitory transmitters, while glutamate and aspartate are excitatory amino acids. Other analogs and amino acids, including some phosphonates, are of importance also. The glutamate analog, DL-2-amino-4-phosphonobutyric acid inhibits the binding of glutamate to "receptor like" proteolipids from locust muscle and antagonizes the excitatory action of glutamate in the locust muscle membrane.[104] However, this compound is probably not useful as an antagonist of aspartate and glutamate transmitter function in mammalian nervous system.[105]

The effects of 2-amino-4-phosphonobutyric acid on responses of the ON and OFF channels of vertebrate retina have also been investigated.[106] The compound blocks all responses in the ON channel but leaves intact the OFF responses. Other homologs, 2-amino-5-phosphonopentanoic acid and 2-amino-7-phosphonoheptanoic acid, are potent and selective antagonists of *N*-methyl D-aspartate. The (−)-D-enantiomer of the phosphonopentanoate derivative is 8 to 10 times more active than the racemate. These data support the hypothesis that *N*-methyl-D-aspartate has a distinct population of receptors in the central nervous system and that aspartate may not be the endogenous ligand for the *N*-methyl-D-aspartate receptor.[107] Gruszecka et al.[108] have synthesized phosphonic acid analogs of α-methyl-aspartic and α-methyl glutamic acids and Tinguy-Moreaud et al.[109] have investigated neuroactive properties of phosphonic analogs of natural guanido compounds.

A series of 4-alkyl derivatives of 2,6,7-trioxa-1-phosphabicyclo [2.2.2] octane, used as flame retardants, vinyl resin stabilizers, and antioxidants, has been found to be highly toxic.[110,111] The toxicity is dependent on the nature of the 4-alkyl group, with the 4-isopropyl derivative having an LD_{50} of 0.18 mg/kg in the mouse or a toxicity some 30 times greater than diisopropylfluorophosphate.[110] The same bicyclic phosphorus ester can be formed by the nonflaming combustion of trimethylol-propane based rigid-urethane foam containing *O,O*-diethyl-*N,N*-bis-(2-hydroxyethyl)aminomethyl phosphonate. The hazard involved is independent of flame contact or carbon monoxide intoxication and is exacerbated by the loss of the behavioral response to escape.[112] The bicyclic phosphate esters have no effect on brain cholinesterase and may be antagonists to the response to γ-aminobutyric acid. The isopropyl and t-butyl derivatives are more potent than bicuculline as antagonists to γ-aminobutyric acid in the isolated frog spinal cord.[111]

4. Antibiotics

The phosphonate antibiotics are primarily of natural occurrence. These antibiotics are produced by various strains of the Streptomycetes and are discussed in Chapter 2, Section II.A. However, there are several compounds of synthetic origin with potential use as antibiotics. L-alanyl-L-1-aminoethylphosphonic acid (alafosfalin) interferes with cell wall synthesis.[113,114] This effect apparently occurs following transport of the peptide and intracellular cleavage by peptidase and is the result of the inhibition of alanine racemase in susceptible organisms.[115]

Another antibiotic of interest here is the organophosphonate, metrifonate (*O,O*-dimethyl[1-hydroxy-2,2,2-trichloroethyl]phosphonate), of which about 5×10^6 doses have been given to patients in the tropics for the treatment of schistosomiasis.[13] Metrifonate (trichlorfon) is transformed to dichlorvos (2,2-dichlorovinyldimethylphosphate) in both aqueous and organic solvents. The latter compound is an active inhibitor of AChE and may be the schistosomicidal agent, as well as being responsible for the production of OPIDN. This material may provide an excellent opportunity to study effects of cholinesterase inhibition and the action of therapeutic agents.[13]

In addition to the antibacterial and antiparasitic compounds, there are two antiviral agents, phosphonoformate and phosphonoacetate which are discussed in Chapter 5.

III. PHOSPHONATE COMPOUNDS WHICH MODIFY LIFE PROCESSES IN PLANTS

A. Herbicides

The problems associated with undesirable plant species are formidable. These include reduced food crop production due to competition for water, sunlight, and space; contamination of harvested materials; disruption of waterways and water flow through irrigation ditches; loss of human and animal life due to poisoning; and human suffering due to contact with toxic or allergenic species. The economic loss alone due to these problems is in the billions of dollars and the control of weeds is an integral part of agriculture, gardening, and even being a homeowner.[116]

The problems are being countered by new approaches to weed control such as improved land management, improved methods of application of herbicides, and development of more efficacious herbicides. Chief among the improvements is the use of herbicides which may have either greater specificity for the target species, effective levels of application below those previously required, activity against weeds which have developed resistance to other herbicides or formulations, or activity against plants which are particularly difficult to control.

The primary phosphonate herbicide is glyphosate[117] (*N*-[phosphonomethyl] glycine, Figure 7). This herbicide is nonselective and is effective against johnson grass, quack grass, and other annual and biennial species of grasses, sedges, and broad leaved weeds. It is primarily used as a postemergence spray on the foliage of vegetation to be controlled. Annual species will normally be controlled by 0.3 to 1.0 lb active ingredient per acre and perennials will

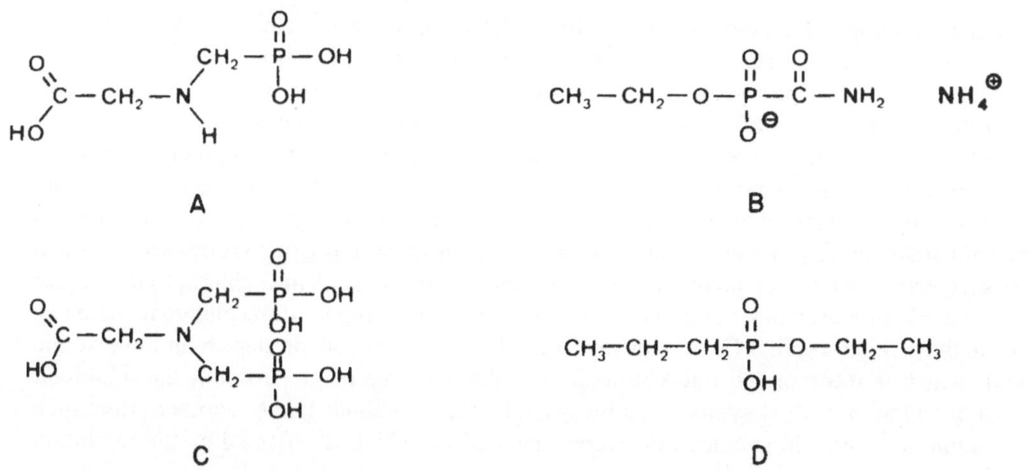

FIGURE 7. Herbicides and plant growth regulators which contain a phosphonate. (A) Glyphosate, *N*-(phosphonomethyl)-glycine; (B) fosamine-ammonium, ammonium ethyl hydrogen carbamoylphosphonate; (C) glyphosine, *N,N*-bis(phosphonomethyl)glycine; (D) ethyl hydrogen 1-propylphosphonate.

be controlled by 1.0 to 4.0 lb active ingredient per acre.[118] A comprehensive bibliography on glyphosate has been compiled and indexed.[119]

Glyphosate is rapidly translocated in the phloem of both annual and perennial weeds to meristematic areas such as shoot and rhizome apices and rhizome nodes.[119-122] The efficacy of a phloem translocated herbicide is related to plant growth and metabolically available herbicide. Humidity and temperature are important in the absorption and translocation,[123] and light will affect translocation.[124] Variability in the efficacy of glyphosate would be expected under varying climatic conditions and has been observed.

The mechanism of action of glyphosate is thought to be associated with the metabolism of aromatic amino acids. Studies by Jaworski[125] indicated that phenylalanine biosynthesis was inhibited by glyphosate at the chorismate mutase or prephenate dehydratase steps and that certain aromatic amino acids could reverse those effects. Further studies[126] using *Escherichia coli* indicated that the above enzymes were not affected but that aromatic amino acids would reverse the phytotoxic effect. Exogenous aromatic amino acids supplied to soybean (Glycine Max [L.] Merr. Hill) seedlings reversed the root growth inhibition slightly (10%) but significantly. Glyphosate had no effect on uptake or incorporation of these amino acids and did not substantially affect shikimate dehydrogenase activity in control or amino acid fed seedlings. Duke and Hoagland[127] concluded that either the mechanism of action of glyphosate is not the inhibition of aromatic amino acid synthesis or that root fed amino acids are compartmentalized differently than endogenous amino acids. Aromatic amino acids will generally reverse effects of glyphosate on unicellular organisms or on cell cultures of higher plants, but have an equivocal effect in higher plants that are intact. Duke and Hoagland[127] suggest that the chelation of divalent metal ion by glyphosate may be important. Ekanayake et al.[128] observed effects of glyphosate on amino acid metabolism but did not examine possible mechanisms.

The first visible effect following application of glyphosate is chlorosis.[129-131] Chlorophyll accumulation has been found to be reduced in corn shoots and etiolated barley grown in the laboratory and in field grown soybeans, indicating that interference with greening may be important in the mechanism of phytotoxicity.[131]

Campbell[129] reported that damage to the chloroplast envelope was induced by glyphosate in quack grass (*Agropyran ripens* [L.] Beariv.). Lee[132] reported that glyphosate completely inhibited chlorophyll accumulation in callus cultures of tobacco stem pith and hypocotyls of soybean seedlings. He found that both synthesis and degradation were involved. In addition, glyphosate promoted chlorophyl degradation in the light but not in the dark in tobacco leaf discs. This effect was not reversible by aromatic amino acids.

Kitchen et al.[133] have investigated the mechanism of this effect by measuring incorporation of ^{14}C precursors into δ-aminolevulinic acid, the formation of which is considered the rate limiting step in chlorophyll synthesis. The incorporation of the precursors glutamate, α-ketoglutarate, and glycine into δ-aminolevulinic acid in etiolated barley shoots was inhibited by glyphosate, while the incorporation of δ-aminolevulinic acid into chlorophyll was not affected. He proposed that the point of inhibition may be common to the glutamate pathway and to the glycine succinyl Co-A pathway. Since the pathways both produce δ-aminolevulinic acid (which is incorporated into chlorophyll in the presence of glyphosate), the inhibition must be in the individual synthetic pathways to δ-aminolevulinic acid or another effect such as inhibition of protein synthesis or energy metabolism which is reflected by the inhibition of these pathways.

The phytotoxicity of glyphosate in soil is very low and may be due either to rapid degradation to aminomethylphosphonic acid or to a high degree of soil binding.[134] This low soil phytotoxicity makes necessary the use of a soil active herbicide to control seedlings for adequate control of perennial weeds, other established vegetation, and seedlings. Thus, the application of combinations of herbicides may be necessary. The effects of various combinations of herbicide with glyphosate have been investigated.[135,136]

A second herbicide of the phosphonate class, fosamine-ammonium (ammonium ethyl carbamoylphosphonate, Figure 7), has primarily been used as a brush control agent.[137] When applied as an aqueous foliar spray at 6 to 12 lb active ingredient per acre it will control or suppress growth of a variety of woody species, although there is poor control in the year of application. The effects on brush have been described.[138-140] There is little immediate effect following application in summer or fall. However, after a normal leaf drop, and a normal winter dormancy,[138] susceptible plants fail to refoliate and then die, apparently as a result of the inhibition of leaf and bud formation.[137] Neihuss and Roediger[139] and Hernandez et al.[138] noted that only sprayed branches of trees are affected, indicating there is little translocation of fosamine. However, studies on the absorption and translocation of fosamine in blackberry[141] demonstrated that fosamine is translocated throughout the plant and accumulates in younger leaves, upper parts of the cane, and in the roots. Apparently, the fosamine is only slowly absorbed (35% in 96 hr) and performance may be improved by more effective surfactants or agents to prevent loss prior to absorption.[141]

A similar finding was reported by Coupland and Peabody[142] with both fosamine and glyphosate, where 47 and 63% of the herbicides, respectively, had been absorbed by field horsetail after 8 days. The patterns of accumulation were similar and the compounds were concentrated in apical regions of both treated and untreated shoots. The accumulation was typical of phloem translocated herbicides and both were accumulated at rhizome nodes and apices. Fosamine accumulated faster in the roots than did the glyphosate which was unexpected due to the earlier reports that fosamine was nonsystemic.[130,139] In addition to the slow absorption, herbicide is lost due to guttation and root exudation and is susceptible to removal by rain, factors which may account for variability in performance in field tests.[142]

Kitchen et al.[143] reported that fosamine is translocated in both woody plant seedlings which are susceptible (black cherry) and those which are tolerant (magnolia and rhododendron). The susceptible seedlings did absorb and translocate significantly greater quantities of the fosamine away from the site of application than did the tolerant species. Thus, the mechanism of resistance is probably not due to physical exclusion. They propose that the phytotoxicity is due to the action of the chemical or a metabolite at or above the point of contact and not in the roots. Field studies have supported this view, since root toxicity would

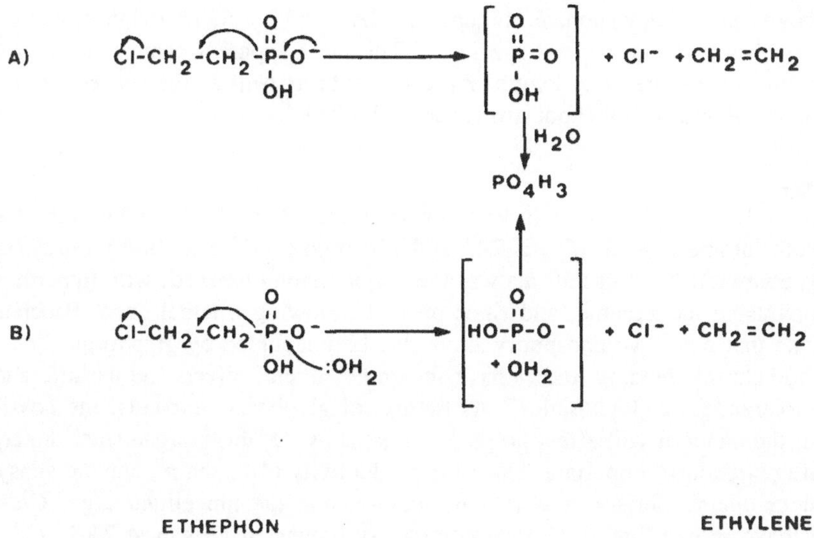

FIGURE 8. Two possible mechanisms for the decomposition of 2-chloroethylphosphonic acid to produce ethylene. (A) An internal fragmentation leading to monomeric metaphosphate, which then reacts with water; (B) a fragmentation initiated by the attack of a water molecule on the phosphonate ion. (From Eto, M., *Organophosphorus Pesticides: Organic and Biological Chemistry*, CRC Press, Boca Raton, Fla., 1979, 77.)

not just kill the treated branches. Morey and Dahl[144] found that growth of mesquite was inhibited for up to 3 years following treatment with fosamine. Their studies indicated that fosamine was effective by inhibiting protein nucleic acid synthesis in both special and lateral meristems in the mesquite.[144]

B. Plant Growth Regulators

Other plant growth effectors are regulators used for inducing flowering, controlling vine formation (pruning), thinning fruit, and for retarding or accelerating ripening. Of these, ethephon, 2-chloroethylphosphonic acid, is used to accelerate fruit ripening, to induce flowering, to promote abscission, to accelerate postharvest ripening, and to stimulate color formation.[118] Ethephon decomposes rapidly in the presence of OH^- to produce ethylene. Two mechanisms have been proposed for this reaction (Figure 8). The plant responses to ethephon are the result of the release of ethylene gas, which has fruit ripening hormone activity. The biochemistry, physiology, and mechanism of action of ethylene have been extensively examined by Abeles.[145]

Glyphosine (*N,N*-bis[phosphonoethyl]glycine, Figure 7) is an analog of glyphosate but is used to accelerate ripening and increase the level of sucrose in sugar cane, *Saccarum officinarum*. The visible effects of its application are cessation of growth, chlorosis, and dessication; however, the mechanism of action is not known.[118,146] At relatively high doses, glyphosine has effects which are similar to those of glyphosate; however, the effects are quantitatively smaller.[146] Glyphosine has been shown to reduce 70-S ribosomes and chloroplastic ribosomal RNA, but to have no effect on photophosphorylation.[147] Potential metabolites of glyphosine (glyphosate, aminomethylphosphonic acid, sarcosine, and glycine) reduced growth but had little effect on reducing dry weight accumulation or on stimulating phenylalanine ammonia-lyase. This indicates that the conversion of glyphosine to glyphosate is not rapid in intact seedlings or that the glyphosate is rapidly metabolized with concomitant decrease in concentrations.[146]

Ethyl hydrogen propylphosphonate, Figure 7, NIA 10637, and propylphosphonic acid will act as growth retardants in plant species.[148] The most apparent change in the growth pattern is a reduction in internode length. Apical growth as well as the subsequent development of subapical and lateral shoot growth are inhibited.

C. Toxicology

The acute oral LD_{50} of glyphosate is 4900 mg/kg in the rat and 1600 mg/kg in the mouse, while the acute intraperitoneal LD_{50} is 240 and 130 mg/kg in the rat and mouse, respectively.[149] Hyperemia in the lungs of rats was the major lesion observed, with hyperthermia, asphyxial convulsive movements, and rigor present following a lethal dose. Biochemical studies indicate that oxidative phosphorylation may be uncoupled by glyphosate.[149]

Subacute and chronic feeding studies have shown no adverse effects and wildlife and fish have a high tolerance for glyphosate.[118] As the use of glyphosate expands, the possibility increases that the material will affect nontarget organisms. Of these organisms, the aquatic microflora are particularly important, since the productivity of the entire aquatic ecosystem is dependent on them. Christy et al.,[150] in studies using the unicellular alga, *Chlorella serokineana,* have shown that glyphosate completely prevented growth at 23.7×10^{-6} *M* and caused a slight reduction in growth at 5.9×10^{-6} *M*.

The acute oral LD_{50} of the fosamine formulation is 24000 mg/kg for rats and 7380 mg/kg for guinea pigs. The acute dermal LD_{50} for rabbits is >4000 mg/kg. This compound is not an irritant to eyes or skin. Rats fed for 90 days at 1000 mg fosamine per kilogram diet suffered no ill effects.[151]

The acute oral LD_{50} of ethephon in rats is 4229 mg/kg and the acute dermal LD_{50} in rabbits is 5730 mg/kg. Chronic toxicity is low; however, the compound is corrosive and can produce corneal injury.[151] Theiss and Shimkin[152] have assessed the carcinogenicity and cocarcinogenicity of 2-chloroethylphosphonic acid in the Strain A mouse pulmonary adenoma bioassay. The plant growth regulator suppressed the spontaneous development of lung tumors and inhibited the development of lung tumors in response to urethane. This effect may be attributable to the ethylene which has been shown to inhibit cell division and DNA synthesis in plants.

D. Discussion

Although the phosphonate herbicides in current use are not particularly specific in their actions, they have properties which make them useful. The rapid translocation of glyphosate has allowed widespread utilization of new methods of application so that only weeds are treated and there is little or no contact of herbicide with low growing food crops. This reduces quantities of herbicide applied, the consequent environmental problems, and costs.[116,153] The rapid translocation also makes glyphosate useful in the control of grasses, such as johnson grass *(Sorghum halepense)* and torpedo grass *(Panicum repens),* which regenerate from rhizomes and are thus difficult to control.[128,154]

The mechanisms of action of the phosphonate herbicides are only partially known. As greater understanding of these mechanisms occurs structure function relationships may allow the development of other herbicides or other uses for existing herbicides.

The utility of the growth regulator 2-chloroethylphosphonic acid is a result of the rapid decomposition of β-haloalkylphosphonic acids to produce the corresponding olefins. The material provides a relatively safe and metabolically available material for the many agricultural uses for ethylene as a plant growth regulator.

IV. SUMMARY

The phosphonate class of the synthetic organophosphate molecules contains many compounds which are extremely reactive in biological systems. In general, either alkyl- or aryl-diesters of phosphonic acid have the potential to inhibit biological esterases. The inhibition of acetylcholinesterase results in accumulation of acetylcholine and severe acute toxicity. However, therapeutic treatment is available and if instituted immediately can significantly reduce acute toxicity. The inhibition of neurotoxic esterase correlates with the occurrence of organophosphorus induced delayed peripheral neuropathy (OPIDN). OPIDN occurs several days to 3 weeks following either a single or multiple exposure to organophosphonates with neurotoxic potential. The paralysis is not reversible to any significant degree and there is no known therapy. The mechanism by which organophosphonates produce OPIDN has not been established unequivocally. The above effects are not specific for phosphonates, but are typical of the triesters of phosphoric acid.

Other phosphonate compounds are used as antibiotics, antivirals, and plant growth effectors. These compounds are usually free phosphonic acids or monoesters of phosphonic acid and have specific metabolic effects as a mechanism of action.

At this time, organophosphonates are most widely used in biological systems as insecticides and as herbicides, although other miscellaneous uses are developing. Phosphonates seem to have wide utility, thus, we should expect very beneficial utilization of synthetic phosphonates as modifiers of biological systems in the future.

REFERENCES

1. **Fisher, E. B. and Van Wazer, J. R.,** Uses of organic phosphorus compounds, in *Phosphorus and Its Compounds,* Vol. II, Van Wazer, J. R., Ed., Wiley-Interscience, New York, 1961, chap. 31.
2. **Chadwick, D. H. and Watt, R. S.,** Manufacture of phosphate esters and organic phosphorus compounds, in *Phosphorus and Its Compounds,* Vol. II, Van Wazer, J. R., Ed., Wiley-Interscience, New York, 1961, 1222 and 1272.
3. **Darlington, W. A., Partos, R. D., and Ratts, K. W.,** Correlation of cholinesterase inhibition and toxicity in insects and mammals. I. Ethylphosphonates, *Toxicol. Appl. Pharmacol.,* 18, 542, 1971.
4. **O'Brien, R. D.,** *Toxic Phosphorus Esters,* Academic Press, New York, 1960, chap. 1.
5. **Murphy, S. D.,** Pesticides, in *Toxicology: The Basic Science of Poisons,* 2nd ed., Doull, J., Klaasen, C. D., and Amdur, M. O., Eds., Macmillan, New York, 1980, chap 16.
6. **Eto, M.,** *Organophosphorus Pesticides: Organic and Biological Chemistry,* CRC Press, Boca Raton, Fla., 1979, 266.
7. **Saunders, B. C.,** *Some Aspects of the Chemistry and Toxic Action of Organic Compounds Containing Phosphorus and Fluorine,* Cambridge University Press, New York, 1957.
8. **Meselson, M. and Robinson, J. P.,** Chemical warfare and chemical disarmament, *Sci. Am.,* 242, 38, 1980.
9. **Lewis, R. J., Sr. and Tatken, R. L.,** Registry of Toxic Effects of Chemical Substances, Vol. 2, DHHS (National Institute for Occupational Safety and Health) Publication No. 80-111, U.S. Government Printing Office, Washington, D.C., 1980, 270.
10. **Fukuto, T. R.,** Relationships between the structure of organophosphorus compounds and their activity as acetylcholinesterase inhibitors, *Bull. W. H. O.,* 44, 31, 1971.
11. **Eto, M.,** *Organophosphorus Pesticides: Organic and Biological Chemistry,* CRC Press, Boca Raton, Fla., 1979, 151.
12. **Eto, M.,** *Organophosphorus Pesticides: Organic and Biological Chemistry,* CRC Press, Boca Raton, Fla., 1979, 73.
13. **Nordgren, I.,** Quantitation of metrifonate and dichlorvos in blood and tissues by gas chromatography — mass spectrometry, *Fundam. Appl. Toxicol.,* 1, 230, 1981.
14. **O'Brien, R. D.,** *Toxic Phosphorus Esters,* Academic Press, New York, 1960, chap. 4.

15. **Fukuto, T. R. and Metcalf, R. L.**, The effect of structure on the reactivity of alkylphosphonate esters, *J. Am. Chem. Soc.*, 81, 372, 1959.
16. **Hansch, C. and Deutsch, E. W.**, The use of substituent constants in the study of structure-activity relationships in cholinesterase inhibitors, *Biochim. Biophys. Acta*, 126, 117, 1966.
17. **Aaron, H. S., Michel, H. O., Witten, B., and Miller, J. I.**, The stereochemistry of asymmetric phosphorus compounds. II. Stereospecificity in the irreversible inactivation of cholinesterases by the enantiomorphs of an organophosphorus inhibitor, *J. Am. Chem. Soc.*, 80, 456, 1957.
18. **Inch, T. D.**, The biological importance of organophosphorus compounds containing a carbon-phosphorus bond, in *Phosphorus in the Environment: Its Chemistry and Biochemistry*, Vol. 1, Elsevier, New York, 1978, 135.
19. **Gallagher, M. J. and Jenkins, I. D.**, Stereochemical aspects of phosphorus chemistry, in *Topics in Stereochemistry*, Vol. 3, Eliel, E. L. and Allinger, N. L., Eds., Wiley-Interscience, New York, 1968, 1.
20. **McEwen, W. E.**, Stereo chemistry of reactions of organophosphorus compounds, in *Topics in Phosphorus Chemistry*, Vol. 2, Grayson, M. and Griffith, E. J., Eds., Wiley-Interscience, New York, 1965, 1.
21. **Ugi, I. and Ramirez, F.**, Stereochemistry of five-coordinate phosphorus, *Chem. Br.*, 8, 198, 1972.
22. **Hassan, A. and Dauterman, W. C.**, Studies on the optically active isomers of *O,O*-diethyl malathion and *O,O*-diethyl malaoxon, *Biochem. Pharmacol.*, 17, 1431, 1968.
23. **Boter, H. L. and van Dijk, C.**, Stereospecificity of hydrolytic enzymes on reaction with asymmetric organophosphorus compounds — III. The inhibition of acetylcholinesterase and butyrylcholinesterase by enantiomeric forms of Sarin, *Biochem. Pharmacol.*, 18, 2403, 1969.
24. **Ooms, A. J. J. and Boter, H. L.**, Stereospecificity of hydrolytic enzymes in their reaction with optically active organophosphorus compounds — I. The reaction of cholinesterase and paraoxonase with *S*-alkyl *p*-nitrophenyl methyphosphonothiolates, *Biochem. Pharmacol.*, 14, 1839, 1965.
25. **Abou Donia, M. B., Graham, D. G., Komeil, A. A., Nomeir, A. A., and Dauterman, W. C.**, Stereospecificity of cholinergic and delayed neurotoxic effects of *O*-ethyl *O*-4-nitrophenyl phenylphosphonthioate optical isomers: effect of daily oral administration to hens, in *Advances in Neurotoxicology*, Manzo, L., Lery, N., LaCasse, Y., and Roche, L., Eds., Pergamon Press, Oxford, 1980, 237.
26. **Benschop, H. P., Berends, F., and de Jong, L. P. A.**, GLC-analysis and pharmacokinetics of the four stereoisomers of Soman, *Fundam. Appl. Toxicol.*, 1, 177, 1981.
27. **Eto., M.**, *Organophosphorus Pesticides: Organic and Biological Chemistry*, CRC Press, Boca Raton, Fla., 1979, 146.
28. **DuBois, K. P., Kinoshita, F. K., and Frawley, J. P.**, Quantitative measurement of inhibition of aliesterases, acylamidase, and cholinesterase by EPN and Delnav, *Toxicol. Appl. Pharmacol.*, 12, 273, 1968.
29. **Hollingworth, R. M., Fukuto, T. R., and Metcalf, R. L.**, The selectivity of Sumithion compared with Methyl Parathion. Metabolism in the white mouse, *J. Agric. Food Chem.*, 15, 242, 1967.
30. **Eto, M.**, *Organophosphorus Pesticides: Organic and Biological Chemistry*, CRC Press, Boca Raton, Fla., 1979, 63.
31. **Menn, J. J., DeBaun, J. R., and McBain, J. R.**, Recent advances in the metabolism of organophosphorus insecticides, *Fed. Proc.*, 35, 2598, 1976.
32. **O'Brien, R. D.**, *Insecticides Action and Metabolism*, Academic Press, New York, 1967.
33. **Dauterman, W.C.**, Biological and nonbiological modifications of organophosphorus compounds, *Bull. W. H. O.*, 44, 133, 1971.
34. **Mayer, S. E.**, Neurohumoral transmission and the autonomic nervous system, in *The Pharmacological Basis of Therapeutics*, 6th ed., Gilman, A. G., Goodman, L. S., and Gilman, A., Eds., Macmillan, New York, 1980, chap. 4.
35. **Taylor, P.**, Anticholinesterase agents, in *The Pharmacological Basis of Therapeutics*, 6th ed., Gilman, A. G., Goodman, L. S., and Gilman, A., Eds., Macmillan, New York, 1980, chap. 6.
36. **Eto, M.**, *Organophosphorus Pesticides: Organic and Biological Chemistry*, CRC Press, Boca Raton, Fla., 1979, 123.
37. **Krupka, R. M.**, Chemical structure and function of the active center of acetylcholinesterase, *Biochemistry*, 5, 1988, 1966.
38. **Epstein, D. J., Berman, H. A., and Taylor, P.**, Ligand-induced conformational changes in acetylcholinesterase investigated with fluorescent phosphonates, *Biochemistry*, 18, 4749, 1979.
39. **Berman, H. A., Becktel, W., and Taylor, P.**, Spectroscopic studies on acetylcholinesterase: influence of peripheral-site occupation on active-center conformation, *Biochemistry*, 20, 4803, 1981.
40. **Sun, M., Chang, Z., Shau, M., Huang, R., and Chou, T.**, The mechanism of ageing of phosphonylated acetylcholinesterase, *Eur. J. Biochem.*, 100, 527, 1979.
41. **Michel, H. O., Hackley, Jr., B. E., Berkowitz, L., List, G., Hackley, E. G., Gillilan, W., and Pankau, M.**, Ageing and dealkylation of soman (pinacolylmethylphosphonofluoridate)-inactivated eel cholinesterase, *Arch. Biochem. Biophys.*, 121, 29, 1967.

42. **Wolthuis, O. L., Berends, F., and Meeter, E.,** Problems in the therapy of soman poisoning, *Fundam. Appl. Toxicol.*, 1, 183, 1981.
43. **Boskovic, B.,** The treatment of soman poisoning and its perspectives, *Fundam. Appl. Toxicol.*, 1, 203, 1981.
44. *Occupational Exposure to Parathion,* U.S. Department of Health, Education and Welfare Publication (NIOSH) 76-190, 1976, 79.
45. **Koelle, G. B.,** Organophosphate poisoning — an overview, *Fundam. Appl. Toxicol.*, 1, 129, 1981.
46. **O'Brien, R. D.,** *Toxic Phosphorus Esters*, Academic Press, New York, 1960, chap 6.
47. **Fonnum, F. and Sterri, S. H.,** Factors modifying the toxicity of organophosphorous compounds including soman and sarin, *Fundam. Appl. Toxicol.*, 1, 143, 1981.
48. **Reiner, E.,** Oxime reactivation of erythrocyte cholinesterase inhibited by ethyl *p*-nitrophenyl ethylphosphonate, *Biochem. J.*, 97, 710, 1965.
49. **Clement, J. G.,** Toxicology and pharmacology of bispyridinium oximes — insight into the mechanism of action vs soman poisoning *in vivo*, *Fundam. Appl. Toxicol.*, 1, 193, 1981.
50. **Szinicz, L. and Weger, N.,** Therapeutic effects of new oximes, benactyzine and atropine in soman poisoning: Part I. Effects of various oximes in soman, sarin, and Vx poisoning in dogs, *Fundam. Appl. Toxicol.*, 1, 162, 1981.
51. **Hauser, W., Kirsch, D., and Weger, N.,** Therapeutic effects of new oximes benactyzine and atropine in soman poisoning: Part II. Effect of HGG12, HGG42, and obidoxime in poisoning with various anticholinesterase agents in beagle dogs, *Fundam. Appl. Toxicol.*, 1, 164, 1981.
52. **Hauser, W., Kirsch, D., and Weger, N.,** Effect of the bispyridinium oximes HGG12 and HGG42 and ganglion blocking agents on synaptic transmission and NAD[P]H fluorescence in the superior cervical ganglion of the rat after soman poisoning *in vitro*, *Fundam. Appl. Toxicol.*, 1, 169, 1981.
53. **Gall, D.,** The use of therapeutic mixtures in the treatment of cholinesterase inhibition, *Fundam. Appl. Toxicol.*, 1, 214, 1981.
54. **Gordon, J. J., Leadbeater, L., and Maidment, M. P.,** The protection of animals against organophosphate poisoning by pretreatment with a carbamate, *Toxicol. Appl. Pharmacol.*, 43, 207, 1978.
55. **Heyl, W. C., Harris, L. W., and Stitcher, D. C.,** Effects of carbamates on whole blood cholinesterase activity: chemical protection against soman, *Drug Chem. Toxicol.*, 3(3), 319, 1980.
56. **Karczmar, A. G. and Ohta, Y.,** Neuromyopharmacology as related to anticholinesterase action, *Fundam. Appl. Toxicol.*, 1, 135, 1981.
57. **Wecker, L., Laskowski, M. B., and Dettbarn, W. D.,** Neuromuscular dysfunction induced by acetylcholinesterase inhibition, *Fed. Proc.*, 37, 2818, 1978.
58. **O'Neill, J. J.,** Non-cholinesterase effects of anticholinesterases, *Fundam. Appl. Toxicol.*, 1, 154, 1981.
59. **Duffy, F. H., Burchfiel, J. L., Bartels, P. H., Gaon, M., and Sim, V. M.,** Long-term effects of an organophosphate upon the human electroencephalogram, *Toxicol. Appl. Pharmacol.*, 47, 161, 1979.
60. **Bignami, G., Rosic, N., Michalek, H., Milosevic, M., and Gatti, G. L.,** Behavioral toxicity of anticholinesterase agents: methodological, neurochemical, and neuropsychological aspects, in *Behavioral Toxicology*, Vol. I, Plenum Press, New York, 1975, chap. 7.
61. **Fleisher, J. H. and Harris, L. W.,** Dealkylation as a mechanism for aging of cholinesterase after poisoning with pinacolyl methyl phosphonofluoridate, *Biochem. Pharmacol.*, 14, 641, 1965.
62. **Deyi, X., Linxiu, W., and Shuaui, P.,** The inhibition and protection of cholinesterase by physostigmine and pyridostigmine against Soman poisoning *in vivo*, *Fundam. Appl. Toxicol.*, 1, 217, 1981.
63. **Smith, M. I. and Elvove, E.,** Pharmacological and chemical studies of the cause of so-called ginger paralysis, *Public Health Rep.*, 45, 1703, 1930.
64. **Kidd, J. G. and Langworthy, O. R.,** Jake paralysis, *Bull. Johns Hopkins Hosp.*, 52, 39, 1933.
65. **Casida, J. E., Eto, M., and Baron, R. L.,** Biological activity of tri-*o*-cresyl phosphate metabolite, *Nature (London)*, 191, 1396, 1961.
66. **Eto, M., Casida, J. E., and Eto, T.,** Hydroxylation and cyclization reactions involved in the metabolism of tri-*o*-cresyl phosphate, *Biochem. Pharmacol.*, 11, 337, 1962.
67. **Baron, R. L., Bennett, D. R., and Casida, J. E.,** Neurotoxic syndrome produced in chickens by a cyclic phosphate metabolite of tri-*o*-cresyl phosphate — a clinical and pathological study, *Br. J. Pharmacol.*, 18, 465, 1962.
68. **Taylor, J. D. and Buttar, H. S.,** Evidence for the presence of 2-(*o*-cresyl)-4H-1:3:2-benzodioxaphosphoran-2-one in cat intestine following tri-*o*-cresyl phosphate administration, *Toxicol. Appl. Pharmacol.*, 11, 529, 1967.
69. **Bidstrup, P. L., Bonnell, J. A., and Beckett, A. G.,** Paralysis following poisoning by a new organic phosphorus insecticide (mipafox), *Br. Med. J.*, 1, 1068, 1953.
70. **Susser, M. and Stein, Z.,** An outbreak of tri-ortho-cresyl phosphate (T.O.C.P.) poisoning in Durban, *Br. J. Ind. Med.*, 14, 111, 1957.

71. **Petty, C. S.,** Organic phosphate insecticide poisoning. Residual effects in two cases, *Am. J. Med.*, 24, 467, 1958.
72. **Smith, H. V. and Spalding, J. M. K.,** Outbreak of paralysis in Morocco due to orthocresyl phosphate poisoning, *Lancet*, 2, 1019, 1959.
73. **Hierons, R. and Johnson, M. K.,** Clinical and toxicological investigations of a case of delayed neuropathy in man after acute poisoning by organophosphorus pesticides, *Arch. Toxicol.,f38 40, 279, 1978.*
74. **Durham, W. F., Gaines, T. B., and Hayes, W. J., Jr.,** Paralytic and related effects of certain organic phosphorus compounds, *Arch. Ind. Health*, 13, 326, 1956.
75. **Johnson, M. K.,** The delayed neuropathy caused by some organophosphorus esters: mechanism and challenge, *Crit. Rev. Toxicol.*, 3, 289, 1975.
76. **Davis, C. S. and Richardson, R. J.,** Organophosphorus compounds, in *Experimental and Clinical Neurotoxicology*, Vol. I, Spencer, P. S. and Schaumburg, H. H., Eds., Williams & Wilkins, Baltimore, 1980, chap. 36.
77. **Abou-Donia, M. B.,** Organophosphorus ester-induced delayed neurotoxicity, *Ann. Rev. Pharmacol. Toxicol.*, 21, 511, 1981.
78. **Abou-Donia, M. B., Othman, M. A., Tantawy, G., Khalil, A. Z., and Shawer, M. F.,** Neurotoxic effect of leptophos, *Experientia*, 30, 63, 1974.
79. **Bondy, H. F., Field, E. J., Worden, A. N., and Hughes, J. P. W.,** A study on the acute toxicity of the tri-aryl phosphates used as plasticizers, *Br. J. Ind. Med.*, 17, 190, 1960.
80. **Johnson, M. K. and Barnes, J. M.,** Age and the sensitivity of chicks to the delayed neurotoxic effects of some organophosphorus compounds, *Biochem. Pharmacol.*, 19, 3045, 1970.
81. **Cavanagh, J. B.,** Peripheral neuropathy caused by chemical agents, *Crit. Rev. Toxicol.*, 2, 365, 1973.
82. **Sheets, L. P. and Norton, S.,** Evidence for leptophos-induced neuropathy in chicks exposed on incubation day 14, *Toxicologist*, 2 (Abstr. 526), 1982.
83. **Johnson, M. K.,** A phosphorylation site in brain and the delayed neurotoxic effect of some organophosphorus compounds, *Biochem. J.*, 111, 487, 1969.
84. **Johnson, M. K.,** Organophosphorus esters causing delayed neurotoxic effects: mechanism of action and structure/activity studies, *Arch. Toxicol.*, 34, 259, 1975.
85. **Johnson, M. K. and Lotti, M.,** Delayed neurotoxicity caused by feeding of organophosphates requires a high-point of inhibition of neurotoxic esterase, *Toxicol. Lett.*, 5, 99, 1980.
86. **Johnson, M. K.,** Structure-activity relationships and substrates and inhibitors of hen neurotoxic esterase, *Biochem. Pharmacol.*, 24, 797, 1975.
87. **Johnson, M. K.,** Organophosphorus and other inhibitiors of brain "neurotoxic esterase" and the development of delayed neurotoxicity in hens, *Biochem. J.*, 120, 523, 1970.
88. **Johnson, M. K.,** The primary biochemical lesion leading to the delayed neurotoxic effects of some organophosphorus esters, *J. Neurochem.*, 23, 785, 1974.
89. **Johnson, M. K. and Lauwerys, R.,** Protection by some carbamates against the delayed neurotoxic effects of di-isopropyl phosphorofluoridate, *Nature (London)*, 222, 1066, 1969.
90. **Johnson, M. K.,** Mechanism of protection against the delayed neurotoxic effects of organophosphorus esters, *Fed. Proc.*, 35, 73, 1976.
91. **Clothier, B. and Johnson, M. K.,** Rapid aging of neurotoxic esterase after inhibition of di-isopropyl phosphorofluoridate, *Biochem. J.*, 177, 549, 1979.
92. **Clothier, B. and Johnson, M. K.,** Reactivation and aging of neurotoxic esterase inhibited by a variety of organophosphorus esters, *Biochem. J.*, 185, 739, 1980.
93. **Davies, D. R., Holland, P., and Rumens, M. J.,** The relationship between the chemical structure and neurotoxicity of alkyl organophosphorus compounds, *Br. J. Pharmacol.*, 15, 271, 1960.
94. **Earl, C. J. and Thompson, R. H. S.,** Cholinesterase levels in the nervous system in tri-ortho-cresyl phosphate poisonings, *Br. J. Pharmacol.*, 8, 212, 1953.
95. **Poulsen, E. and Aldridge, W. N.,** Studies on esterases in the chicken central nervous system, *Biochem. J.*, 90, 182, 1964.
96. **Aldridge, W. N. and Barnes, J. M.,** Esterases and neurotoxicity of some organophosphorus compounds, *Biochem. Pharmacol.*, 15, 549, 1966.
97. **Joel, C. D., Moser, H. W., Majno, G., and Karnovsky, M. L.,** Effects of bis-(monoisopropylamino)-fluorophosphine oxide (Mipafox) and starvation on the lipids in the central nervous system of the hen, *J. Neurochem.*, 14, 497, 1967.
98. **Foil, L. D., Chambers, H. W., Stinson, R. S., and Glick, B.,** Immunological aspects of tri-*o*-tolyl-phosphate-induced delayed neurotoxicity in chickens, *Toxicol. Appl. Pharmacol.*, 56, 259, 1980.
99. **Olajos, E. J., Caprio, A. P., and Rosenblum, I.,** Central and peripheral neurotoxic esterase activity and dose-response relationship in adult hens after acute and chronic oral administration of diisopropyl fluorophosphate, *Ecotoxicol. Environ. Safety*, 2, 383, 1978.

100. **Davies, D. R. and Holland, P.**, Effect of oximes and atropine upon the development of delayed neurotoxic signs in chickens following poisoning by DFP and Sarin, *Biochem. Pharmacol.*, 21, 3145, 1972.
101. **Lotti, M. and Johnson, M. K.**, Neurotoxicity of organophosphorus pesticides: predictions can be based on *in vitro* studies with hen and human enzymes, *Arch. Toxicol.*, 41, 215, 1978.
102. **Dudek, B. R., Barth, M., Gephart, L., Huggins, J., and Richardson, R. J.**, Correlation of brain and lymphocyte neurotoxic esterase inhibition in the adult hen following dosing with neurotoxic compounds, *Toxicol. Appl. Pharmacol.*, 48 (A198), 397, 1979.
103. **Johnston, G. A. R.**, Neuropharmacology of amino acid inhibitory transmitters, *Ann. Rev. Pharmacol. Toxicol.*, 18, 269, 1978.
104. **Cull-Candy, S. G., Donnellan, J. F., James, R. W., and Lunt, G. G.**, 2-Amino-4-phosphonobutyric acid as a glutamate antagonist on locust muscle, *Nature (London)*, 262, 1976.
105. **Watkins, J. C., Curtis, D. R., and Brand, S. S.**, Phosphonic analogues as antagonists of aminoacid excitants, *J. Pharm. Pharmacol.*, 29, 324, 1977.
106. **Slaughter, M. M. and Miller, R. F.**, 2-Amino-4-phosphonobutyric acid: a new pharmacological tool for retina research, *Science*, 211, 182, 1981.
107. **Perkins, M. N., Stone, T. W., Collins, J. F., and Curry, K.**, Phosphonate analogues of carboxylic acids as amino acid antagonists on rat cortical neurones, *Neurosci. Lett.*, 23, 333, 1981.
108. **Gruszecka, E., Soroka, M., and Mastalerz, P.**, Phosphonic analogs of α-methylaspartic and α-methylglutamic acids, *Pol. J. Chem.*, 53, 2327, 1979.
109. **de Tinguy-Moreaud, E., Bioulac, B., Vincent, J.-D., and Neuzil, E.**, Neuroactive properties of some phosphonic analogs of natural guanido compounds, *Gen. Pharmacol.*, 11, 513, 1980.
110. **Bellet, E. M. and Casida, J. E.**, Bicyclic phosphorus esters: high toxicity without cholinesterase inhibition, *Science*, 183, 1135, 1973.
111. **Petajan, J. H., Voorhees, K. J., Packham, S. C., Baldwin, R. C., Einhorn, I. N., Grunnet, M. L., Dinger, B. G., and Birky, M. M.**, Extreme toxicity from combustion products of a fire-retarded polyurethane foam, *Science*, 198, 742, 1975.
112. **Bowery, N. G., Collins, J. F., and Hill, R. G.**, Bicyclic phosphorus esters that are potent convulsants and GABA antagonists, *Nature (London)*, 261, 601, 1976.
113. **Atherton, F. R., Hall, M. J., Hassall, C. H., Holmes, S. W., Lambert, R. W., Lloyd, W. J., and Ringrose, P. S.**, Phosphonopeptide antibacterial agents related to alafosfalin: design, synthesis and structure-activity relationships, *Antimicrob. Agents Chemother.*, 18, 897, 1980.
114. **Allen, J. G., Havas, L., Leicht, E., Lennox-Smith, I., and Nisbet, L. J.**, Phosphonopeptides as antibacterial agents: metabolism and pharmacokinetics of alafosfalin in animals and humans, *Antimicrob. Agents Chemother.*, 16(3), 306, 1979.
115. **Atherton, F. R., Hall, M. J., Hassall, C. H., Lambert, R. W., Lloyd, W. J., and Ringrose, P. S.**, Phosphonopeptides as antibacterial agents: mechanism of action of alaphosphin, *Antimicrob. Agents Chemother.*, 15(5), 696, 1979.
116. **Sanders, H. J.**, Herbicides, *Chem. Eng. News*, 20, 1981.
117. **Baird, D. D., Upchurch, R. P., Homesley, W. G., and Franz, J. E.**, Introduction of a new broad spectrum postemergence herbicide class with utility for herbaceous perennial weed control, *Proc. North Cent. Weed Control Conf.*, 26, 64, 1971.
118. *Herbicide Handbook*, 4th ed., Weed Science Society of America, Champaign, Ill., 1979.
119. **Chykaliuk, P. B., Abernathy, J. R., and Gipson, J. R.**, *Bibliography of Glyphosate*, The Texas Agricultural Experiment Station, Lubbock, 1980.
120. **Sprankle, P., Meggitt, W. F., and Penner, D.**, Adsorption action and translocation of glyphosate, *Weed Sci.*, 23, 235, 1975.
121. **Sandberg, C. L., Meggitt, W. F., and Penner, D.**, Adsorption, translocation and metabolism of ^{14}C-glyphosate in several weed species, *Weed Res.*, 20, 195, 1980.
122. **Lolas, P. C. and Coble, H. D.**, Translocation of ^{14}C-glyphosate in johnsongrass (*Sorghum Halepense* L. Bers.) as affected by growth stage and rhizome length, *Weed Res.*, 20, 267, 1980.
123. **Sherp, L. A.**, A review of some factors affecting the performance of glyphosate, *Proc. N. Z. Weed Pest Control Conf.*, 28, 165, 1975.
124. **Kells, J. J. and Rieck, C. E.**, Accumulation of ^{14}C-glyphosate in johnsongrass, *Proc. South. Weed Sci. Soc.*, 31, 243, 1978.
125. **Jaworski, E. G.**, Mode of action of *N*-phosphonomethylglycine inhibition of aromatic amino acid biosynthesis, *J. Agric. Fd. Chem.*, 20, 1195, 1972.
126. **Roisch, U. and Lingens, F.**, Effect of the herbicide *N*-phosphonomethylglycine on the biosynthesis of aromatic amino acids, *Angew. Chem.*, 13, 400, 1974.
127. **Duke, S. O. and Hoagland, R. E.**, Effects of glyphosate on the metabolism of phenolic compounds: VII. Root-fed amino acids and glyphosate toxicity in soybean *(Glycine max)* seedlings, *Weed Sci.*, 29, 297, 1981.

128. **Ekanayake, A., Wickremasinghe, R. L., and Liyanage, H. D. S.**, Studies on the mechanism of herbicidal action of *N*-(phosphonomethyl)glycine, *Weed Res.*, 19, 39, 1979.
129. **Campbell, W. F., Evans, S. O., and Reed, S. C.**, Effects of glyphosate on chlorophyll ultrastructure of quackgrass mesophyll cells, *Weed Sci.*, 24, 22, 1976.
130. **Hull, H. M., Bleckmann, C. A., and Morton, H. L.**, Some effects of glyphosate on the foliar ultrastructure of velvet mesquite seedlings, *Proc. West. Weed Sci. Soc.*, 30, 18, 1977.
131. **Kitchen, L. M., Witt, W. W., and Rieck, C. E.**, Inhibition of chlorophyll accumulation by glyphosate, *Weed Sci.*, 29, 513, 1981.
132. **Lee, T. T.**, Effects of glyphosate on synthesis and degradation of chlorophyll in soybean and tobacco cells, *Weed Res.*, 21, 161, 1981.
133. **Kitchen, L. M., Witt, W. W., and Rieck, C. E.** Inhibition of δ-aminolevulinic acid synthesis by glyphosate, *Weed Sci.*, 29, 571, 1981.
134. **Nomura, N. S. and Hilton, H. W.**, The adsorption and degradation of glyphosate in five Hawaiian sugercane soils, *Weed Res.*, 17, 113, 1977.
135. **Selleck, G. W. and Baird, D. D.**, Antagonism with glyphosate and residual herbicide combinations, *Weed Sci.*, 29, 185, 1981.
136. **O'Sullivan, P. A. and O'Donovan, J. T.**, Interaction between glyphosate and various herbicides for broad leaved weed control, *Weed Res.*, 20, 255, 1980.
137. **Zoebisch, O. C., Rushing, T. T., and Barrier, G. E.**, A new brush control agent, *Proc. Northeastern Weed Sci. Soc.*, 28, 347, 1974.
138. **Hernandez, T. J., Hudson, W. H., and Gonzalez, F. E.**, A progress report on Krenite brush control agent, *Proc. South. Weed Sci. Soc.*, 28, 261, 1975.
139. **Neihuss, M. H. and Roediger, K. J.**, Ammonium ethylcarbamoylphosphonate, a new plant growth regulator for the control of undesirable brushwood species, *Proc. Br. Weed Control Conf.*, 3, 1015, 1974.
140. **Welch, A. W., Weed, M. B., and Bingeman, C. W.**, Control of brush with a new growth regulator, *Proc. South. Weed Sci. Soc.*, 27, 245, 1974.
141. **Richardson, R. G.**, Foliar absorption and translocation of fosamine and 2,4,5-T in blackberry (*Rubus procerus* P. J. Muell.), *Weed Res.*, 20, 159, 1980.
142. **Coupland, D. and Peabody, D. V.**, Absorption, translocation, and exudation of glyphosate, fosamine, and amitrole in field horsetail *(Equisetum arvense)*, *Weed Sci.*, 29, 556, 1981.
143. **Kitchen, L. M., Rieck, C. E., and Witt, W. W.**, Absorption and translocation of ^{14}C-fosamine by three woody plant species, *Weed Res.*, 20, 285, 1980.
144. **Morey, P. R. and Dahl, B. E.**, Inhibition of mesquite (*Prosopsis juliflora* var. *glandulosa*) growth by fosamine, *Weed Sci.*, 28, 251, 1980.
145. **Abeles, F. B.**, Biosynthesis and mechanism of action of ethylene, *Ann. Rev. Plant Physiol.*, 23, 259, 1972.
146. **Hoagland, R. E.**, Effects of glyphosate on metabolism of phenolic compounds: VI. Effects of glyphosine and glyphosate metabolites on phenylalanine ammonia-lyase activity, growth, and protein, chlorophyll, and anthocyanin levels in soybean *(Glycine max)* seedlings, *Weed Sci.*, 28, 393, 1980.
147. **Croft, S. M., Arntzen, C. J., Vanderhoef, L. N., and Zettinger, C. S.**, Inhibition of chloroplast ribosome formation by *N,N*-bis(phosphonomethyl)-glycine, *Biochim. Biophys. Acta*, 335, 211, 1974.
148. **Dollwet, H. H. A. and Kumamoto, J. J.**, Ethylene production of ethyl propylphosphonate, NIA 10637, *Plant Physiol.*, 46, 786, 1970.
149. **Bababunmi, E. A., Olorunsogo, O. O., and Bassir, O.**, Toxicology of glyphosate in rats and mice, *Toxicol. Appl. Pharmacol.*, 45, 319, 1978.
150. **Christy, S. L., Karlander, E. P., and Parochetti, J. V.**, Effects of glyphosate on the growth rate of *Chlorella*, *Weed Sci.*, 29, 5, 1981.
151. **Worthing, C. R.**, *The Pesticide Manual*, 6th ed., BCPC Publications, Croyden, England, 1979, 286.
152. **Theiss, J. C. and Shimkin, M. B.**, Effect of the plant growth regulator, 2-chloroethylphosphonic acid, on spontaneous and chemically-induced lung tumorigenesis in strain A mice, *Food Cosmet. Toxicol.*, 18, 129, 1980.
153. **McWhorter, C. G.**, Weed control in soybeans with glyphosate applied in the recirculating sprayer, *Weed Sci.*, 25, 135, 1977.
154. **Monaghan, N.**, The biology of johnsongrass *(Sorghum halepense)*, *Weed Res.*, 19, 261, 1979.
155. **Frear, D. E. H.**, *Pesticide Index*, 4th ed., College Science Publishers, State College, Pa., 1969.
156. **Aldridge, W. N. and Barnes, J. M.**, Further observations on the neurotoxicity of organophosphorus compounds, *Biochem. Pharmacol.*, 15, 541, 1966.
157. **Abou-Donia, M. B. and Graham, D. G.**, Delayed neurotoxicity of *O*-ethyl*O*-4-nitrophenyl phenylphosphonothioate: subchronic (90 days) oral administration in hens, *Toxicol. Appl. Pharmacol.*, 45, 685, 1978.
158. **Sanborn, J. R., Metcalf, R. L., and Hansen, L. G.**, The neurotoxicity of *O*-(2,5-dichlorophenyl)*O*-methyl phenylphosphonothioate, an impurity and photo-product of leptophos (Phosvel) insecticide, *Pestic. Biochem. Physiol.*, 7, 142, 1977.

159. **Hollingshaus, J. G., Abu-El-Haj, S., and Fukuto, T. R.,** Delayed neurotoxicity of *O*-alkyl *O*-aryl phenylphosphonothioate analogues related to leptophos administered orally to the hen, *J. Agric. Food Chem.*, 27, 1197, 1979.
160. **Abou-Donia, M. B.,** Delayed neurotoxicity of phenylphosphonothioate esters, *Science*, 205, 713, 1979.
161. **Abou-Donia, M. B., Graham, D. G., Ashry, M. A., and Timmons, P. R.,** Delayed neurotoxicity of leptophos and related compounds: differential effects of subchronic oral administration of pure, technical grade and degeneration products on the hen, *Toxicol. Appl. Pharmacol.*, 53, 150, 1980.
162. **Abou-Donia, M. B., Graham, D. G., and Komeil, A. A.,** Delayed neurotoxicity of *O*-ethyl *O*-2,4-dichlorophenyl phenylphosphonothioate: effects of a single oral dose on hens, *Toxicol. Appl. Pharmacol.*, 49, 293, 1979.
163. **Abou-Donia, M. B.,** Delayed neurotoxicity of *O*-(2,4-dichlorophenyl) *O*-ethyl phenylphosphonothioate (S-Seven), *Toxicol. Lett.*, 3, 61, 1979.
164. **Abou-Donia, M. B. and Graham, D. G.,** Delayed neurotoxicity of a single oral dose of *O*-ethyl *O*-4-cyanophenyl phenylphosphonothioate in the hen, *Neurotoxicology*, 2, 449, 1979.

Chapter 7

INDUSTRIAL USES OF PHOSPHONATES

George L. Drake, Jr. and Timothy A. Calamari, Jr.

TABLE OF CONTENTS

I. INTRODUCTION

Although organic phosphonates offer many advantages over organic phosphates in numerous commercial applications, extensive use of these agents has been hindered by their higher price compared to organic phosphates and their limited commercial availability. Nonetheless, organic phosphonates have slowly begun to establish a position in the commercial marketplace. Phosphates and phosphonates are similar both in structure and in their physical properties, but they differ in that the phosphonates have a C–P bond, one less oxygen atom in the analogous molecule, corresponding lower molecular weights and boiling points, higher refractive indexes, different solubilities, and greater hydrolytic stability. Many users of organo phosphorous compounds are taking advantage of these property differences by substituting phosphonates into formulations that previously used phosphates. Even so, the commercial potential of the organo phosphonates has only barely been tapped.

II. USES

Organic phosphonates have been recommended for use in a number of applications some of which are listed below.

A. List of Applications Covered

1. Adhesives[1-5]
2. Agents for extraction, concentration, and purification of uranium, thorium, and plutonium[6-8]
3. Antioxidants[9-12]
4. Antistatic agents[13]
5. Blowing agents[14]
6. Catalysts[15-19]
7. Corrosion inhibitors[20-34]
8. Coupling agents[35]
9. Crystallization inhibitors[36-37]
10. Dentifrice compositions[38-40]
11. Deodorants[41]
12. Detergent additives[42-47]
13. Detergents for cleaning metal surfaces[48]
14. Dye modifiers[49-50]
15. Flame retardant polymers[51-53]
16. Flame retardants for textiles[54-96]
17. Fire retardants for synthetic fibers[97-103]
18. Flotation agents[104-105]
19. Fuel additives[106]
20. Gelling agents[107-110]
21. Hardening oil composites[111]
22. Heat and light stabilizers[112-129]
23. Hydraulic fluid additives[130-133]
24. Ion exchange resins[134-135]
25. Lubricants[136-162]
26. Photography[163]
27. Plasticizers[164]
28. Polyester, polyethylene, and polycarbonate discoloration inhibitors[165-168]
29. Polyurethan additives[169-195]

30. Rayon additives[196-202]
31. Resin and plastic additives[203-224]
32. Scale inhibitors[225-231]
33. Settling retardants[232]
34. Sequestering agents[234-236]
35. Solvent extraction[233]
36. Suspending agents[237]
37. Synthetic fiber preparation[238]
38. Viscosity modifiers[239]
39. Wood fireproofing agents[240-241]

With the exception of insecticides and herbicides, the two principle industrial uses of phosphonates are as lubricant additives, and as fire retardant agents. There are so many other areas in which phosphonates can cause a vast improvement in an existing product or process, that some consideration should be given to the most promising of these many and diverse applications.

B. Major Applications

1. Adhesives

Adhesives are substances capable of holding materials together by surface attachment. They can attach to surfaces and develop the internal or cohesive strength necessary to hold materials together while cooling from the liquid to the solid state, while losing solvent, or during chemical reaction. Pressure sensitive adhesives generally do not undergo a phase change in order to hold surfaces together. Present day technology in adhesives has made this a very significant area of growth.

Improved pressure-sensitive adhesives with increased resistance to hot bar creep of an aged sample result when rubbery elastomers containing neutral tackifiers are mixed with a dialkyl phosphonate and either a phenolic or secondary-amine antioxidant.[1]

Improved adhesives for joints with good strength and heat resistance result when diglycidylmethylphosphonate is mixed with an epoxy resin and polyethylenepolyamine.[2]

Pressure sensitive adhesives for plastics can be prepared by mixing vinyl carboxylatedialkyl vinyl phosphonate copolymer with a natural or synthetic resin. Such polymers bond polyethylene, polypropylene, polyterephthalate, rigid and plasticized poly(vinyl chloride), superpolyamide, cellulose triacetate, and poly(vinyl chloride)-coated cellulosic film to glazed and bleached craft paper. Bonding remains effective over 6 months at 50°C.[3]

The compound bis(2-chloroethyl)vinyl phosphonate is also useful in making adhesive compositions.[4] Dialkyl alkyl phosphonates are used in compositions as antifoaming and dust inhibitors for proteinaceous adhesives.[5]

2. Agents for Extraction, Concentration, and Purification of Uranium, Thorium, and Plutonium

Some phosphonates have found application as extractive solvents in the concentration and purification of uranium and thorium for use in atomic energy applications. Some examples of these compounds are di-butylphenylphosphonate and di-sec-butylphenylphosphonate in aromatic and aliphatic solvents in the presence of HNO_3. The di-sec-butylphenylphosphonate is best for extractions from HNO_3 systems where U/Th separation is required. Either of the two above phosphonates could reduce chemical costs to alleviate HNO_3 recovery from waste solutions. Although these phosphorus compounds are beneficial to the industry, the market, as yet, has not been large enough to justify production of such phosphonates on a large scale.[6-8] Siddall[8] gives a good discussion of the nomenclature and types of organophosphorus compounds that could be useful for aqueous processing of irradiated fuels and by-product isotopes.[7-8]

3. Antioxidants

(Bis[3,5-di-tert-butyl-4-hydroxybenzyl]methyl)-phosphonates when used in conjunction with dilauryl thiodipropionate are good antioxidants for polypropylene and nylon 66. When polypropylene is stabilized with such an antioxidant it has a stability of 525 hr at 150°C as compared with 3 hr when no stabilizer is used.[9] These compounds are also good antioxidants for rubbers, petroleum waxes,[10] and motor oils.[11]

Phosphonates such as *O,O*-di-*n*-octadecyl-3,5-di-tert-butyl-4-hydroxybenzylphosphonate when used in polyester molding compositions as an antioxidant, produces molded articles with improved impact strength and impact strength retention.[12]

4. Catalysts

Several U.S. patents have been issued in which phosphonates have been shown to act as catalysts.[15-19] For example, $(CH_3)_2CHP(O)(ONa)_2$ was used as catalyst in the preparation of poly(arylene sulfides) from an alkali sulfide, and a *p*-dihalobenzene giving polymers with high inherent viscosity and low melt index.

Bis(beta-chloroethyl) vinylphosphonate is reported to be a good catalyst for the condensation of isocyanates to carbodiimides.[17] It is also effective for coupling with aluminum alkyls as catalysts for olefin synthesis.[18] See the section on phosphonate chemistry for additional uses of phosphonates as catalysts.

5. Corrosion Inhibitors

The presence of dissolved oxygen in water accelerates corrosion. Few of the inhibitors used for alleviation of anaerobic attack will function in the presence of oxygen. Phosphonates however, can be used in the treatment of cooling waters to control oxygen corrosion.[31] They have properties which make them more desirable than the polyphosphates which are frequently used for control of oxygen attack in waters.

Amino methylenephosphonates protect against corrosion especially when used in conjunction with zinc salts. Inhibition is similar to that of the polyphosphates but show some significant differences. For instance, the hydrolytic stability of the aminomethylene phosphonates is very attractive when prolonged exposure to high water temperatures is involved. Their use also eliminates deposition of $Ca_3(PO_4)_2$, one of the major problems encountered in cooling towers. The protective film formed with the aminophosphonates is more tolerant of alkaline pH than the phosphates.[20-24,26,29,31,34]

Lindeblad and Pousette[30] have reported new methods for the treatment of cooling tower water to prevent scaling and corrosion. They describe the use of complexing agents with various phosphonates to prevent corrosion.

Use of phosphonate-phosphate in the treatment of cooling water systems avoids the pollution of blowdown streams that results when chromates are used.[25,28,29]

Decyl (2,5-di-tert-amyl-4-hydroxyphenyl) phosphonate has been shown to be a good inhibitor of metal corrosion on both aluminum and cast iron.[27,32,33]

6. Detergent Additives

Most of the built detergents sold today contain about 50% by weight of sodium-phosphate and/or sodium pyrophosphate. Quimby[42] has described a number of new triphosphonates for use in built detergent compositions that have good sequestering properties, such as complexing >2 mol of Ca per mole of sequestrant.

The triphosphonates are efficient in treatment of solutions containing Ca, Fe, Mg, Zn, or Cu. The ethane triphosphonates are 3 to 4 times as efficient as sodium tripolyphosphate and can be used in photography to avoid sludges and turbidity in developing and fixing solutions. They are also used in industrial cleaning and in textile finishing. As builders in detergents, the tripolyphosphonates are stable against deterioration during storage and use. Detergents containing them perform equally well with tapwater at room temperature or in water heated to 140 to 160°F or higher. These are useful in granular or liquid form.[42-43] Alkylenediphosphonate detergent builders such as tetra-sodium methylenediphosphonate are reported to be equal to and in some cases, better in hot water than sodium tripolyphosphate, a widely used builder.[44] Hexasodium diaminotris(methylenephosphonate) is useful in compositions for making textile nonyellowing detergents.[45]

U.S. patents have been issued covering antiseptic detergent compositions. These comprise soap or synthetic detergents containing 50% by weight of a dialkyl ester of a long chain alkyl phosphonate such as di-ethyl dodecylphosphonate as a bacteriostatic agent.[46] Other phosphonates have been used in formulations as lime soap dispersants.[47]

7. *Flame Retardant Polymers*

A fire retardant composition for polyethylene, polypropylene, polyester resins, phenolic resins, and polystyrene is produced by reacting an arylhaloalkyl phosphonate, such as cresyl-2-bromo 3 ethylmethyl phosphonate with a chlorinated paraffin. Antimony oxide, (Sb_2O_3), is usually necessary in formulations of this type to produce flame retardance. This composition is effective without the use of Sb_2O_3 in the formulation.[51-52]

Polyesters prepared from aromatic dicarboxylate esters, diaryl phosphonates, and aromatic diols have good fire resistance, mechanical properties (especially heat deformation resistance), and processability. They are useful in the preparation of moldings.[53]

8. *Flame Retardants for Textiles*

a. Background

The flame retarding of textiles is important for improvement of the safety characteristics of apparel, bedding, protective clothing, tenting materials, carpets, home furnishings, aircraft and automobile interior fabrics, and industrial fabrics which may be woven, knitted, tufted, and nonwoven.

Of the various chemical agents used for the flame retardant treatment of textiles, phosphonates have been among the more successful.

These compounds have been used in three ways: (1) in the preparation of polymers that are applied to the surface of the fibers, (2) for making derivatives of cellulose, and (3) as monomers that react with cellulose and polymerize in the fiber.

b. Allyl Phosphonates

Polymers for use on the surface of fibers have been based on chloromethyl diallyl phosphonate[54] and cyanoethyl diallyl phosphonate[55-56]

$$(CH_2=CHCH_2O)_2\overset{O}{\overset{\|}{P}}{-}CH_2Cl \qquad (CH_2=CHCH_2O)_2\overset{O}{\overset{\|}{P}}{-}CH_2CH_2CN$$

These monomers may be polymerized with a benzoyl peroxide catalyst at about 85 to 100°C. Better flame retardants are prepared; however, when the monomers are first partially brominated and then polymerized. About 25% add-on of the partially brominated polymer product is required for an 8 oz cotton fabric to provide sufficient flame resistance to pass the standard vertical flame test. The relatively high add-ons are probably accounted for by the high percentage of fuel (carbon and hydrogen) that is added per unit of phosphorus and bromine. In general, these polymers are about equivalent in flame retardancy to the bromoform adduct of triallyl phosphate. The phosphonate polymers are somewhat more impervious to alkaline laundering than polymers made from phosphates. The phosphonates are applied to fabric from an organic solvent or a mixture of solvents, (such as methanol and ethylene dichloride), using a pad-dry-cure procedure. Curing at an elevated temperature is not required. The flame-retardant finishes withstand several accelerated laundering cycles.

c. Phosphonomethyl Cellulose

Phosphonomethyl cellulose is prepared by reacting the disodium salt of chloromethyl phosphonate with cotton in the presence of a strong base.[57] The preferred method for carrying out this reaction is to pad the fabric with a solution containing 25% NaOH and 15% of the sodium salt of chloromethyl phosphonate, dry the fabric, then heat for 5 min at 140°C. The wet pickup for most cotton fabrics is about 125 to 140%. Under these conditions, nearly 2% phosphorus can be attached to the cotton fiber in a single processing. The phosphonomethyl cellulose produced in this manner exists as the sodium salt and is not very flame resistant. It is necessary to exchange the sodium ions for hydrogen or ammonium ions to obtain a high degree of flame resistance. In the ammonium-salt form only 0.7% phosphorus is required in a cotton sheeting to pass the standard vertical flame test.[57] About 1% phosphorus is required to pass the 180° angle strip match test.[58] This treatment generally does not reduce breaking strength and reduces tearing strength only slightly.[59] Stiffness is increased severalfold. Phosphonomethyl cellulose is very resistant to both acid and base hydrolysis, however, flame resistance is lost when the material forms a salt with either group I or II metal ions. Tretment of the phosphonomethyl cellulose with a dimethylol-type cross-linking agent reduces the stiffness.[59]

d. Phosphonopropionamides

The *N*-methylol derivatives of several dimethyl, diethyl, and di(chloroethyl) phosphonopropionamides have been studied as flame retardants for cellulosic fabrics.[60,87]

The dimethylphosphonopropionamide is by far the most useful and is synthesized as follows:

$$PCl_3 + 3\,CH_3OH \longrightarrow P(OCH_3)_3 + 3\,HCl \qquad (1)$$

$$P(OCH_3)_3 + HCl \longrightarrow HP(O)(OCH_3)_2 + CH_3Cl \qquad (2)$$

The phosphonate adds across the activated double bond on reaction with acrylamide according to the following reaction:

$$(CH_3O)_2\overset{\overset{O}{\|}}{P}—H + CH_2=CH—\underset{\underset{O}{\|}}{C}—NH_2 \longrightarrow$$

$$(CH_3O)_2P(O)CH_2CH_2—\underset{\underset{O}{\|}}{C}—NH_2 \qquad (3)$$

The hydroxymethyl derivative is made by reaction with formaldehyde under acidic conditions:

$$(CH_3O)_2\underset{\underset{O}{\|}}{P}—CH_2CH_2\underset{\underset{O}{\|}}{C}NH_2 + CH_2O \xrightarrow{H^+}$$

$$(CH_3O)_2\underset{\underset{O}{\|}}{P}—CH_2CH_2—\underset{\underset{O}{\|}}{C}—NHCH_2OH \qquad (4)$$

Dimethylphosphonopropionamide has been developed by the Ciba Geigy Company and the commercial product is sold under the trade name Pyrovatex C.P.® A typical treating solution contains 5% trimethylolmelamine (or comparable type reactive amine), 1% urea,

0.1% of a suitable stabilizing agent, 0.5% polyethylene softener, 0.4% amine hydrochloride catalyst (or other acid catalysts such as H_3PO_4), and 30% of the *N*-methyloldialkyl phosphonopropionamide. The add-on required for an 8 oz cotton fabric to pass the standard vertical flame test[60a] is about 25%. Fabrics are padded in the treating solution, dried, and then cured for 1 to 2 min at 175°C or 5 min at 160°C. Treated fabrics undergo ion exchange when subjected to acid sours or hard water because of the acidic hydroxyl groups generated by partial hydrolysis of the P–OCH_3 bonds in the dimethylphosphonopropionamide. There is also some build-up of volatilized matter in the ovens during curing. Although this finish has been used in the U.S., it is presently used mainly in Europe in hospitals, hotels, public buildings, local authority buildings, aerospace, and the shipping industry. Presently there are approximately 30 finishers in the U.K. who can finish fabric with Pyrovatex C.P.,® of which 12 units are in large scale production.[61] The contract flame retardant furnishing market in the U.K. is currently thought to be over 10 million m of fabric annually.[61]

Tesoro et al.[62] prepared cellulosic derivatives closely resembling those based on the dialkylphosphonopropionamide. Fabric was treated with three *N*-hydroxymethylhaloacetamides (chloro, bromo, and iodo) from dimethylformamide (DMF) solution by a pad-dry-cure technique using zinc nitrate as the catalyst. These derivatives were then reacted in solvent with trimethyl phosphite at about 140 to 150°C. The reactions are as follows:

$$\text{CellOH} + \text{HOCH}_2\text{NHCOCH}_2\text{I} \xrightarrow{\text{DMF}} \text{CellOCH}_2\text{NHCOCH}_2\text{I} \tag{5}$$

$$\text{CellOCH}_2\text{NHCOCH}_2\text{I} + (\text{CH}_3\text{O})_3\text{P} \longrightarrow \text{CellOCH}_2\text{NHCOCH}_2\text{P(O)(OCH}_3)_2 + \text{CH}_3\text{I} \tag{6}$$

Phosphorus contents above 1.5% provided good flame resistance which was durable to laundering. Loss of fabric strength was not significant. These authors also verified that the nitrogen and phosphorus act synergistically when the phosphonopropionamide is used in conjunction with trimethylolmelamine to impart flame retardancy to cotton or rayon.[63]

e. Vinyl Phosphonates

Vinyl phosphonates undergo homopolymerization or copolymerization with *N*-hydroxymethyl acrylamide in the presence of a free radical catalyst like potassium persulfate.[64] Stauffer Chemical Company has developed vinyl phosphonate flame retardants for cellulose. The trade name of their commercial product is Fyrol® 76, represented by the following formula:

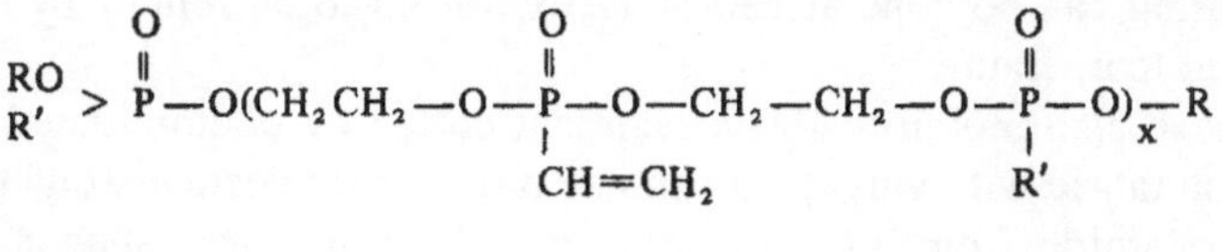

It is prepared as follows:

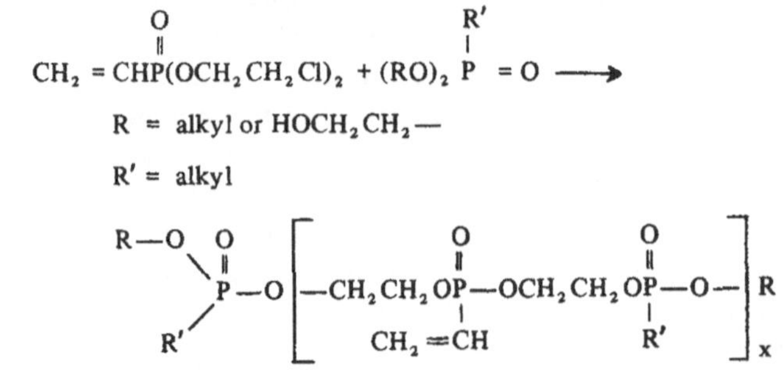

(7)

It is a clear water soluble liquid containing 22.5% phosphorus. The homopolymer of vinyl phosphonate or copolymer of vinyl phosphonate and *N*-hydroxymethyl acrylamide is deposited within the interstitial spaces of the cellulose matrix.

Since there is no reaction with cellulose the tensile and tearing strength and hand of the treated fabric is not affected greatly. There is some degradation of the cellulose due to oxidation by the persulfate catalyst. A typical Fyrol® 76 formulation is as follows:

Water	52.99%
Triton® X-100	0.01%
Chelon® 100	0.50%
Softener (25%)	4.00%
Fyrol® 76	20.00%
Methylolacrylamide	22.00%
Potassium persulfate	0.50%

The finish is applied by a pad-dry-cure technique and a soft polymeric finish durable to laundering is obtained. With a 30% add-on this product has been shown to be useful in treating polyester cotton blends. Addition of bromine compounds to the formulation enhances the fire retardancy of polyester/cotton blends. Fyrol® 76 can also be cured by electron beam radiation techniques to achieve durable flame retardant properties.[65-67b] Potential advantages using this method are softer hand, good physical properties, very fast processing speeds, and favorable economics.

f. Bis(beta-chloroethyl)vinyl Phosphonate

Bis(beta-chloroethyl)vinyl phosphonate manufactured under the trade name Fyrol® Bis Beta has also been shown to impart durable flame resistance to polypropylene by applying a 5:1 ratio of Fyrol® Bis Beta, and *N*-methylol-acrylamide using 2% ammonium persulfate as the catalyst. Curing can be done at 120 to 125°C for 45 to 60 min or by electron beam radiation at ambient temperature.[68]

Harris et al.[69] have also produced flame retardant cotton by photoinitiated graft copolymerization of cotton fabric with vinyl phosphonate oligimer (molecular weight 500 to 1000) and *N*-methylolacrylamide from aqueous solution. The finish was shown to be mainly concentrated between and on the surfaces of the fibers.[69]

Inagaki and Katsuura[70] reacted cellulose phosphonate and *N*-vinyl-2-pyrrolidone in ethanol in the presence of sodium ethoxide to form graft polymerization of *N*-vinyl-2-pyrrolidone at P–H sites in cellulose phosphonate with an average chain length of the graft polymer of about 5 units of vinyl pyrrolidone. These grafts improve the flame retardant properties of cellulose fabrics.

Arney and Kuryla[71] have studied the effects of alkyl phosphates, phosphonates, and phosphites on cellulose flammability.

Other phosphonates which have been reported to be flame retardants for textiles, but which have not been used in quantity are as follows:

1. Dialkyl(carbamoylethyl) phosphonates[71a]
2. Monoammonium methyl phosphonate[71b]
3. Chloromethylphosphonic acid-urea derivatives[72]
4. (Hydroxymethyl) phosphonic acid-urea[73]
5. Monoalkyl or arylphosphonate-urea[74]
6. Dialkyl 1-amino-1-deoxyglucitylphosphonates[75]
7. Bis(2-chloropropyl) propenyl phosphonate, bis(2-bromoethyl) vinyl phosphonate[76]
8. 3-(Dialkylphosphono) propionamides[77]
9. $(CH_3O)_2P(O)CH_2CH_2CONHCH(OH)CH(OH)\text{-}NHCOCH_2CH_2P(O)(OCH_3)_2$, $(CH_3O)_2P(O)CH_2CH_2CONHCH(OH)CH(OH)N(CH_2OH)\text{-}COCH_2CH_2P(O)(OCH_3)_2$[78]
10. *N*-(hydroxymethyl)-3-phosphonoproionamide, *N*-(hydroxymethyl)-3-(diethylphosphono)-propionamide, *N*-(hydroxymethyl)-3-(dimethylphosphono)-propionamide, *N*-(hydroxymethyl)-3-[bis(2-chloroethyl)-phosphono]propionamide *N*-(hydroxymethyl)triazines containing phosphonate substituents[79]
11. Alkenylphosphonates[80-81]
12. Methyldiethylphosphonopropionate[82]
13. *N*-containing polyphosphonates[83]
14. $HOCH_2NHCOCH_2CH_2PO(OR)_2$[63,84]
15. 2,4-Diamino-6-(*O,O'*-diethylphosphonomethyl)-*s*-triazine[85]
16. Trisphosphonates[86]
17. $(CH_3O)_2P(O)CH_2CH_2CONHCH_2OH$ (Pyrovatex C.P.®), *N*-methylol-3-(diethylphosphono)propionamide, *N*-methylol-3-(dimethylphosphono)-2-methyl-propionamide[60,87]
18. Amino phosphonates[88]
19. 3-phosphonopropionamide, 3-(dimethylphosphono)propionamide, *N*-(hydroxymethyl)-3-(dimethylphosphono)propionamide[89]
21. 3-(Dialkylphosphono)propionamide condensation with urea and HCHO[90]

22.

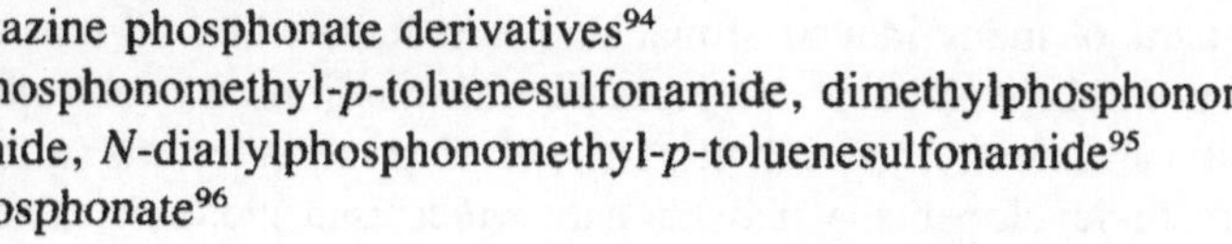

23. Ureidophosphonates[92]
24. Triazinylaminoalkylphosphonates[93]
25. Hexahydrotriazine phosphonate derivatives[94]
26. *N*-dimethylphosphonomethyl-*p*-toluenesulfonamide, dimethylphosphonomethyl methane sulfonamide, *N*-diallylphosphonomethyl-*p*-toluenesulfonamide[95]
27. Cellulose phosphonate[96]

9. Gelling Agents

Phosphonates can be used in compositions that can be pumped or sprayed as a ramjet fuel or a missile propellant. The thixotropic fuel mix can be prepared by combining powdered metals or metallic salts, liquid hydrocarbon fuels, metallic phosphonate thickening agents, oxidizers, and dispersion agents.[107] One example of the thixotropic gelling agent is the aluminum salt of monobutyldecylthiomethylphosphonate. Liquid hydrocarbon compositions are improved with respect to their wear-inhibiting and extreme pressure properties. They can be transformed into stable gels or greases by incorporation of small amounts of the oil soluble metal salt of the above phosphonate. These can be used in natural and synthetic lubricating oils, fuels, (gasoline, kerosene, gas oil, or burner oil) and slushing, metalworking, drawing, quenching, textile, and hydraulic oils in concentrations of 1 to 2%.[108]

The homopolymers and copolymers of *O,O*-bis-(methacryloyloxyethyl)phosphonates are also useful as gelling agents.[109] Mono-butyl 2-thiadodecylphosphonate is another compound which can be used in compositions to form firm, smooth, nonbleeding gels.[110]

10. Heat and Light Stabilizers

Heat stabilizers are chemical additives used to protect certain polymers especially those that contain regular, repeating units against degradation resulting from heat or exposure to UV radiation. By virtue of their chemical composition and modes of function, many heat stabilizers also function as light stabilizing agents.

There are numerous patents in the literature in which phosphonates of various types play an important part in formulations useful for heat and light stabilizers especially for polyvinyl chloride,[112-115] polyethylene,[113] nylon and polyolefins,[117-119] polyesters,[120-124] polycarbonate molding compositions.[125-126] Examples of phosphonates used in formulations as heat stabilizers are as follows:

1. Di-ethyl 3,5-di-tert-butyl-4-hydroxybenzylphosphonate[125]
2. Dibenzylphosphonate[112]
3. Metal *P*-hydrocarbonphosphonate[113,127]
4. Ba, Ca, Zn, Mg, Sr, Cd, Ni, K, Sb, Sn, or Pb organophosphonates[114]
5. Bis(4-hydroxy-3,5-di-tert butylbenzyl) phosphonate[116]
6. Nickel bis(O-C_2H_5-3,5-di-tert-butyl-4-hydroxybenzylphosphonate)[117]
7. Dimethyl 2-(methoxycarbonyl)ethylphosphonate[120]
8. Diallyl chloromethylphosphonate[122]
9. Dimethyl(2-methoxycarbonylethane)phosphonate[123]
10. Polyoxyalkaline alkenylthiophosphonate[128]
11. Di-C_2H_5-ethoxycarbonylethyl phosphonate[124]

Examples of phosphonates used in compositions as light stabilizers are as follows:

1. Diphenyl isodecanephosphonate, diisodecyl isodecanephosphonate[115]
2. Metal alkyl 3-acetyl-4-hydroxybenzylphosphonates[110]
3. *O,O*-diisopropyl 1-hydroxy-1-methyl-2-phenyl-2-nitroethyl phosphonate[129]

11. Hydraulic Fluid Additives

The moving parts of many industrial machines are actuated by oil that is under pressure. Lubricating oil not only is suitable for pressure transmission and controlled flow, but also minimizes friction and wear of moving parts as well as protects ferrous surfaces from rusting.

Diphenyl 1-methylcyclopentane phosphonate when used alone or mixed with other additives in hydraulic fluids imparts improved fire resistance, lubricity, wear resistance, and viscosity-temperature properties.[130,131]

Incorporating phosphonates with the following structures, $[R'P(O)(OR)O]_2M$ (where M is Na; R is CH_3; R′ is polyisobutenyl having 10 to 30 atoms), into lubricating oils imparts antiwear-detergent properties.[132]

The formation of deposits by "functional fluids," such as heat-exchange fluids and hydraulic fluids is prevented by addition of phosphonates such as $CH_3CH(OH)P(O)P(O)(OCH_3)_2$.[133]

12. Ion Exchange Resins

Ion exchange is defined as the reversible interchange of ions between a solid and a liquid in which there is no permanent change in the solid, which is the ion exchange material. The largest use of ion exchange resins is in the field of water conditioning, however, more recently broader applications are being found in the field of chemical processing, hydrometallurgy, purification, product isolation, concentration, catalysis, food processing, and a variety of other areas.

Phosphonates have been used in the synthesis of ion exchangers by copolymerization of sodium vinylphosphonate with acrylic acid, and cross-linking with *N,N'*methylenediacrylamide. The resulting resin had an ion exchange capacity of 10 to 11 meq/g, and had a high selectivity for Cu^{+2} ions.[134]

Amphoteric ion-exchange resins are made by the condensation of poly(ethylenimine) with di-Na chloromethylphosphonate. The resulting resin had an ion exchange capacity of 8.2 meq/g for acids and 7.8 meq/g for bases.[135]

Phosphonomethyl cellulose prepared by the reaction of sodacellulose with the sodium salt of chloromethylphosphonic acid has an excellent ion exchange capacity. One of the good features of this material is that the ion exchange process can be done on a continuous basis. A belt can be made with agents of this type such that ions can be picked up by the belt by passing it first through a solution containing the undesired ion and then into a regenerating bath. This entire process can be carried out continuously.

13. Lubricants

A lubricant is any material that can be placed between surfaces to lessen friction. The purpose of lubrication is to reduce the frictional resistance between two contacting surfaces forced to slide over one another and to minimize wear and prevent corrosion.

The production of lubricant additives has grown to a billion dollar segment of the chemical industry. Although the bulk of additives are used in automotive lubricants, they also are used in oils for turbines, diesel and aircraft engines, two-cycle engines, hydraulic equipment, gears, and metal working.

Aminoethanephosphonate copolymers were prepared for imparting dispersant and antirust properties, pour point depressing action, and improvements in viscosity to lubricating and fuel compositions. Antirust agents are surface active materials that are preferentially adsorbed as a film onto iron and steel surfaces to protect them from attack by moisture.[136]

Antiwear agents produce a surface film by either a chemical or physical adsorption mechanism to minimize friction and wear under boundary lubrication conditions. An anticorrosion and antiwear additive for synthetic liquid lubricants, such as used in aircraft engines, is made from a mixture of primary and tertiary alkyl amines[18-22] and dinonylphenylphosphonate. These can be mixed with pentaerythrithyl ester lubricants.[137] Other phosphonates used as antiabrasive oil additives are the dibutyl(trichloromethyl)phosphonates.[138] Dimethyltetradecanephosphonate is used in a composition to make water-based lubricants with good pump-wear characteristics. Thermally stable antiwear additives for lubricants were prepared from $P(OCH_2CHClR)_3$ and alkyl phenols to give $RCHClCH_2PO(OC_6H_4R'\text{-}P)_2$, where $R' =$ octyl and $R = CH_2CH_2Cl$.[141]

Improved diester lubricants for aircraft gas turbines are produced by adding 1 to 5% of an aminophosphonate to a diester lubricant. This mixture improves load characteristics, anticorrosive properties, and viscosity indices.[142] Phosphonates are also used as lubricants for two-stroke engines.[143-144]

A number of phosphonates are used in compositions as extreme pressure lubricants. Some examples of such materials are dialkyl trichloromethylphosphonate-olefin adducts,[145] dialkylphosphonate-alkyl tallate and sulfur,[146] aminophosphonates,[147] potassium chloromethylphosphonate-potassium mercaptide,[148] and bis(2-chloroethyl) (1-hydroxyundecanyl) phosphonate.[149]

Dioctadecyl and dioleyl phosphonate form extreme high pressure lubricants in combination with active organic sulfur compounds of animal or vegetable oils, fats, and mineral oil. These oils contain 10% tightly bound, and 10 to 70% loosely bound, sulfur (polysulfurized C_2-C_6 olefins, such as the polysulfides of $CH_2 = CH_2$, $CH_3CH{=}CH_2$, and $(CH_3)_2C{=}CH_2$ with 35 to 50% S). The phosphonates mentioned above have a synergistic effect on lubricating oils when added in concentrations of 0.5 to 5% of each component to give a final mixture which contained 0.1% P and 0.9% S.[150]

Easily washable lubricants for the working of aluminum or aluminum alloys contain polyethylene glycol sorbitan monolaurate and dilauryl phosphonate.[151]

Some research has been done on the development of lubricating oils suitable for use with liquid oxidizers.[152-153] Dialkyl phosphonates have been incorporated in formulations to inhibit oxidation and stabilize motor fuels, hydrocarbon-base lubricating oils, greases, and hydrocarbon polymers.[154]

Compositions containing di-butyl mono and trichloromethylphosphonates have been prepared and used in compounded motor oils for heavy duty gear lubricants.[155-158]

Friction in internal combustion engines is reduced by using a combination of graphite and di-methyloctadecylphosphonate.[156,159]

A transaxle lubricant containing esters of mercapto acids and a phosphonate of the formula $(RP(O)(OR)_2$, (where R is a $C_{2\text{-}20}$ alkyl group), is suitable for use in the automatic transmission and the rear axle systems of automobiles.[160]

Phosphonates have also been used in the areas of ash-free lubricating compositions[161] and as flame retardant lubricating oil additives.[162]

14. Polyester, Polyethylene, and Polycarbonate Discoloration Inhibitors

When polyesters are prepared with Ce or La catalysts, they tend to turn yellow, especially in the second step of the polycondensation. This can be prevented by using phosphorus or phosphoric acid. These materials are seldom used because of their strong influence on the activity of the catalysts. When the Ce and La catalysts are used in combination with a phosphonate such as di-phenylbeta-phenylethane phosphonate, colorless products are formed with no reduction in the catalytic activity.[165]

Linear polyesters are stabilized against discoloration by the addition of a phosphite-polyphosphonate, $(CH_3CHCl\text{-}CH_2O)_2POCHCH_3P(O)(OCH_2CHClCH_3)OCHCH_3P(O)\text{-}(OCH_2CHClCH_3)_2$, to the polyester either during or after polymer preparation. Polyesters in which the stabilizer was used had an initial whiteness of 95. Those not stabilized had a whiteness value of 45.[166]

Alkylphosphonates such as $(C_4H_9O)_2P(O)C_4H_9$ and $(C_8H_{17}O)_2P(O)C_8H_{17}$ inhibited color formation in high density polyethylene but not in low density polymers.[167]

Dialkyl alkylphosphonates such as butyl $P(O)(O\ butyl)_2$ are useful in making polycarbonates resistant to opacification and discoloration in high temperature molding.[168]

15. Polyurethan Additives

Over 6 million lb of organo phosphorus-containing flame retardants were used in 1966 to impart flame resistance to polyurethans. A typical reactive flame retardant for the above use is $(C_2H_5O)_2P(O)CH_2N(CH_2CH_2OH)_2$, trade named Fyrol® 6 by Stauffer Chemical Company.[168a] This compound which is made from diethanolamine, formaldehyde, and diethylphosphonate, acts as a polyol reagent in the polymerization of rigid urethan foams. The following flame retardant, Phosgard C22R®, is an additive type retardant for imparting flame resistance to urethan foams.[168b] The formula for this product is as follows:

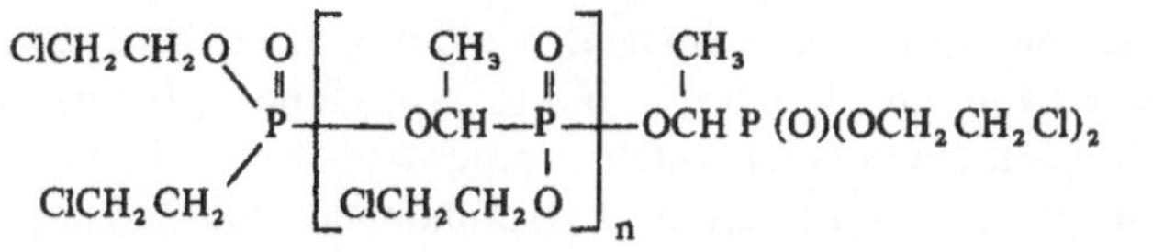

Where n is usually about 0, 1, 2, and 3. This product is prepared by reacting phosphorus trichloride with ethylene oxide, next treating with acetaldehyde, followed by thermal rearrangement. By substituting the chlorine atoms on the above compound with amino groups such as $-NHCH_2CH_2NH_2$, a product suitable for flame retarding cotton is obtained.

As is well known, one of the most common methods of forming polyurethans is by the reaction of di- or polyfunctional compounds, such as hydroxyl-terminated polyesters or polyethers, with di- or polyfunctional isocyanates. The general structure of a polyurethan from a dihydroxy compound, HOROH, and a diisocyanate, OCN-R′-NCO, is as follows:

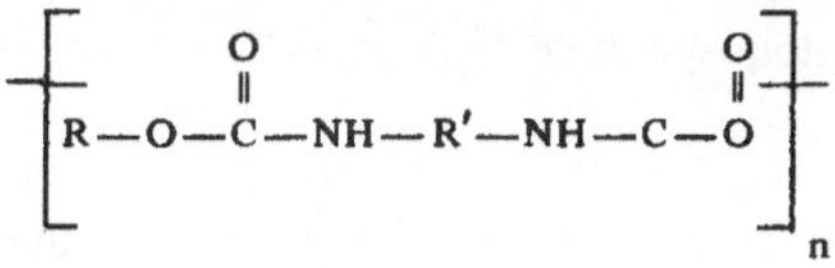

Markets for these type polymers have grown tremendously especially in the furniture industry, and in the automotive industry for seat cushioning, instrument panel trim, safety pads, arm rests, floor mats, sun visors, underlays, roof insulation, weatherstripping, air filters, etc. It is also being used as backing for fabrics and carpets, for refrigerator insulation, in the building industry, in boats, and numerous other end uses. Since urethans burn, there is the necessity to impart flame resistance to these products that are used in all of the above items. Numerous references appear in the literature in which phosphonates have been incorporated into polyurethan or polyester formulations to form structurally bound components of the resulting polymers. Polymers containing these organo phosphonates are strong, flame retardant, and self-extinguishing. They have good hydrolytic stability, good mechanical strength, and insulating characteristics.[169] Addition of phosphonates to urethan foams usually increases smoke development.[169a] Foams with good cell structure, filaments, and coatings with good flame resistance were formed from polyurethans which were the reaction products of an organic polyisocyanate and a phosphonate of the formula $(RO)_2P(O)R'P(O)(OR)_2$, where R is hydroxyalkyl, hydroxy alkoxyalkyl, or hydroxy polyalkoxy alkyl, and R′ is a $C_{2\text{-}10}$ bivalent alkane or alkene radical. Examples are tetrakis(dipropylene glycol)2-butene-1,4-diyldiphosphonate, pentaerythritol diallyldiphosphonate, and bis(dipropylene glycol) allyl phosphonate.[170] Other phosphonatesused to impart flame retardancy to polyurethans are as follows:

1. Cyclic phosphonates[171-172]
2. *N,N*-diethanolaminomethyl phosphonate[173]
3. Bis(2,3-dichloropropyl)2,3-dichloropropyl phosphonate[174]
4. Diethyl *p*-[*N,N*-bis(2-hydroxyethyl)amino]benzyl phosphonate[175-176]
5. Diethyl *p*-[*N,N*-bis(2-hydroxypropyl)amino]benzyl phosphonate[175]
6. Triacyl decapropoxyacetodiphosphonate[177]
7. Dialkanolaminoalkylphosphonates[178]
8. β-Hydroxyethyl α-hydroxyethanephosphonate[179]
9. Mannich condensate phosphonates such as: bis(dipropylene glycol) diethylaminomethanephosphonate, bis(dipropylene glycol)diallylaminomethanephosphonate, bis(diethylene glycol)diethylaminomethanephosphonate, bis(triethylene glycol)bis-(2-

hydroxyethyl)aminomethanephosphonate, tris(dipropylene glycol)bis[bis(2-hydroxyethyl)aminomethanephosphonate], bis-(dipropylene glycol) bis(2-hydroxyethyl)aminofurylmethanephosphonate, piperazinomethylbisphosphonates, and bis(dipropylene glycol) bis(2-hydroxyethyl)aminobutanephosphonate, bis(dipropylene glycol) bis(2-hydroxyethyl)aminomethanephosphonate[180]

10. Halogen containing phosphonates[181-185]
11. Hydroxyalkylaminoalkylphosphonates[186]
12. Carbamoylphosphonates[187-189]
13. Benzyl phosphonates[190]
14. Bis(oxypropylated pentaerythritol)oxypropylated pentaerythritol phosphonates, bis(tripentaerythritol)tripentaerythritol phosphonate, bis(pentaerythritol)butane phosphonate[191]
15. Dialkoxyphosphono-substituted alkanols[192]
16. 3-(Dialkoxyphosphonyl)propionamides[193]
17. Diethylphosphonate[194]
18. Poly(oxyethylene)phenylphosphonate[195]

16. Rayon

Many phosphonate derivatives or compositions have been useful for imparting flame retardance to viscose and acetate rayon. In many cases, the rayon has been processed by wetting the fabrics with the flame retardant followed by drying and curing to impart flame resistance.[196-197] In some cases, the rayon is shrinkproofed also by treatment with dimethylolurea.[198] In other cases, the flame retardant has been first added to the viscose or acetate solution. The viscose is then spun into an acid solution producing a fire retardant regenerated cellulose.[187,188,199-201] Bis(2-bromoisopropyl)2-bromopropyl phosphonate, bis(2-chloro-3-bromopropyl)2-chloro-3-bromopropylphosphonate, and bis(2,3-dibromopropyl)vinyl phosphonate are all examples of the type phosphonates used.[202]

17. Scale Inhibitors

A combination of a water soluble phosphonate and sulfonate polymer produces a synergistic effect in reducing the formation of boiler scale especially in hard water.[225,226]

Phosphonates like tetra-Na 1-hydroxyethane-1,1-diphosphonate prevent scale buildup in suspension polymerization of vinyl chloride.[227-229]

Scales such as Ca, Ba, and Mg, carbonates, sulfates, and silicates which form in water systems are inhibited by the addition of minor amounts of water soluble poly(phosphonomethylated)poly(ethylene amine) compounds containing from 3 to 15 amine groups.[230] Aminophosphonates of the formula $RN[CH_2P(O)(OM)_2]$ where R is $CH_2P(O)(OM)_2(CH_2)\infty N\ [CH_2P(O)(OM)_2]_2$, n = 1 to 3, and M is H, NH_4, or an alkali metal is used for the prevention of scale formation in salt water evaporators.[231]

18. Solvent Extraction

Close boiling hydrocarbons such as paraffins, monoolefins, diolefins, naphthenes, and aromatic hydrocarbons can be separated by extractive distillation. An extracting agent such as the α-oxophosphonates of the general formula $(RO)_2P(O)COR'$, where R is a C_{1-10} alkyl radical and R′ is a hydrocarbon, a C_{1-6} alkoxy radical or an aryloxy and alkyl substituted aryloxy radical containing C_{6-10} aryl C and C_{1-3} alkyl groups, is excellent for this purpose. With the extractive distillation, the more volatile paraffins are withdrawn as an overhead vapor fraction and the less volatile hydrocarbons are withdrawn as a liquid bottom product dissolved in the phosphonate. Conventional means are then used to separate the fractions from the phosphonate solvents.[233]

C. Miscellaneous Phosphonate Applications

In addition to those phosphonate applications discussed above, there are a number of other uses mentioned in the literature which should be pointed out.

Antistatic agents — Phosphonium phosphonates and phosphonium xylenesulfonates when added to polyamides to increase their electrical conductance make them suitable as cores in antistatic sheath-core composite fibers.[13]

Blowing agents — Phosphonates can be used as blowing agents in various formulations to produce flexible foam sheets which exhibit good fire retardant properties.[14]

Coupling agents — Organophosphonates such as diethylaminobenzylphosphonate or diethyl bis(2-hydroxyethyl)aminoethylphosphonate are used to impregnate glass cloth to prevent water absorption and to act as a coupling agent between the glass and an epoxy resin in glass reinforced plastics.[35]

Crystallization inhibitors — Damage caused by salt deposits on oil field equipment can be minimized by the addition of an inhibitor consisting of diethylcyclohexylaminomethylphosphonate in an organic solvent to the normal lubricants employed.[36-37]

Dentrifice compositions — Some polyphosphonate-containing compositions are useful as anticalculus agents.[38] Di-Na-1-hydroxy-1,1-ethanediphosphonate, when used in combination with NaF for treatment of molar surfaces, reduced calculus formation and F retention was enhanced.[39-40]

Deodorants — Esters of alkanediphosphonic acids of the general structure $(RO)_2P(O)(CH_2)\infty P(O)(OR)_2$, where R = C_{1-6} alkyl; n = 1 to 6 have been used in deodorant compositions in 0.1 to 10% concentrations.[41] These compositions are preferably prepared as aerosols. The esters are usually odorless, high boiling liquids which can be easily used in all types of cosmetic preparations.

Detergents for cleaning metal surfaces — Phosphonates suitable for cleaning metallic surfaces are prepared by adding 2-ethylene glycol alkylphosphonate-diethanolamine salt to phosphate containing detergents. Addition of this phosphonate to detergents decreases the foaming and minimizes environmental pollution.[48]

Dye modifiers — Phosphonates such as diethylethanephosphonate have been used in tone-in-tone dyeing.[49] Polyacrylonitrile fibers are made more receptive to basic dyes by addition of phosphonate anionic surfactant salts, which are insoluble in water and polyacrylonitrile solvents.[50]

Fire retardants for synthetic fibers — Polyesters for film and for fibers which have improved dye receptivity and resistance to pilling in woven form are produced by adding a phosphonate which is neutral and has no esterifiable groups as one of the ingredients in the polyester process.[97] Phosphonates used alone or in combination with other agents to impart flame resistance to paper, wood, cotton, polyester, or polyester cotton blends are as follows:

1. Bis(phosphonomethyl) chloromethylphosphonate, bis(phosphonomethyl) phenylmethylphosphonate, bis(diethylphosphonomethyl) phenylphosphonate[98]
2. 3-(dimethylphosphono)-*N*-methylolpropionamide[99]
3. Dimethylmethanephosphonate[100]

Phosphonates used for imparting fire resistance to acrylic fibers:

1. Dimethylphosphonate, dibutylphosphonate, diallylphosphonate[101]
2. Bis(3-chloro-2-bromopropyl)-3-chloro-2-bromopropylphosphonate, allyl 2,3-dichloropropyl 2,3-dichloropropylphosphonate[102]
3. Bis(2,3-dibromopropyl)2,3-dibromopropylphosphonate[102-103]

Flotation — Diethyl 2-[2(glycidyloxy)ethoxy] ethyl phosphonate is used as the frothing agent for the flotation of complex ores.[104-105]

Fuel additives — Gasoline engine manufacturers were allowed to go to higher compression ratios when tetraethyl lead (TEL) was added to the gasoline to prevent knocking. The effectiveness of TEL can be enhanced by the addition of dimethyl methylphosphonate to leaded fuels. The phosphonate reduces preignition and fouling of the spark plugs in spark-ignition combustion engines.[106]

Hardening oil composites — Alkyl phosphonates such as $(C_{12}H_{25}O)_2P(O)R$ or $(C_4H_9O)_2P(O)R$ are useful as hardening oil composites.[111]

Photography — *O*-alkylphosphonates such as $PhCH_2P(O)[O(CH_2)_4CH_3]OH$ are used in the preparation of photographic emulsions.[163]

Plasticizers — Dialkyl (dialkoxyallyl) phosphonates are useful as plastic additives.[164]

Resin and plastic additives — Phosphonates that have been reported to be useful in producing flame retardant resins are as follows:

1. Bromophosphonates having 3 aliphatic substituents one or more of which is halogen-substituted and contains one or more bromine atoms[203-204]
2. Diallyl *p*-chlorobenzylphosphonate[205]
3. Bis(dibromopropyl) dibromopropylphosphonate[206-207]
4. Bis(2-chloroethyl) naphthylphosphonate[208]
5. Bis(2-bromoethyl) 2-bromoethylphosphonate[209]
6. α-hydroxyphosphonate silicates[210]
7. Unsaturated fatty acid phosphonates[211]
8. Bis(beta chloroethyl) vinylphosphonate[212]
9. Imidomethylphosphonates[213]
10. Polymeric pentaerythrityl phosphonates[214-215]
11. Chloroalkyl chloroalkylphosphonates[216]
12. Cyclic phosphonates[217]
13. Benzyl neopentylene phosphonate[218]
14. Mono(2-methacryloxy-3-chloropropyl)phosphonate[219]
15. Polyphosphonates[220-221]
16. Dialkyl alkyl phosphonates such as dimethyl methyl phosphonate[222]
17. Bis(β-chloroisopropyl)α,β-dibromo-α-methylethyl phosphonate[223]
18. Dimethyl(β,β,β,trichloro-α-hydroxyethyl)phosphonate[224]

Settling retardants — Sodium ethylenediaminetetramethylene phosphonate has been used as a settling retardant for gypsum or gypsum plaster, in the manufacture of molds.[232]

Sequestering agents — Phosphonates such as triammonium tridecylidenediphosphonate have been reported to have been used as surfactants.[234] Phosphonates like tetra-Na methylenediphosphonate have been used with chlorine forming agents with sequestering properties.[235-236]

Suspending agents — Aliphatic phosphonates are useful as suspending agents in the production of polyvinyl chloride which exhibits good colloidability, plasticizer absorption, heat stability, and color.[237]

Synthetic fiber preparation — Polyesters prepared in the presence of various organic phosphonates such as hydroxypoly (oxyethylene) hydroxymethylphosphonate, hydroxypoly(oxyethylene) trichloromethyl phosphonate, as well as certain bromo containing phosphonates, reduce fouling of dyes. The color of the products and their dielectric properties are improved.[238]

Viscosity modifiers — Henkel and Cie[239] report in the French patent literature the use of sodium ethylenediamine-*N,N,N',N'*-tetrakis (methylene)phosphonate for reducing the viscosity of aqueous mineral suspensions.

Wood fireproofing agents — Alkyl and aryl phosphonates have been used to impart flame retardancy to wood by impregnation followed by heating to 100 to 200°C to fix the compounds in the wood. One example of such a phosphonate is bis(2-chloroethyl) chloroethyl phosphonate.[240]

Wood impregnated with unsaturated organophosphorus compounds such as dimethylmethallylphosphonate or diallylallylphosphonate, then irradiated, exhibits flame resistance durable to over 14 days immersion in water.[241]

III. TOXICITY

A. Toxicity of Some Selected Phosphonates

Toxicity is defined as the ability of a chemical molecule or compound to produce injury once it reaches a susceptible site in or on the body. In recent years, there has been a vast increase in the number of phosphonates made and used by mankind to benefit his living standards, but risks from their use have not always been considered. In fact, many have probably been produced with essentially no toxicological data. The producers of most of the phosphonates used industrially furnish product safety information which is very helpful to the safety of the users. Toxicity data related to many of the phosphonates follows:

Dialkyl alkylphosphonates [diethyl ethyl, dibutyl butyl, and bis(2-ethyl hexyl) 2-ethylhexyl] — Studies of the toxicity of the above phosphonates indicate that they are quite low, with the ethyl ester being slightly more toxic than the butyl and 2-ethylhexyl esters. Acute oral toxicities to male white rats are 1.2 to 1.4 g/kg body weight for the ethyl ester and greater than 2 g/kg body weight for the butyl and 2-ethylhexyl esters. No unusual difficulties were reported in the handling of these products.[242]

Dimethyl methylphosphonate[243] — Animal studies indicate that this material is practically nontoxic. It is only mildly irritating to the skin; however, no human effects have been reported at this time from exposure to this product. Based on animal studies, it is likely that ingestion of large quantities may produce symptoms of nonspecific irritation of the gastrointestinal tract accompanied by nausea, vomiting, cramps, and diarrhea. Studies on rabbits indicate the product is not an eye irritant. At this time no threshold limit value has been established by the American Conference of Governmental Industrial Hygienists.

[Bis(beta-chloroethyl) vinylphosphonate (Fyrol® Bis Beta)] — The acute oral toxicity shows an LD_{50} 937 mg/kg (rat-oral). It is an irritant to the eyes and tests on rabbits showed some signs of corneal damage. It has no immediate effects on the skin, but prolonged exposure may lead to absorption and dermatitis. The acute dermal LD_{50} is above 4640 mg/kg (rabbit).[244]

Fyrol® 58 [aliphatic phosphate phosphonate oligomer (mixture with isopropyl alcohol)] — Animal studies indicate the material is essentially nontoxic. Ingestion may produce abdominal discomfort and local irritation may result from contact with mucous membranes or upper respiratory tract. It is nonirritating to the eyes and is only a mild skin irritant. No ill effects are known to occur at this time from inhalation of the vapors. Inhalation of high concentrations of sprays or mists may result in nonspecific irritation of the upper respiratory tract. No threshold limit value has been established by the American Conference of Governmental Industrial Hygienists.[245]

Antiblaze® 19 — Acute oral LD_{50} was greater than 10 g/kg body weight and can be considered nontoxic. Dermal toxicity studies showed the acute dermal LD_{50} to be greater than 15 g/kg body weight. Primary dermal irritation studies made on the abraded skin of rabbits revealed no evidence of edema, erythema, and eschar formation. The primary Dermal Index was calculated to be 0.0/8.0. There was no irritation or apparent toxic effects observed after 48 hr in unwashed rabbit eyes. Studies also showed that the product was not mutagenic[246] and the combustion products of Antiblaze® 19 flame retardant and combinations with rigid urethane foams did not generate any unusual toxicity hazard. Results of patch tests and freshwater fish toxicity studies showed no evidence of skin sensitization and that these products were essentially nontoxic to aquatic life.[247]

Chlorinated phosphonate ester (Antiblaze® 78)[248] — Acute oral toxicity (rats) LD_{50} was determined to be greater than 8 mℓ/kg (12 g/kg body weight). Primary Dermal Irritation studies revealed no scoreable evidence of edema, erythema, and eschar formation (rabbit) and no irritation or apparent toxic effects observed after 48 hr in the unwashed rabbit eyes. The Antiblaze® 78 is not considered mutagenic (Ames test).[246] Human Skin Irritation studies showed none of the subjects exhibited irritation or sensitization to the product under semi-occlusive conditions after 2 days of continuous contact.

Fyrol® 76[249] — Fyrol® 76 has an acute oral LD_{50} (male rats) of >4640 mg/kg, is nonirritating to the skin or eye. The Delayed Neurotoxicity of 420 mg/kg (chicken) is negative. The acute Dermal LD_{50} (rabbit) >4640 mg/kg, Acute Inhalation LC_{50} > 4.0 mg/ℓ/hr and the Ames test was also negative.[250] Cotton and rayon fabrics treated with Fyrol® 76 have been tested and found to be nonmutagenic.

Pyrovatex C.P.® (dialkylphosphonopropionamide)[251] — The Ames test[246] shows that fabric treated with Pyrovatex C.P.® is not mutagenic, according to Ciba-Geigy. Similarly, there have been no reports of any other health hazards associated with cotton fabric which has been so treated with Pyrovatex C.P.®[250]

B. Epilogue

Many of the phosphonates in commercial use today are in the field of pesticides, insecticides, and herbicides, as well as other areas. Articles related to their toxicity[60,251] will be covered elsewhere throughout the book.

Although a great deal has been published on the toxicity of phosphonates, there is still room for studies to shed more light on this important group of compounds.

REFERENCES

1. **Samour, C. M.,** U.S. Patent 3,162,610, 1964.
2. **Zuev, B. M., Gubanov, E. F., Rizpolozhenskii, N. I., Kudryavtsev, B. V., Stepashkina, L. V., and Dikolenko, E. P.,** U.S.S.R. 794,054, 1981; *Chem. Abstr.*, 94, 193199, 1981.
3. **Farbwerke Hoechst, A. G.,** British Patent 1,115,719, 1968.
4. **Lewis, A. F.,** U.S. Patent 3,291,861, 1966.
5. **Stephan, J. T.,** U.S. Patent 3,056,687, 1962.
6. **Blake, C. A., Jr., Gresky, A. T., Schmitt, J. M., and Mansfield, R. G.,** Comparison of dialkyl phenylphosphonates with tributyl phosphate in nitrate systems. Extraction properties, stability, and effect of diluent on the recovery of U and Th from spent fuels, U.S. Atomic Energy Commission ORNL-3374, Oak Ridge, Tenn., 1963.
7. **Siddal, T. H., III,** Organophosphorus compounds, other than tributyl phosphate (TBP) for processing irradiated fuels and by-products, *Aqueous Reprocess. Chem. Irridiat. Fuels, Symp. 57,* Brussels, 1963.
8. **Siddal, T. H., III,** Organophosphorus compounds other than TBP for processing of irradiated fuels and by-products, U.S. Atomic Energy Commission TID-18299, Oak Ridge, Tenn., 1963.
9. **Spivack, J. D.,** German Offen., 2,023,973, 1970.
10. **Meltsner, B. R.,** U.S. Patent 3,565,855, 1971.
11. **Novoselova, L. V., Babel, V. G., Proskuryakov, V. A., Novitskii, K. I., and Ginzburg, O. F.,** Synergism of mixtures of dialkyl phosphonates and alkylphenol in the oxidation of MK-8 motor oil, *Okislenie Uglevodorodov, Ikh. Proizvod. Bitumov,* 9, 51, 1971; *Chem. Abstr.*, 79, 21271k, 1973.
12. **Charles, J. J.,** U.S. Patent, 4,277,391, 1981.
13. **Singh, G. and von Rutte, R. E.,** U.S. Patent 4,005,057, 1977.
14. **Herweh, J. E. and Poshkus, A. C.,** U.S. Patent 3,423,485, 1969.
15. **Idel, K. and Merten, J.,** German Patent 2,930,797, 1981.

16. Monsanto Co. Neth., Appl. 6,510,412, 1966.
17. **Begunov, A. V., Rutkovsky, G. V., and Kuznetsov, S. G.,** Micellar catalysis. II. Effect of the nature of surface active agents on the alkaline hydrolysis of isobutyl *p*-nitrophenylmethylphosphonate, *Zh. Org. Khim.*, 17(8), 1668, 1981; *Chem. Abstr.*, 95, 186272e, 1981.
18. **Price, J. A.,** U.S. Patent 3,216,987, 1965.
19. **Monagle, J. J.,** U.S. Patent 3,056,835, 1962.
20. **Kuznetsov, Y. I., Trunov, E. A., Rosenfeld, I. L., and Belik, R. V.,** Study of phosphorus containing compounds in inhibitors in industrial water supply systems (hard water), *Kovroz. Zashch. Neftegazov. Promsti*, (2), 6, 1981; *Chem. Abstr.*, 95, 103087z, 1981.
21. **Panacek, F., Kocica, J., and Eliasek, J.,** The use of phosphonates for the treatment of cooling waters, *Sb. Vys. Sk. Chem.-Technol. Praze, Technol. Paliv*, D41, 5, 1980; *Chem. Abstr.*, 95, 155007t, 1981.
22. **Verma, K. M., Gupta, M. P., Sinha, B. B., Rai, J. S., and Oswal, D. R.,** New trends in cooling water treatment, *Chem. Age India*, 31(11), 1137, 1980.
23. **Khullar, M., Chakraborti, S., and Allard, B.,** *Chem. Age India*, 31(11), 815, 1980.
24. **King, T. M. and Mitchell, R. S.,** U.S. Patent 3,714,067, 1973.
25. **Von Koeppen, A., Emerle, G. A., Nishio, K., and Metz, B. A.,** Pollution control technology applied to cooling water treatment, *Mater. Prof. Performance*, 12(7), 23, 1973.
26. **Thoren, A.,** A comparison of electrochemical methods for evaluation of phosphonate-zinc inhibitors in cooling water, *Ann. Univ. Ferrara Sez. V*, 3(Suppl. 7), 909, 1980.
27. **Orudzheva, I. M. and Dzhavadova, A. A.,** Phosphorus acid esters as inhibitors of metal corrosion, *Prisadki Smaz. Maslam*, 5, 171, 1978; *Chem. Abstr.*, 92, 62738e, 1980.
28. **Rue, J. R.,** Non-chromate treatment of cooling water, *Ammonia Plant Saf.*, 19, 133, 1977.
29. **Gesick, J. A.,** A comparative study of non-chromate cooling water corrosion inhibitors, *Proc. Int. Water Conf., Eng. Soc. West. Pa.*, 35, 155, 1974.
30. **Lindeblad, B. and Pousette, R.,** Treatment of cooling tower water-new methods, *Kem. Tidskr.*, 90(12), 62, 1978; *Chem. Abstr.*, 90, 92186, 1979.
31. **Hatch, G. B. and Ralston, P. H.,** Zincaminomethylenephosphonates for control of oxygen corrosion, *Oil Field Chem., Natur. Gas Liquids Chem. Ind., Symp.*, 159th Natl. ACS Meet., Houston, Tex., 1970, 69.
32. **Suzuki, T. and Kubo, S.,** Japanese Patent 15,220, 1975.
33. **Suzuki, T.,** *Jpn. Kokai*, 73, 34, 742, 1973; *Chem. Abstr.*, 79, 60894k, 1973.
34. **Ralston, P. H.,** U.S. Patent 3,393,150, 1968.
35. **Schrader, M. E. and Lerner, I.,** U.S. Patent 3,420,726, 1969.
36. **Dytyuk, L. T., Samakaev, R. K., Narozhnyi, G. A., Zaripov, R. K., and Tsekhmistrov, A. N.,** U.S.S.R. 850,612, 1981; *Chem. Abstr.*, 95, 189843c, 1981.
37. **Driker, B. N., Prostakov, S. M., Rempel, S. I., Vakulenko, V. A., and Samborskii, I. V.,** *Zh. Prikl. Khim.*, 54(5), 1006, 1981.
38. Proctor and Gamble Co., British Patent 1,201,984, 1970.
39. **Regolati, B. and Muehlemann, H. R.,** Effects of diphosphonate and fluoride on carries, fluorine content, and dissolution of rat molars, *Helv. Odontol. Acta*, 14(1), 37, 1970; *Chem. Abstr.*, 73, 54412m, 1970.
40. **Medcalf, R. F., Jr.,** German Offen. 1,908,067, 1969.
41. **Osberhaus, R. and Blum, H.,** German Offen. 2,607,225, 1977.
42. **Quimby, O. T.,** U.S. Patent 3,551,339, 1970.
43. **Budnick, E. G.,** U.S. Patent, 3,471,406, 1969.
44. Monsanto Co., British Patent 1,024,387, 1966.
45. **Henkel and Cie GmbH.,** French Patent 1,516,223, 1968; *Chem. Abstr.*, 70, 98165a, 1969.
46. **Fitch, S. J. and Irani, R. R.,** U.S. Patent 3,312,623, 1967.
47. **Irani, R. R. and Moedritzer, K.,** U.S. Patent 3,298,956, 1967.
48. **Karaseva, A. D., Kruchinina, R. N., Doroshenko, V. G., Vedenov, G. N., Bromberg, L. D., and Nazarenko, V. T.,** U.S.S.R. 857,253, 1981; *Chem. Abstr.*, 95, 221721z, 1981.
49. **Meyer, E. and Lenz, J.,** German Patent 1,260,429, 1968.
50. **Ogani, K., Teranishi, N., Haioka, H., and Okimoto, M.,** Japanese Patent 16,670, 1965.
51. **Hindersinn, R. R. and Porter, J. F.,** U.S. Patent 3,598,733, 1971.
52. **Benghiat, I.,** U.S. Patent 3,787,528, 1974.
53. **Schmidt, M., Freitag, D., Bottenbruch, L., Reinking, K., Rohr, H., and Medem, H.,** Eur. Pat. Appl. 21,213, 1981; *Chem. Abstr.*, 94, 176071d, 1981.
54. **Ward, F.,** The principles and practice of the flameproofing of textiles, *J. Soc. Dyers Colorists*, 71, 569, 1955.
55. **Wyld, O.,** British Patent, 551, 1735.
56. **Toy, A. D. F. and Rattenburg, K. H.,** U.S. Patent 2,735,789, 1956; **Costello, J. R., Jr. and Traise, T. P.,** U.S. Patent 2,841,507, 1958; **Toy, A. D. F. and Cooper, R. S.,** U.S. Patent 2,867,547, 2,867,548, 1959.

57. **Drake, G. L., Jr., Reeves, W. A., and Guthrie, J. D.,** Phosphonomethylation of cotton, *Textile Res. J.*, 29, 270, 1959.
58. **Reeves, W. A., McMillan, O. J., Jr., and Guthrie, J. D.,** Chemical and physical properties of aminized cotton, *Textile Res. J.*, 23, 527, 1953.
59. **Hobart, S. R., Drake, G. L., Jr., and Guthrie, J. D.,** Changes in the properties of partially phosphonomethylated cotton caused by crosslinking prior to phosphonomethylation, *Textile Res. J.*, 29, 884, 1959.
60. **Aenishanlin, R., Guth, C., Hoffman, P., Maeder, A., and Nachbur, H.,** A new chemical approach to durable flame retardant cotton fabrics, *Textile Res. J.*, 39, 375, 1969.
60a. U.S. Federal Supply Service, Textile Test Methods, Federal Specification CCC-T-191b, U.S. Government Printing Office, Washington, D.C., Method 5902, 1951.
61. **Taylor, G. F.,** Growth of flame-retardant contract fabrics in the U.K., *Int. Dyer, Text. Printer, Bleacher Finish.*, 26, 1980.
62. **Tesoro, G. C., Sello, S. B., and Willard, J. J.,** Flame-retardant properties of phosphonate derivatives of cotton cellulose, *Textile Res. J.*, 38, 245, 1968.
63. **Tesoro, G. C., Sello, S. B., and Willard, J. J.,** Nitrogen-phosphorus synergism in flame-retardant cellulose, *Textile Res. J.*, 39, 180, 1969.
64. **Bruce, J. P., Brunner, R. H., Eisenberg, B. J., and Weil, E. D.,** *Proc. 1974 Symp. Text. Flammability*, LeBlanc, R. B., Ed., LeBlanc Research Corp., E. Greenwich, R. I., 1974, 47.
65. **Tripp, E. P. and Nablo, S. V.,** Electron-cured flame retardant cotton, paper presented at 45th Annu. Res. Technol. Conf. Text. Res. Inst., New York, March 1975.
66. **Walsh, W. K. and Wieder, A.,** to be published in thesis of **Wieder, A.,** North Carolina State University, School of Textiles, Raleigh, 1975.
67. **Walsh, W. K.,** High Energy Radiation for Textiles, Assessment of a new technology 1st Semiannual Report under NSR-Rann Program, School of Textiles, North Carolina State University, Raleigh, November 1974.
67a. **Choi, J. H., Lee, J. K., Kong, Y. K., and Clang, H. S.,** Radiation grafting of flame retardants onto polyester/cotton blend fabric by electron beams, Korea Atomic Energy Research Institute, Seoul, S. Korea, Rep. 1980, KAERI-396/RR-129/80.
67b. **Kim, Y. H. and Ko, S. W.,** Phosphorus-nitrogen synergism and the thermal behavior of vinyl phosphonate oligomer-NMA treated cotton fabrics, *Hanguk Sumyu Konghakhoe Chi*, 18(2), 88, 1981.
68. **Adler, A. and Brenner, W.,** New durable flame retardant finish for polypropylene textiles, *Nature (London)*, 225, 60, 1970.
69. **Harris, J. A., Arthur, J. C., and Goynes, W. R.,** Flame-resistant cotton by photoinitiated graft polymerization, *J. Appl. Poly. Sci.*, 24, 201, 1979.
70. **Inagaki, N. and Katsuura, K.,** Reaction of cellulose phosphonate with *N*-vinyl-2-pyrrolidone and flame-retardant properties of the product, *J. Poly. Sci., Poly. Chem. Ed.*, 18, 441, 1980.
71. **Arney, W. C., Jr. and Kuryla, W. C.,** Structure-property relations in flame retardant systems. Relative effects of alkyl phosphates, phosphonates, and phosphites on cellulose flammability, *J. Fire Flammability*, 3, 183, 1972.
71a. **Peterson, H.,** German Offen. 1,938,101, 1971.
71b. **Nachbar, H. and Rohringer, P.,** U.S. Patent 4,288,489, 1981.
72. **Schiffner, R. and Lange, G.,** German Patent 1,150,651, 1963.
73. **Schiffner, R. and Lange, G.,** Flameproof finishing of cellulosic textiles. IV. Effect of (hydroxymethyl)phosphonic acid and urea on cellulosic fabrics, *Faserforsch. Textiltech.*, 14(9), 375, 1963.
74. **Petrov, K. A., Andreev, L. N., and Sopikova, I. I.,** U.S.S.R. 215,483, 1968.
75. **Lew, B. W.,** German Offen. 2,020,909, 1970.
76. **Yoshizawa, M., Kobayashi, Y., Higuichi, T., and Iida, A.,** Japan Patent 68 20,999, 1968.
77. Ciba Ltd., French Patent 1,560,824, 1969.
78. **Nachbar, H., Stark, B. P., and Maeder, A.,** S. African Patent 68 04,706, 1968; *Chem. Abstr.*, 71, 71881q, 1969.
79. **Tesoro, G. C.,** German Offen. 1,913,137, 1969.
80. **Sello, S. B. and Tesoro, G. C.,** Flame retardant cellulose, *Textilveredlung*, 5(5) 391, 1970.
81. **Spaulding, D. C.,** French Patent 1,580,005, 1969; *Chem. Abstr.*, 73, 5161v, 1970.
82. Compagnie Francaise de Produits Industriels, British Patent 1,558,082, 1979.
83. **Tesoro, G. C.,** Flame retardants for cotton fabrics, *Textilveredlung*, 2(7), 435, 1967.
84. **Tesoro, G. C.,** Private communication.
85. **Tesoro, G. C., Sello, S. B., Moore, D. R., and Wurster, R. F.,** U.S. Patent 3,551,422, 1970.
86. J. P. Stevens and Co., Inc., French Patent 1,534,719, 1968.
87. Ciba Ltd., French Patent 1,395,178, 1965.
88. **Domovs, K. B., Lee, W. K., and Tesoro, G. C.,** British Patent 1,176,495, 1970.
89. **Nachbur, H. and Maeder, A.,** German Offen. 1,933,479, 1970; *Chem. Abstr.*, 72, 80286x, 1970.

90. **Nachbur, H. and Maeder, A.,** German Patent 1,964,757, 1970.
91. **Nachbur, H., Kern, J., and Maeder, A.,** German Offen. 1, 961,884, 1970.
92. **Nachbur, H. and Maeder, A.,** German Offen. 1,964,963, 1970.
93. **Weil, E. D. and Fearing, R. B.,** German Offen. 2,131,040, 1972.
94. **Weil, E. D.,** U.S. Patent 3,906,136, 1975.
95. **Golborn, P. and Duffy, J. J.,** U.S. Patent 3,959,551.
96. **Inagaki, N. and Katsuura, K.,** Reaction of cellulose with *N*-vinyl-2-Pyrrolidone and flame retardant properties of the product, *J. Polymer Sci.*, 18, 441, 1980.
97. Teikoku Rayon Co., Ltd., Belgian Patent 619,210, 1962.
98. **Temin, S. C.,** U.S. Patent 3,179,522, 1965.
99. **Nachbur, H., Kern, J., and Maeder, A.,** German Offen. 2,013,665, 1969.
100. **Hendricks, U. W., Walz, K., and Schaefer, K.,** German Offen. 2,544,550, 1977; *Chem. Abstr.*, 87, 54486c, 1977.
101. **Okada, K., Mitsumura, K., Otani, K., Ogawa, Y., and Hisada, H.,** Japanese Patent 72 02,774, 1972.
102. **Hiraoka, S., Mitsumura, K., Hirajima, M., Ogawa, Y., and Hisada, H.,** Japanese Patent 72 02,773, 1972; *Chem. Abstr.*, 77, 141304d, 1972.
103. **Okada, K., Hirajima, M., Otani, T., Ogawa, Y., and Hisada, H.,** Japanese Patent 77 02,775, 1977.
104. **Leonov, S. B., Roginskaya, L. V., and Khilko, M. Y.,** U.S.S.R. 766,649, 1980; *Chem. Abstr.*, 94, 87816h, 1981.
105. **Wottgen, E. and Dietze, U.,** Accumulation of phosphonic acids on cassiterite in flotation, *Z. Anorg. Allg. Chem.*, 369(1-2) 64, 1969.
106. **Miller, C. O.,** German Patent 1,163,080, 1964.
107. **Calhoun, G. M. and Toombs, A. J. L.,** U.S. Patent 3,105,747, 1963.
108. **Calhoun, G. M.,** U.S. Patent 3,114,710, 1963.
109. **Bykova, L. V., Mikhailova, Z. V., Li, P. Z., Rubtsova, I. K., and Osadchuk, O. S.,** Homopolymers and copolymers of *O,O*-bis(methacryloxyethyl)phosphonates and phosphates, *Plast. Massy*, 6, 31, 1967.
110. **Fuller, G. and Toombs, A. J. L.,** U.S. Patent 3,567,531, 1971.
111. **Hattori, S.,** Japanese Patent 74 32,288, 1974.
112. **Maruyama, T., Yukutomi, M., Osada, M., Hanaue, K., and Uejima, T.,** Jpn. Kokai 77, 72,746, 1977; *Chem. Abstr.*, 87, 136876u, 1977.
113. **Minagawa, M., Sekiguchi, T., Kurita, N., and Sugawara, Y.,** British Pat. Appl. 2,037,777, 1980; *Chem. Abstr.*, 94, 85140j, 1981.
114. Argus Chemical N.V., French Demande 2,449,705, 1980; *Chem. Abstr.*, 95, 26146p, 1981.
115. **O'Konski, T. R. and Stretanski, J. A.,** French Patent 1,519,364, 1968.
116. **Saito, I., Ishizuka, S., Komatsu, Y., and Nakamura, K.,** Jpn. Kokai 75 65,691, 1975; *Chem. Abstr.*, 83, 195115k, 1975.
117. **Wishman, M., Anthony, P., Leininger, J. C.,** Eur. Pat. Appl. 27,926, 1981; *Chem. Abstr.*, 95, 99251d, 1981.
118. **Popov, L. G., Tokareva, L. G., and Mikhailov, N. V.,** Light stabilization of polypropylene fibers, *Khim. Volokna*, 6, 26, 1966; *Chem. Abstr.*, 66, 47218e, 1967.
119. **Fenyes, J. G. E. and Flanagan, K. J.,** U.S. Patent 3,767,735, 1973.
120. Toyobo Co. Ltd. Japan. Kokai Tokkyo Koho 81 43,321, 1981; *Chem. Abstr.*, 95, 99248h, 1981.
121. **Zhdanov, A. A., Grechkin, E. F. and Kalabina, A. V.,** British Patent 1,022,899, 1966.
122. **Zaska, O. C., Fintelmann, C. H., and Igdaloff, H. B.,** French Patent 1,584,020, 1969.
123. **Ehara, M., Endo, S., Osako, A., and Ikegami, T.,** Japan. Kokai 77 19,625, 1977; *Chem. Abstr.*, 88, 106228n, 1978.
124. Toyobo Co. Ltd., Japan. Kokai Tokkyo Koho 80 155,021, 1980, *Chem. Abstr.*, 94, 157557a, 1981.
125. Teijin Chemicals, Ltd., Japan. Kokai Tokkyo Koho 80 123,646, 1980; *Chem. Abstr.*, 94, 48328f, 1981.
126. **Schmidt, M., Freitag, D., and Bottenbruch, L.,** German Offen. 2,944,093, 1981.
127. **Mukmeneva, N. A., Sabirova, L. K., Kadyrova, V. K., and Kirpichnikov, P. A.,** Color and heat stability of stabilized polyethylene in the presence of phosphites, *Plast. Massy*, 1, 51, 1975.
128. **Steinmec, F., Patzau, S., Montewski, W., and Zajezierska, A.,** Polish Patent 103,351, 1979; *Chem. Abstr.*, 92, 131,876n, 1980.
129. **Zhidkova, L. A., Baranov, G. M., Kutimova, G. V., Efimov, A. A., and Tyutereva, A. F.,** New class of light stabilizers of impact resistant polystyrene, *Plast. Massy*, 10, 42, 1980; *Chem. Abstr.*, 94, 48199q, 1981.
130. **Peterson, G. H.,** U.S. Patent 3,514,505, 1970.
131. **Nowotny, K. A.,** U.S. Patent 3,383,318, 1965.
132. **Mann, J. T.,** U.S. Patent 4,179,384, 1979.
133. **Sullivan, J. D. and Duncker, C. P.,** French Patent 1,566,028, 1969; *Chem. Abstr.*, 72, 45790k, 1970.

134. **Efendiev, A. A., Kasimova, F. S., Abbasova, B. G., and Nazirov, B.,** Synthesis of ion exchangers by copolymerization of sodium vinyl phosphonate and acrylic acid, *Dokl. Akad. Nauk Az. SSR*, 36(10), 37, 1980; *Chem. Abstr.*, 94, 192993m, 1981.
135. **Manecke, G. and Heller, H.,** German Patent 1,168,078, 1964.
136. **Sims, H. J., Bauer, L. N., and Preuss, A. F., Jr.,** U.S. Patent 3,268,450, 1966.
137. **Bieber, H. E. and Hepplewhite, H. L.,** French Patent 1,557,534, 1969.
138. **Vinogradova, I. E. and Alekseeva, E. A.,** Thermographic investigation of anti-abrasive oil additives, *Khim. i Tekhnol Topliv i Masel*, 6(7), 56, 1961; *Chem. Abstr.*, 60, 5251g, 1964.
139. **Fein, R. S. and Powers, C. T.,** U.S. Patent 4,260,499, 1981.
140. **Papay, A. G. and O'Brien, J. P.,** U.S. Patent 4,246,125, 1981.
141. **Tammik, K. D., Slobodkin, Y. M., Afanas'ev, I. D., Grigor'eva, I. S., and Gorozhankina, V. A.,** Preparation of thermally stable antiwear additives [for lubricants], *Tr. Vses. Nauch.-Tekh. Soveshch. Prisadkam Miner. Maslam*, 2, 143, 1966; *Chem. Abstr.*, 71, 14776k, 1969.
142. **Todd, A. R. and Blanchard, P. M.,** German Patent 1,126,056, 1962.
143. **Ratner, H., Townsend, M. E., and Gauley, R. W.,** French Patent 1,433,354, 1966.
144. **Peterson, J. M., Reed, D. D., and Kluge, H. D.,** U.S. Patent 3,256,193, 1966.
145. **Fields, E. K.,** U.S. Patent 3,193,570, 1965.
146. **Hotten, B. W.,** U.S. Patent 3,899,475, 1975.
147. British Petroleum Co. Ltd., Belgian Patent 666,662, 1966.
148. **Mastschappij, N. V.,** British Patent 904,968, 1962.
149. **Birum, G. H.,** U.S. Patent 3,243,370, 1966.
150. **Meyers, H.,** German Offen. 1,937,243, 1970.
151. Showa Aluminum K. K., Japan. Kokai Tokkyo Koho 81 26,997, 1981.
152. **Breed, L. W., Wiley, J. C., Jr., and Elliott, R. L.,** Developoment of lubricating oils suitable for use with liquid oxidizers, NASA Contract Rep. NASA-CR-98480, 1969, from *Sci. Tech. Aerosp. Rep.*, 7(16), 2966, 1969; *Chem. Abstr.*, 71, 126849t, 1969.
153. Shell Internationale Research Maatschappij, Neamloze Vennootschap, Nether. Appl. 6,506,779, 1965.
154. **Knapp, G. C. and Orloff, H. D.,** German Patent 1,217,382, 1966.
155. **Vinogradova, I. E. and Petyakina, E. I.,** Compounded motor oils for heavy duty gear driving, *Freiberger Forschungsh.*, A296, 117, 1963.
156. **Papay, A. G.,** U.S. Patent 4,228,020, 1980.
157. **Tada, F., Koga, T., Inaba, S., Nabata, S., Hatanaka, T., and Sakata, K.,** Japan. Kokai 75 111,106, 1975.
158. **Papay, A. G. and O'Brien, J. P.,** U.S. Patent 4,293,432, 1981.
159. **Papay, A. G. and O'Brien, J. P.,** U.S. Patent 4,225, 449, 1980.
160. **O'Halloran, R.,** U.S. Patent 3,206,401, 1965.
161. **Fuller, G. and Henderson, B. M.,** U.S. Patent 3,243,371, 1966.
162. **Lewis, M.,** U.S. Patent 3,534,073, 1970.
163. Ciba Ltd., Belgian Patent 657,853, 1965.
164. **Gross, H. and Costisella, B.,** German (East) Patent 77,712, 1970.
165. Teijin Ltd., Nether. Appl. 6,611,742, 1967.
166. Monsanto Co., Nether. Appl. 6,500,381, 1965.
167. **Mukmeneva, N. A., Akhmadullina, A. G., and Kirpichnikov, P. A.,** Color stabilizing action of phosphites during the bisphenol inhibited thermal oxidation of high density polyethylene, *Vysokomol. Soedin., Ser. B*, 16(12), 867, 1974, *Chem. Abstr.*, 82, 140926z, 1975.
168. Teijin Chemicals, Ltd. Japan. Kokai Tokkyo Koho 81 70,053, 1981, *Chem. Abstr.*, 95, 133883y, 1981.
168a. **Papa, A. J.,** Reactive flame retardants for polyurethane foams, *Ind. Eng. Chem., Prod. Res. Dev.*, 9(4), 478, 1970.
168b. **Granzow, A.,** Flame retardation by phosphorus compounds, *Acc. Chem. Res.*, 11(5), 177, 1978.
169. **Walsh, E. N., Uhing, E. N., and Beck, T. M.,** Flame retardant polyurethan and polyester resins, *Am. Chem. Soc., Div. Org. Coatings, Plastics Chem., Preprints*, 23(1), 1, 1963.
169a. **Einhorn, I. N., Mickelson, R. W., Shah, B., and Craig, L.,** Smoke development in urethane foams, *J. Cell. Plast.*, 4(5), 188, 1968.
170. **Friedman, L.,** U.S. Patent 3,225,010 1965.
171. Hooker Chemical Corp., Nether. Appl. 6,502,541, 1965.
172. **Friedman, L.,** U.S. Patent 3,330,888, 1967.
173. Stauffer Chemical Co., Nether. Appl. 6,613,257, 1967; *Chem. Abstr.*, 67, 65056a, 1967.
174. **Coates, H. and Cooksley, M. V.,** British Patent 1,069,524, 1967.
175. Upjohn Co., British Patent 1,062,869, 1967.
176. **Hindersinn, R. R. and Iliopulos, M. I.,** U.S. Patent 3,385,914, 1968.
177. **Jacques, J. K.,** French Patent 1,455,978, 1966.

178. **Duke, R., Mayo, H. P., and Molinario, L.**, French Patent 1,470,166, 1967.
179. **Clampitt, R. B. and Anderson, R. M.**, U.S. Patent 3,332,893, 1967.
180. **Friedman, L.**, U.S. Patent 3,309,342, 1967.
181. **Cook, J. W. and Haggis, G. A.**, Behavior of halogen-containing phosphates and phosphonates in urethane foams, *J. Cell Plast.*, 5(2), 119, 1969.
182. **Parrish, D. B. and Pruitt, R. M.**, Thermal stability of flame flexible urethane foam, *J. Cell. Plast.*, 5(6), 348, 1969.
183. **Nakano, T.**, Japan. Kokai 74 27,542, 1974; *Chem. Abstr.*, 81, 26509v, 1974.
184. **Studnicki, M. and Karminski, W.**, Self-extinguishing polyurethane foams based on chloroalkyl phosphonates, *Polimery*, 23(11), 405, 1978.
185. **Fulmer, G. E. and Murch, R. M.**, U.S. Patent 4,237,182, 1980.
186. **Bernhart, D. N.**, German Offen. 1,928,265, 1969.
187. **Sayigh, A. A. R. and Tilley, J. N.**, U.S. Patent 3,498,937, 1970.
188. **Walsh, E. N.**, French Patent 1,578,091, 1969.
189. **Nachbur, H. and Maeder, A.**, German Offen. 2,104,729, 1971.
190. **Bright, J. H. and Cannelongo, J. F.**, U.S. Patent 3,547,842, 1970.
191. **Baranauckas, C. F. and Gordon, I.**, U.S. Patent 3,538,196, 1963.
192. **Boyer, N. E. and Weaver, M. P.**, German Offen. 2,023,788, 1970.
193. **Nachbur, H. and Maeder, A.**, German Offen. 2,012,137, 1970.
194. **Biranowski, J.**, U.S. Patent 3,956,200, 1976.
195. **Rosenberg, H. and Cooper, R. S.**, French Patent 1,388,667, 1965.
196. **Kawai, A., Kaneiwa, M., and Tanaka, H.**, Japanese Patent 72 04,756, 1972; *Chem. Abstr.* 77, 153830k, 1972.
197. **Takatani, T., Date, M., Fukuoka, S.**, Japanese Patent 74 01,680, 1974; *Chem. Abstr.*, 81, 92995b, 1974.
198. **Uno, H., Urata, K., and Tsuji, T.**, Japan. Kokai 74 117,799, 1974.
199. **Tsuji, T.**, Japanese Patent 73 42,125, 1973.
200. **Nagai, K., Okada, H., and Takeuchi, I.**, Japan. Kokai 73 91,312, 1973; *Chem. Abstr.*, 80, 146919e, 1974.
201. **Nagy, G.**, French Patent 1,559,000, 1969.
202. **Mohney, W. K.**, U.S. Patent 3,321,330, 1967.
203. **Jahn, A. K.**, British Patent 959,419, 1964.
204. **Dow Chemical Co.**, French Patent 1,332,588, 1963.
205. **D'Alelio, G. F.**, U.S. Patent 3,325,569, 1967.
206. **Ando, T., Osuga, H., Kotera, H., Matsubara, H., Imamura, K., and Kikuchi, T.**, Japanese Patent 68 28,179, 1969; *Chem. Abstr.*, 72, 101377v, 1970.
207. **Lowes, F. J., Jr.**, U.S. Patent 3,242,124, 1966.
208. **Savides, C.**, German Offen. 1,933,425, 1970.
209. **Jann, A. K.**, German Patent 1,495,419, 1970.
210. **Kodama, Y., Kodama, T., Nakabayashi, M., and Komatu, S.**, German Offen. 2,013,547, 1970.
211. **Nakazato, S. and Higuchi, K.**, Japanese Patent 71 42,995, 1971.
212. **Wu, M. Y. and Rogovin, Z. A.**, Synthesis of graft copolymers of cellulose and vinylphosphonate esters, *Vysokomolekul. Soedin*, 5(5), 706, 1963; *Chem. Abstr.*, 59, 10336e, 1963.
213. **Duffy, J. J. and Golborn, P.**, German Offen. 2,315,829, 1973.
214. **Granzow, A. H.**, U.S. Patent 4,278,591, 1981.
215. **Granzow, A. H.**, U.S. Patent 4,257,931, 1981.
216. **Vlasova, L. A. and Kuznetsov, E. V.**, Chloroalkylchloroalkylphosphonates as modifiers for epoxy resins, *Plast. Massy*, 9, 39, 1980.
217. **Saito, T. and Ohishi, H.**, German Offen. 3,010,375, 1980.
218. **Granzow, A. H. and Savides, C.**, U.S. Patent 4,255,324, 1981.
219. **Koehler, W.**, German Patent 1,134,836, 1962.
220. **Schmidt, M. and Reinking, K.**, German Offen. 3,002,792, 1981, *Chem. Abstr.*, 95, 133901c, 1981.
221. Rohm and Haas Co., Nether. Appl. 6,603,629, 1966; *Chem. Abstr.*, 66, 38653r, 1967.
222. **Hoffman, P. and Rohringer, P.**, German Offen. 2,231,633, 1973.
223. **Schroll, G. E.**, U.S. Patent 3,250,827, 1966.
224. Pierrefitte-Kalaa Djerda, Societe Generale d'Engrais et Produits Chemiques, French Patent 1,298,491, 1962.
225. **Greaves, B. and Ingham, P.**, German Offen. 3,039,356, 1981.
226. **Ralston, P. H.**, Scale control with aminomethylenephosphonates, *J. Petrol. Technol.*, 21, 1029, 1969.
227. Lion Corp., Japan. Kokai Tokkyo Koho 80 106,201, 1980; *Chem. Abstr.*, 94, 16551g, 1981.
228. **Ralston, P. H.**, Eur. Pat. Appl. 24,247, 1981; *Chem Abstr.*, 94, 16551g, 1981.

229. **Gevecker, J. R.,** Predicting and preventing oilfield scales, *Proc. Annu. Southwest Pet. Short Course*, 23, 117, 1976.
230. **Ralston, P. H.,** U.S. Patent 3,434,969, 1969.
231. Calgon Corp., Nether. Appl. 6,604,037, 1966; *Chem. Abstr.*, 67, 98413x, 1967.
232. Daiichi Kogyo Seiyaku Co. Ltd. Japan. Kokai Tokkyo Koho 81 88,856,1981; *Chem. Abstr.*, 95, 208512n, 1981.
233. **Cornell, D. and Birum, G. H.,** U.S. Patent 3,155,597, 1964.
234. **Irani, R. R. and Fitch, S. J.,** French Patent 1,394,386, 1965.
235. Monsanto Co., Nether. Appl. 6,407,365, 1965.
236. **Blaser, B., Germscheid, H. G., and Worms, K. H.,** U.S. Patent 3,303,139, 1965.
237. **Kearney, J. J.,** U.S. Patent 3,026,308, 1962.
238. **Eaton, E. E. and Small, J. R.,** Belgian Patent 632,324, 1963.
239. **Henkel and Cie GmbH,** French Patent 1,509,449, 1968; *Chem. Abstr.*, 70, 40405n, 1969.
240. **Gurgiolo, A. E.,** U.S. Patent 3,501,339, 1970.
241. **Raff, R. A. V., Herrick, I. W., and Adams, M. F.,** Flame-retardant wood by *in situ* polymerization of organophosphorus monomers, *Forest Prod. J.*, 16(2), 43, 1966.
242. Virginia-Carolina Chemical Corp., Dialkyl Alkylphosphonates, Richmond, Va., 1953.
243. Stauffer Chemical Co. Speciality Chemical Div., Dimethyl Methylphosphonate, Product Safety Inf., Westport, Conn., 1973.
244. Stauffer Chemical Co. Speciality Chemical Div., Fyrol® Bis-Beta Bul., F-1567A-5 Westport, Conn., 1970.
245. Stauffer Chemical Co. Speciality Chemical Div., Fyrol® 58, Product Safety Inf., Westport, Conn. 1975.
246. **Ames, B. N., McCann, J., and Yamasaki, E.,** Methods for detecting carcinogens and mutagens with Salmonella/mammalian microsome mutagenicity test, *Mutat. Res.*, 31, 347, 1975.
247. Mobil Chemical Co., Phosphorus Div., Antiblaze® 19 flame retardant, Prod. Inf. Bul., FRD-3F, 1976.
248. Mobil Chemical Co., Phosphorus Div., Antiblaze® 78 flame retardant, Bul., FRB-1E, 1976.
249. Cotton Inc., Cotton fabrics treated to meet Federal FR regulations are safe, Raleigh, N.C., 1978.
250. **Brusick, D.,** Mutagenic evaluation of Fyrol® 76, Final Report, LB Project 2547 Litton Biometrics Inc. Kensington, Md., 1976.
251. **Menn, J. J. and McBain, J. B.,** New aspects of organophosphous pesticides. IV. Newer aspects of the metabolism of phosphonate insecticides, *Residue Rev.*, 53, 35, 1974.

INDEX

A

B

C

D

E

F

G

H

I

J

K

L

M

N

O

P

Q

R

S

T

U

V

W

X

Z